D0927049

CONCEPTIONS OF COSMOS

Conceptions of Cosmos

From Myths to the Accelerating Universe: A History of Cosmology

HELGE S. KRAGH

History of Science and Technology
University of Aarhus

OXFORD
UNIVERSITY PRESS

Great Clarendon Street, Oxford OX2 6DP

Oxford University Press is a department of the University of Oxford.
It furthers the University's objective of excellence in research, scholarship,
and education by publishing worldwide in

Oxford New York

Auckland Cape Town Dar es Salaam Hong Kong Karachi
Kuala Lumpur Madrid Melbourne Mexico City Nairobi
New Delhi Shanghai Taipei Toronto

With offices in

Argentina Austria Brazil Chile Czech Republic France Greece
Guatemala Hungary Italy Japan Poland Portugal Singapore
South Korea Switzerland Thailand Turkey Ukraine Vietnam

Oxford is a registered trade mark of Oxford University Press
in the UK and in certain other countries

Published in the United States
by Oxford University Press Inc., New York

First published 2007

British Library Cataloguing in Publication Data
Data available

Library of Congress Cataloging in Publication Data

Kragh, Helge, 1944–
Conceptions of cosmos : from myths to the accelerating universe : a history
of cosmology / Helge S. Kragh.
p. cm.
Includes bibliographical references and index.
ISBN-13: 978–0–19–920916–3 (acid-free paper)
ISBN-10: 0–19–920916–2 (acid-free paper)
1. Cosmology—History. I. Title.
QB981.K729 2006 2007
523.109—dc22 2006027692

Typeset by Newgen Imaging Systems (P) Ltd., Chennai, India
Printed in Great Britain
on acid-free paper by
Biddles Ltd, www.biddles.co.uk

ISBN 0–19–920916–2 978–0–19–920916–3

10 9 8 7 6 5 4 3 2 1

PREFACE

Cosmology is likely to be the branch of the physical sciences that can boast of having the greatest popular appeal. Compared with the wealth of books written by astronomers, physicists, and science writers, many of them with a historical dimension, the history of cosmology has not received much attention by historians of science. This is the case for modern cosmology in particular, and also for comprehensive works dealing with the entire history of how philosophers and scientists have studied the universe. The present work offers a full account of the history of cosmology from the ancients to the beginning of the twenty-first century, although of course an account which is far from complete. The book is written with a diverse audience in mind, students of the history of science and ideas as well as the general reader interested in how the picture of the world has developed over nearly three thousand years. Although not a technical work, it will hopefully be of interest also to astronomers, physicists, and other scientists working with, or teaching, cosmology. I would like to express my gratitude to Ole Bjælde of the University of Aarhus, who kindly read part of the manuscript and corrected some mistakes.

Helge S. Kragh

CONTENTS

Introduction **1**

1 From myths to the Copernican universe **6**
1.1 Ancient cosmological thought 6
1.2 The Greek cosmos 18
1.3 Medieval cosmology 32
1.4 The Copernican revolution 46

2 The Newtonian era **67**
2.1 Newton's infinite universe 67
2.2 Enlightenment cosmologies 75
2.3 Astrophysics and the nebulae 89
2.4 Thermodynamics and gravitation 100
2.5 The Via Lacta 110

3 Foundations of modern cosmology **125**
3.1 Early relativistic models 125
3.2 The expanding universe 139
3.3 Towards a finite-age universe 149
3.4 Alternative cosmologies 163

4 The hot Big Bang **177**
4.1 Cosmology—a branch of nuclear physics? 177
4.2 The steady-state challenge 184
4.3 Relativistic standard cosmology 200

5 New horizons **221**
5.1 Early-universe cosmology 221
5.2 Observational surprises 228
5.3 Anthropic and other speculations 235
5.4 The problem of creation 240
5.5 Cosmology in perspective 242

References **253**

Index **267**

INTRODUCTION

The term 'cosmology' derives from Greek, essentially meaning the rational or scientific understanding of the *cosmos*, a word which to the ancient Greeks carried connotations such as 'order', 'regular behaviour' and 'beauty' (it is no accident that the words 'cosmology' and 'cosmetology', or 'cosmos' and 'cosmetics', are so similar). The wildly ambitious claim that the universe can be described rationally—that it is a cosmos, not a *chaos*—had its origin in ancient Greek natural philosophy, which consequently must occupy a central place in any comprehensive history of cosmology. Although in Chapter 1 I refer briefly to the cosmological views of non-Western cultures, the present book is concerned with the development of the scientific understanding of the universe, which effectively means that it is a contribution to the history of science in the European cultural tradition. Incidentally, although attempts to understand the universe in scientific terms go back to the very birth of science, until the twentieth century the word 'cosmology' was rarely used in a scientific context. The first books that carried the word in their titles date from the 1730s. As will become clear, cosmology did not have a professional identity until after the Second World War. Strictly speaking, there were no 'cosmologists' before that time, only scientists who ocassionally dealt with questions of a cosmological nature. Although it is a bit anachronistic to refer to these scientists as 'cosmologists', it is a convenient label and I have made no particular effort to avoid it.

The domain of cosmology is a frightening concept, the *universe* or the cosmos in the sense of everything that has (or has had, or will have) a physical existence, whether matter, energy, space, or time. I use the two words 'cosmos' and 'universe' synonymously, and also do not distinguish them from the word *world*. In German and the Scandinavian languages this all-encompassing concept is sometimes known as 'all'; compare the German *Weltall*. Cosmology in the traditional sense refers principally to the study of the structure of the universe, what in the seventeenth century was often known as *cosmography*, a term which stresses the mapping of the universe and which could also refer to what we would consider as geography today. Indeed, when Ptolemy's famous geographical work (*Geographia*) was first translated into Latin in 1406, it carried the title *Cosmographia*. Whereas cosmology and cosmography were sciences dealing with a static world, *cosmogony* means literally the study of how the universe came to be what it is and so includes a temporal dimension. However, the term is not widely used any longer, and today the evolutionary aspects of the universe, including its so-called creation, are included under the label 'cosmology'.

Confusingly, cosmogony and cosmography often referred to the planetary system (its formation and description, respectively) rather than the universe as a whole, as may be exemplified by Petrus Apianus' *Cosmographia* of 1524 and Henri Poincaré's *Hypothèses cosmogoniques* of 1913. Neither of these works was about cosmology, in the present meaning of the term. *Cosmophysics* may come closer, but this was originally a name employed for a mixture of astrophysics, meteorology, and geophysics, with little concern for the universe at large. The term may first have been used by the German Johannes

Müller, the author of *Lehrbuch der kosmischen Physik* (1856), and in 1903 the Swedish chemist Svante Arrhenius published a massive work in two volumes with the same title. Neither Müller nor Arrhenius had much to say about physical cosmology as we understand the subject today.

It should, further, be pointed out that the word 'cosmology' is sometimes used in a sense very different from the scientific study of the universe. One may, for instance, speak of communist cosmology, romantic cosmology, or the cosmology of Australian aboriginals, in which case one refers to the world view of the corresponding group or era (in German, *Weltanschauung* rather than *Weltbild*). The world views of individuals, periods, or societies may be related to the more narrow, astronomically oriented meaning of cosmology, but this is not generally the case. For instance, the philosopher Stephen Toulmin published in 1982 a book with the title *The Return to Cosmology*, an analysis of leading intellectual 'cosmologists' such as Arthur Koestler, Teilhard de Chardin, and Jacques Monod, none of whom have contributed to the study of the physical universe. Likewise, the fact that Alfred Whitehead's *Process and Reality* of 1929 was subtitled *An Essay in Cosmology* does not make it relevant to astronomers and physicists trying to understand the world. Nor is that the case with the Russian philosopher Peter D. Ouspensky's *A New Model of the Universe*, first published in 1914.

In a wider historical perspective, cosmology as a world view or an ideology cannot be cleanly separated from cosmology as a science. Indeed, the latter largely grew out of the former, and consequently the historian has to deal with both. Even when focusing on the scientific aspects of cosmology, as I do, one cannot ignore the philosophical and religious dimensions, which for a long time were inextricably connected with scientists' efforts to unravel the secrets of the universe. This connection was particularly strong in the old days, especially before the Enlightenment period, after which it weakened. However, it never disappeared completely and probably never will. (Those who believe that cosmology has nowadays severed its links to philosophy and religion should consider the anthropic principle and so-called physical eschatology, topics which will be discussed in Section 5.3.)

From an epistemic and sociological point of view, cosmology is a peculiar science, unlike any other. Historians have traditionally investigated its development either from the perspective of the history of ideas and culture or as part of the history of astronomy. There certainly is a tight connection between astronomy and cosmology, but in my view it would be a mistake to look at cosmology as merely a subfield of the astronomical sciences. This is not the case today, and it was not the case in the past. In fact, for long periods of time astronomers wanted to have as little as possible to do with questions of cosmology and cosmogony, fields they were happy to leave to the philosophers. It has been my intention to write a history of how scientists—or, until fairly recently, natural philosophers—explored and thought of the universe and how they, in the process, changed the very meaning of it. Astronomers always played a most important role in this development, but they were not alone. Contrary to most other histories of cosmology, I pay close attention to reasoning based on physics and chemistry. Mathematical modelling compared with astronomical observations may have been the single most important approach to the study of the universe, but there have always been people who considered the heavens in material terms, as something chemists and physicists could investigate. 'Physical cosmology' is generally believed to be an invention of the second half of the twentieth century, something only made possible by the discovery in 1965 of the cosmic microwave background, but this view I believe is contradicted by history.

There is no one way to write the history of cosmology, just as there is no one way to write the history of any other field of science (or branch of history in general). I have chosen to present the development rather broadly, to include physical and philosophical perspectives alongside the unavoidable astronomical perspectives. As far as chronology is concerned, I highlight the twentieth century, which is given as much space as the entire previous development. I believe this is justified for at least two reasons. For one thing, the history of pre-twentieth-century cosmology is well covered by the existing literature. For another thing, and more importantly, scientific cosmology has changed drastically since the early part of the twentieth century (more precisely, since 1917), which marked a new and revolutionary phase in the age-old study of the universe. The development since Einstein's breakthrough during the First World War, and even more so since the 1960s, has been so remarkable—and so sketchily covered by historians—that it needs to be given high priority. It goes without saying that this is no easy job and that my account can undoubtedly be criticized. It is an especially difficult task to cover recent developments, not only because they are so varied and confusing, but also because it is difficult to judge their historical significance. But this is a general problem for any kind of recent historiography.[1] In any case, I feel that a somewhat inadequate and objectionable historical account of modern cosmology is better than no history. It is about time that historians of science discovered the immense richness of modern cosmology, and it is my hope that this book may be a modest contribution to changing the state of affairs.

The structure of the book is, by and large, chronologically organized. The earliest cosmological views we know of, those of the Mesopotamian and Egyptian cultures, were cosmogonies rather than cosmographies. They were mythical tales of how the world and the gods came into existence, to be followed by the first humans. This is dealt with in Chapter 1, which proceeds to consider the Greek cosmos, first in its speculative–philosophical version and next as developed into a scientific model by Eudoxus, Aristotle, Hipparchus, Ptolemy, and others. The Aristotelian-Ptolemaic picture of the world was, in a Christianized version, adopted by the theologian–philosophers of the Middle Ages, who turned it into a pillar not only of knowledge but also of faith. The stable medieval world picture was, however, challenged by Copernicus' heliocentric system of 1543, an innovation which heralded the coming of a new age. The Copernican universe was immensely larger than the traditional universe, yet the two rival world systems had much in common, including their shared conception of the stellar system as a huge spherical shell populated with countless stars. Also, both systems presupposed that the universe had a centre and that the heavenly bodies moved uniformly in circles, a view which was finally abandoned a couple of decades after Kepler introduced ellipses as the true planetary orbits.

Chapter 2 describes some of the advances in astronomical and cosmological knowledge from Newton in the 1680s to Hubble in the 1920s, a long period in which progress occurred by astronomical observations rather than theoretical innovations. Newton's universal law of gravitation became the cornerstone of theoretical astronomy and the basis of the first scientific (or scientific-looking) cosmogonies in the magnificent style of Kant and Lambert. In the second half of the eighteenth century an evolutionary perspective made its entry into cosmology, a trend which continued in the nebular world view of the following century. Until that time theology had been part of the cosmological tradition, but from about 1820 it becomes rare to find references to God in scientific works on the cosmos.

The invention of spectroscopy in 1860 introduced for the first time a physical (and chemical) dimension to cosmology, providing new and fruitful ways to deal with the riddle of the nebulae. At the same time, the laws of thermodynamics were used to discuss the long-term development of the universe, its fate in the far future, and its possible origin in an unknown past. These discussions of a more speculative nature were not of great concern to the astronomers, who preferred to use their telescopes to obtain positive knowledge about the universe in its present state. By the turn of the century, one of the great questions concerned the size of the Milky Way and the distribution of the nebulae. These difficult problems, epitomized in the 'Great Debate' of 1920, were solved when it became possible to determine the distances to some of the nebulae. It turned out that they were at vast distances, island universes majestically floating around in the vast sea of space.

The work done by observational astronomers was of little relevance to Einstein's development of the general theory of relativity and its subsequent transformation into a theory of a closed universe. As we can see today, but which was far from obvious at the time, Einstein's work marked a watershed in the history of cosmology, easily comparable to the Copernican revolution. The main part of Chapters 3 and 4 deal with aspects of the amazing consequences of Einstein's cosmological field equations. The static nature of the universe had a paradigmatic status in early relativistic cosmology, to the extent that the first theories of an evolving universe were ignored. Only in 1930, when Hubble's observations were combined with the theoretical insights of Friedmann and Lemaître, did the expanding universe become part of mainstream cosmology. We may be tempted to identify the expansion of the universe with relativistic cosmology, and also to think that it led automatically to the notion of a finite-age universe, but history shows otherwise. Cosmologists could favour a universe with an origin in time without subscribing to general relativity; and those in favour of the relativistic theory of the expanding universe could deny that it had a definite age.

The emergence and development of the Big Bang theory of the universe, from the mid-1940s to the late 1970s, forms the main part of Chapter 4. In the early 1950s, Gamow and his collaborators had developed a sophisticated model of the early universe based on nuclear physics, the first version of hot Big Bang cosmology. The theory came to a halt, though, and it took more than a decade until it was developed further and became generally accepted. An important reason for the non-linear development of cosmology in that period was the emergence of a strong rival theory of the universe in the form of the steady-state cosmology of Bondi, Hoyle, and Gold. The controversy between this theory and relativistic evolutionary theories is a classical case in the history of cosmology, described in greater detail in my *Cosmology and Controversy* of 1996. New observations, in particular the discovery of the cosmic microwave background radiation in 1965, killed the steady-state theory, which by 1970 was no longer taken to be a serious alternative by the majority of astronomers and physicists. The hot Big Bang theory quickly became the paradigm of the new cosmology, a field which for the first time emerged as a scientific discipline with its own standards and rules for solving problems. It short, cosmology became a scientific profession.

Chapter 5 summarizes the most important developments since about 1980. On the theoretical side, the inflationary scenario of the very early universe led to a minor revolution which further strengthened the already strong links between particle physics and early-universe cosmology. Even more important was that the standard Big Bang model of the 1970s began to lose its status as observations indicated that the universe was in a state of accelerated expansion. It was believed for theoretical reasons that the energy-mass density was critical, but even when the large amounts of hypothetical dark matter were taken into account it was not enough. By the end of the millennium many cosmologists believed that the main part of the universe consisted of a 'dark energy', which was possibly a form of quantum vacuum energy. Most remarkably, in this way Einstein's controversial cosmological constant made a dramatic comeback on the cosmological scene. Progress in cosmology during the last couple of decades has been mainly observation-driven, yet at the same time interest in highly theoretical and in part speculative areas of cosmology has flourished. In the final sections I offer a characterization of some of the more speculative areas which, whatever their scientific merits, have greatly appealed to the public. They have helped make modern cosmology a fashionable science far beyond the world of research cosmologists.

It goes without saying that the book covers the development of cosmology incompletely. There are many names, events, and themes that are not included, and some that are mentioned only too briefly. At the end of the book I take up a few themes which are best treated in a broad, non-chronological perspective, such as the importance of technological innovations for the progress of cosmological knowledge. I also comment on various questions of a more philosophical nature, not in order to 'philosophize' about cosmology but because they have been recurrent themes in the historical development of cosmology. In 1996, after having been in the business of cosmology for some thirty years, Stephen Hawking wrote:

> Cosmology used to be considered a pseudoscience and the preserve of physicists who might have done useful work in their earlier years, but who had gone mystic in their dotage. . . . However, in recent years the range and quality of cosmological observations has improved enormously with developments in technology. So this objection against regarding cosmology as a science, that it doesn't have an observational basis, is no longer valid.[2]

Hawking was right about the last part—observations of cosmological relevance have improved enormously—but his appreciation reveals an inadequate understanding of the history of cosmology, to put it gently. As this book demonstrates, cosmology as a science dates back much farther in time than the 'recent years' Hawking talked about. I see no reason why Aristotle's cosmology, or that of later researchers such as Copernicus, Newton, William Herschel, and Hugo von Seeliger, was not 'scientific'. Granted, their cosmologies were not very scientific by our standards, but then, how will cosmologists five hundred years from now look upon the current relativistic Big Bang theory of the universe?

Notes

1. On the problems and promises associated with writing the history of contemporary science, see Söderqvist 1997.
2. Hawking and Penrose 1996, p. 75.

1

FROM MYTHS TO THE COPERNICAN UNIVERSE

1.1 Ancient cosmological thought

Cosmology, in the elementary sense of an interest in the natural world and the heavenly phenomena, predates science and can be traced back several thousand years before humans learned to write and read. The cave dwellers knew how to communicate by means of pictures, as we know from the fascinating artwork found in the Lascaux caves in France and the Altamira caves in Spain, for example. Some of this cave art possibly had an astronomical significance. There are drawings that may symbolize the Sun and others that have been interpreted as depictions of the phases of the Moon. If so, they provide evidence that *Homo sapiens* had a sense of wonder about the universe more than 10 000 years ago.

Evidence of a different kind, and relating to a later period in pre-literary culture, comes from the arrangements of large stones—megaliths—that are found many places in Europe, most notably in Great Britain, and which date back to around 3500 BC. The most famous of these impressive megalithic documents is undoubtedly Stonehenge in southern England. For what purpose was the enigmatic Stonehenge projected and constructed? Nobody knows for sure, but today it is widely accepted that it partly served astronomical purposes, that it was a huge megalithic observatory or 'an astronomical temple', as John Smith suggested as early as 1771. More than a century later, the idea appealed to the prominent astrophysicist Norman Lockyer, who was convinced that the Egyptian pyramids had astronomical orientations and saw no reason why that shouldn't be the case with Stonehenge as well. In 1906 he argued his case in a book titled *Stonehenge and Other British Monuments Astronomically Considered*, but the book failed to convince the majority of astronomers and archaeologists. Lockyer may be considered the father of archaeoastronomy, but it was only in the 1960s that the field took off, revived in particular by the British–American astronomer Gerald Hawkins. Appropriately, his classic papers of 1963 and 1964, 'Stonehenge decoded' and 'Stonehenge: a Neolithic computer', appeared in *Nature*, the journal that Lockyer had founded nearly a century earlier.

Hawkins's arguments in favour of British archaeoastronomical activities aroused a good deal of controversy but also attracted positive responses and helped to create an interest in the field. Among the early supporters of archaeoastronomy was Fred Hoyle, the eminent astrophysicist and cosmologist, who entered the debate in 1966, and in 1977 gave a full presentation of his ideas in his book *On Stonehenge*. During the last couple of decades, archaeoastronomy has flourished, and claims that at least some of the megalithic monuments were observatories of a kind are today generally accepted.[1] It seems that humans, even in pre-literary times, had a keen interest in astronomical phenomena and constructed sophisticated tools to study celestial motions. Unfortunately, archaeoastronomy tells us little about the cosmological views of Neolithic man, his conception of the structure of the universe, and how it came into being.

1.1.1 *Cosmo-mythologies*

The ancient Egyptians thought of the world as consisting of three parts. The flat Earth, situated in the middle, was divided by the Nile and surrounded by a great ocean; above the Earth, where the atmosphere ended, the sky was held in its position by four supports, sometimes represented by poles or mountains. Beneath the Earth was the underworld, called Duat. This dark region contained all things which were absent from the visible world, whether deceased people, stars extinguished at dawn, or the Sun after having sunk below the horizon. During the night, the Sun was thought to travel through the underground region, to reappear in the east next morning.

Although the universe of the Egyptians was static and essentially timeless, apparently they imagined that the world had not always existed in the form in which they knew it. Theirs was a created world, the creation being described in cosmogonies, of which there existed at least three different versions.[2] Common to them is that they start with a state of primeval waters, a boundless, dark, and infinite mass of water which had existed since the beginning of time and which would continue to exist in all of the future. Although the gods, the Earth, and its myriads of inhabitants were all products of the primeval waters, these waters were still around, enveloping the world on every side, above the sky, and beneath the underworld.

To the Egyptians, the universe and all its components were living entities, some of them represented as persons. The original watery state of chaos was personified as the god Nun, who, in one of the cosmogonies associated with Heliopolis ('the city of the sun'), gave rise to Atum; according to other versions, Atum emerged out of the primeval waters, as a hill or standing upon a hill. Atum was the true creator-god, and he created out of himself—by masturbation, according to one source—two new gods, one personified as Shu, god of the air, and the other as Tefenet, goddess of rain and moisture. A passage from the Book of the Dead expresses the first creation as follows: 'I am Atum when I was alone in Nun; I am Re in his [first] appearances when he began to rule that which he had made,. . .[meaning that] Re began to appear as a king, as one who existed before Shu had lifted [heaven from Earth], when he [Re] was on the primeval hillock which was in Hermopolis.' The Earth and the sky came next, represented by the deities Geb and Nut, respectively. However, the Earth and the sky had not yet been created as separate parts, for initially they were locked closely together in a unity. It was only when Shu raised the body of Nut high above himself that the heavens came into existence; at the same time Geb became free and formed the Earth. The creation story continues with the emergence of a variety of new gods, but what has been said is enough to give an impression of the nature of the Egyptian cosmo-myths.

Another text, dating from the old kingdom in Memphis (about 2700–2200 BC), likewise includes Nun as the original god of the waters, but it differs from the other cosmogonies by speaking of an even more original god or spirit, Ptah, who is described more abstractly as a cosmic eternal mind, the maker of everything. Ptah was the one god, a cosmic intelligence and creator who was responsible for all order in the universe, physical as well as moral. Atum and the other gods were said to emerge from Ptah, or be contained in him, Atum being the heart and tongue of Ptah. According to the text, 'Creation took place through the heart and tongue as an image of Atum. But greatest is Ptah, who supplied all gods and their faculties with [life] through his heart and tongue—the heart and tongue through which Horus and Thoth took origin as Ptah.'[3]

Fig. 1.1 The Egyptian creation story. Shu, the god of air, separates Earth and heaven, personified by Geb and the goddess Nut, who are dressed in leaves and stars, respectively. The daily journey of the Sun is represented by a god in a boat traversing the sky from east to west. Reproduced from J. Norman Lockyer, *The Dawn of Astronomy* (London: Cassell and Co., 1894), p. 35.

Many of the features of Egyptian cosmology, as sketched here, can be found in other ancient cosmologies, both in the Near East and elsewhere. Generally, these depict the universe as a dynamic entity, something which was created and is full of life, change and activity—cosmology and cosmogony were intertwined and parts of the same story. On the other hand, the extant texts have disappointingly little to say about the geometry of the universe, an aspect which came to be central in scientific cosmology but which was of no importance to the Egyptians or other ancient cultures. For example, the Egyptian texts tell us nothing about the spatial location of Duat, except that it is symmetric to the visible world. As has already become clear, the universe of the ancients was thoroughly mythological and it would be a grave mistake to try interpreting it in scientific terms. Gods, often personified, were the central players in the cosmologies of the ancient world. Not only were objects, forces, and places associated with such gods, the same could also be the case with abstract concepts such as time and victory (which in Greek mythology were identified with the god Chronos and the goddess Nike).

Mesopotamian cosmology was essentially a mythological tale, and the tale has some similarities with that told in Egypt. The universe was ruled by three gods, each with their separate domain. Heaven was ruled by Anu, the Earth and the waters around and below it were the domain of Ea, and Enlil was the ruler of the air in between (the names here are those used by the Babylonians; the earlier Sumerian names were different). Although Anu was thought of as a kind of father god, he only ruled the universe as part of a triumvirate, together with Ea and Enlil. Not unlike the Egyptian cosmology, the gods were descended

Fig. 1.2 The Sun (Samas), the Moon (Sin), and Venus (Ishtar), placed in the centre of a Babylonian monument from the twelfth century BC. The three celestial bodies are surrounded by a heavenly army of animals. Reproduced from Schiaparelli 1905, p. 80.

from a primeval chaos of waters, in this case a mingling of salt water and sweet, associated with the goddess Tiamat and the god Apsu, respectively. Again in conformity with the Egyptian myths, what came to be the domains of Anu and Ea were originally tied together and only became separated after Enlil moved heaven away from the Earth. The Mesopotamian universe also included an underworld, ruled by a god or a goddess.

It is well known that the Mesopotamian civilizations came to include a sophisticated scientific astronomy, more highly developed than that of the Egyptians. In view of this, it is remarkable that the world picture of the Babylonians remained mythological and that their mathematical astronomy had almost no impact at all on their cosmology. The clay tablets do not discuss the shape of the Earth, but it was evidently thought to be a flat disc. There are only a few glimpses of astronomical knowledge in the creation myth known as *Enuma Elish*, the earliest known version of which was composed around the middle of the second millennium BC but is based on material going further back in time. One of these glimpses relates to the Moon as a timekeeping device. The Moon is portrayed as a god wearing a crown which changes in shape through the month, corresponding to the lunar phases. The young warrior god Marduk, city god of Babylon, not only organized the calendar, but also 'bade the Moon come forth; entrusted night to her.' He 'made her a creature of the dark, to measure time; and every month, unfailingly, adorned her with a crown. "At the beginning of the month, when rising over land, thy shining horns six days shall measure; on the seventh day let half [thy] crown [appear]. At full Moon thou shalt face the Sun. . . .[But] when the Sun starts gaining on thee in the depth of heaven, decrease thy radiance, reverse its growth." '[4]

Finally, it is worth pointing out that the world picture of the Jewish people, as reconstructed from various passages in the Bible, was essentially the same as that of the Egyptians and the Babylonians. According to the Italian astronomer Giovanni Schiaparelli, who in 1903 published a book on the subject,[5] it can be summarized in a drawing such as that given in Fig. 1.3. The flat, disc-shaped Earth is surrounded by a sea; beneath the Earth, there are wells and fountains connected with the upper part of the Earth as well as with the great deep, called Tehom. The Earth rests on pillars, and above it is the sky or firmament. Waters are to be found not only on the Earth or beneath it, but also above the firmament. After all, on the second day of creation, God commanded, ' "Let there be a dome to divide the water and to keep it in two separate places"—and it was done. So God made a dome, and it separated the water under it from the water above it. He named the dome "Sky" ' (Genesis 1:6–7). It is from this heavenly water that the rain, formed by water in the clouds, comes. The Jews' equivalent to the Egyptian Duat, the underworld and abode of the dead, was called Sheol. The only difference is that Sheol includes a deep cave which houses a kind of hell for those who have lived a particularly immoral life. On the other hand, the writers of the Bible did not think in diagrams or pictures, and one should not try to make a definite world view out of the Old Testament. That is just not what the Bible is about.

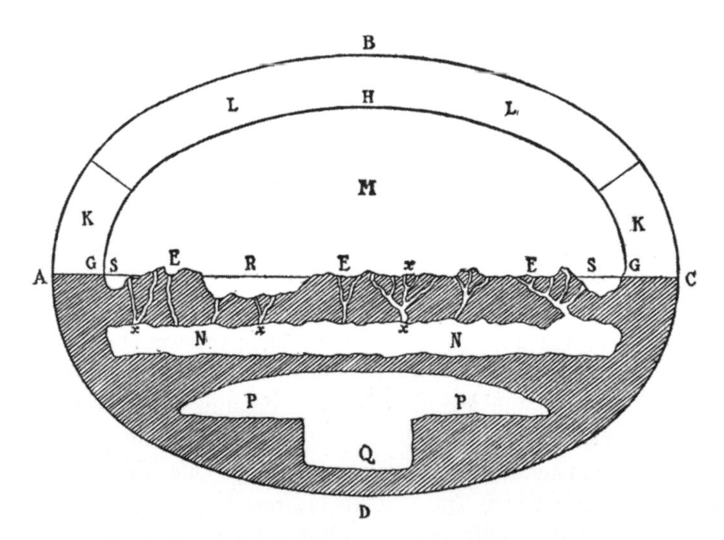

Fig. 1.3 Schiaparelli's reconstruction of the universe of the the Old Testament. The Earth (EEE) is surrounded by a sea (SS), and its surface is connected by streams to a large underground water deposit (NN). Above the Earth is the heavenly tent (ABC), supported by a solid vault (GHG). The space LL contains the waters in heaven, the source of rain. Beneath is the underworld Sheol (PQP), the land of the dead. From Schiaparelli 1905, p. 33.

Egyptian and Babylonian astronomy and cosmology influenced to some extent Greek thought and thereby became linked to the European tradition, out of which scientific cosmology would eventually emerge. But there were other old cultures and these, too, had their conceptions of the universe. In ancient India there were four distinct traditions of cosmology, each of them exceedingly complicated, often fantastic, and rarely exhibiting much consistency. The reason may have been that there was no tradition in India of abandoning a theory or idea just because a new one had been accepted; rather, new ideas were added to the existing belief system by aggregation and inclusion. The complexity of the cosmological traditions makes it impossible to describe them in brief.[6] Suffice it to mention a theme from the *Rig-Veda*, written around 1500 BC, which also occurs in the later Vedic literature, namely, that the world started when fire and water came to meet. In one of the hymns of the *Rig-Veda*, it is said that originally there was nothing, neither existence nor non-existence. 'Darkness was in the beginning hidden by darkness; indistinguishable, all this was water. That which, coming into being, was covered with the void, that One arose through the power of heat.'[7]

Early Chinese astronomy differed in several respects from that of ancient Greece, notably by admitting that the heavens could change. New stars and novae were duly recognized when they appeared. Astronomical inscriptions related to divination practices have been found on a large number of animal bones and shells, dating back to about 1400 BC. Some of these records refer to solar and lunar eclipses, some to comets, and others to stars. One of these oracle bones records a 'guest star' near the star Antares. As to the world picture, the Kai Thien school of the third century BC conceived of the Earth as an inverted bowl lying within a similar but larger bowl, representing the heavens. The two bowls shared the same axis, around which the celestial bodies revolved. At the bases of the bowls, the space between them was filled with water. Remarkably, this model was supplied with precise dimensions (for instance, the distance between the two concentric domes—or between the Earth and the heavens—was about 43 000 km).

According to the later Hun Tian cosmology, the heavens formed a system of celestial spheres, essentially a cosmological model of the type that had been developed by Greek astronomers. A book from the first century AD says that the world is like a hen's egg, with the central Earth like the yolk of the egg. Of more interest is the somewhat later Xuan Ye school, according to which the celestial bodies floated freely around in an infinite space. The world view of this school included a kind of physical cosmology with a certain affinity to the ideas of the Greek atomists. A description of the Xuan Ye school from the early fourth century AD gives this account:

> They said that the heavens were entirely empty and void of substance. When we look up at them, we can see they are immensely high and far away, without any bounds. . . . The Sun, Moon, and company of stars float freely in empty space, moving or standing still, and all of them are nothing but condensed vapour. . . . The speed of the luminaries [the Sun, Moon, and the five brighter planets] depends on their individual natures, which shows they are not attached to anything, for if they were fastened to the body of heaven, this could not be so.[8]

1.1.2 *Cosmogonies and theogonies*

As we have seen, the ancient cosmologies were primarily concerned with how the world and its inhabitants (gods and humans) came into existence. They were cosmogonies and, because of the crucial role played by the gods, at the same time theogonies. Most of the cosmologies share some common features, among them a starting point in some undifferentiated, perhaps chaotic state which separates into two or three opposites, such as the Earth, air, and sky. A variant of this theme is included in the opening verse of Genesis:

In the beginning, when God created the universe, the Earth was formless and desolate. . . . Then God commanded, 'Let there be light'—and light appeared. . . . Then He separated the light from the darkness . . . Then God commanded, 'Let the water below the sky come together in one place, so that the land will appear'—and it was done. He named the land 'Earth' and the water which had come together he named 'Sea'.

However, this scenario differs from most other cosmogonies not only in its monotheism, but also in the absence of any primordial state. The initial elements, such as 'darkness' and 'water', are not pre-existing and active principles, but God's creations. It should also be noted that Genesis does not speak of nothingness as the 'state' out of which God created the universe. In later Christianity, *creatio ex nihilo* would become a dogma, but the notion is not mentioned explicitly in the Bible; it was only introduced in the second half of the second century by the church, which wanted to emphasize in this way God's absolute sovereignty.[9]

Dating from roughly the same time as the Book of Genesis, Hesiod's *Theogony* offers the Greek version of how the world and the gods came into existence. Unsurprisingly, there are considerable similarities with the cosmogonies known from Mesotopamia and Egypt. Although the *Theogony* is mainly a cosmogony, it also comprises a picture of the structure of the universe, however crude. The Earth (flat, of course) is surrounded by a river or ocean, and above the Earth there is a hemispherical heaven, separated from the Earth by a gap which is bright by day and dark by night; beneath the Earth, dark and gloomy Tartaros is located, an underworld symmetric to heaven. We are informed about the size of the universe, as measured by the distance from the Earth to Tartaros, in the following way: 'A bronze anvil falling nine days and nights from heaven would reach Earth on the tenth. And a bronze anvil falling nine days and nights from Earth would reach Tartaros on the tenth.' Hesiod's account of the earliest gods is, at the same time, an account of how the main components of the physical world were created. Contrary to the Jewish creation story, there is no creator, but an initial state of chaos, which, without any cause or explanation, develops into a heaven and a sky. First night (Nyx) and darkness (Erebos) are produced, and then day (Hemera) and bright air (Aither):

First of all Chaos came into being. Next came broad-breasted Gaia [Earth],. . . and murky Tartaros in a recess of the broad-roaded earth, and Eros . . . From Chaos there came into being Erebos and black night. From night, Aither and Hemera came into being . . . Gaia first brought forth starry Ouranos [heaven] equal to herself, to cover her all about in order to be a secure dwelling place forever for the blessed gods.[10]

Note that the original chaos has not always been there, but 'came into being'. Hesiod's text provides no answer to the question of what the chaos emerged from or how it happened. Historians believe that in the *Theogony* Chaos does not refer to a formless and structureless fluidum, as in the later meaning of the term, but to the gap between the Earth and the sky.

Apart from giving an account of the origin of the universe, many of the ancient cosmo-mythologies included accounts of the end of the world, typically by some cosmic catastrophe represented by a huge battle between the good and evil forces of nature. The catastrophe did not necessarily imply an absolute end of the world, though, for some of the cosmogonies were recurring cosmogonies, with a new world arising out of the ashes of the old. In some cultures, notably in India, the process was thought to go on endlessly, with an eternal change between creative and destructive phases. This is the archetypical conception of the cyclical universe, an idea which has fascinated humans throughout history and can be found in mythical as well as scientific cosmologies right up to the present.

The Hindus had a predilection for large numbers, which they used in elaborating cosmic cycles of vast proportions. A single cycle of the universe, called a *mahayuga*, consisted of 12 000 divine years, each of a duration of 360 solar years, thus totalling 4.32 million years. Two thousand such cosmic cycles made up one day of Brahma, called a *kalpa*. The life of Brahma, corresponding to 100 Brahma years, was also the lifetime of the lower part of the universe, 311 trillion (3.11×10^{14}) years—or 23 000 times the age of the modern Big Bang universe. And Hindu cosmology operated with even larger numbers.

Cyclical conceptions of the universe can be found in many other civilizations. Although there was no unique Greek idea of time, the notion of cyclical time, or a cyclical universe, was entertained by several Greek philosophers. According to a Greek historian, writing about 40 BC, there were two competing views with respect to time and the universe: 'One school, premising that the cosmos is ungenerated and indestructable, declares that the human race has always existed, and there was no time when it began to reproduce itself. The other holds that the cosmos has been generated and may be destroyed, and that men similarly first came into existence at a definite time.' [11]

1.1.3 *Ionian natural philosophy*

Historians sometime speak of the period between 600 BC and 450 BC as 'the first scientific revolution'. By this grand name they refer to the emergence of a group of Greek (or Ionian) thinkers who initiated a paradigmatic change in humanity's understanding of the natural world: they approached nature in a new way, asked different questions than previously, and provided different kinds of answers. These Ionians and Milesians were were not only philosophers, they were also natural philosophers. They believed that the world could be understood rationally or, rather, naturalistically, that it could become the subject of human reasoning. The Olympian gods were still there, but they were no longer held responsible for natural phenomena. The Ionian philosophers, also known as the Presocratics, thought of the world as a cosmos, a structure of matter and forces bound together by law-like connections into a harmonious whole. It followed that they endeavoured to explain natural phenomena as instances of general patterns of explanation, not as individual phenomena, each with its own explanation.

According to tradition, the first of the natural philosophers—the first 'physicist' if one likes—was Thales of Miletus. He allegedly predicted a solar eclipse in 585 BC, although this is undoubtedly more myth than historical reality. This wise man had a high reputation among the Greeks, who said that he thought hard about how to explain celestial phenomena. Perhaps he thought too hard, for Aristotle reports that once while Thales was studying the heavens, he fell into a well. 'A clever and delightful Thracian serving-girl is said to have made fun of him, since he was eager to know the things in the heavens but failed to notice what was in front of him and right next to his feet.'[12]

A Milesian follower of Thales, Anaximander, postulated an eternal and spatially unlimited principle or medium, an indefinite something called *apeiron*, out of which the present world order grew by a process of separation. He wanted to explain how the diversity of the world had emerged out of the undifferentiated and indeterminate *apeiron*; characteristically for the new spirit of enquiry, he refrained from invoking the intervention of the gods. Anaximander's explanation may appear obscure and unconvincing, but his question—how can the formation of a complex world out of an originally simple state be understood?—would remain central to cosmological thinking. Indeed, it is still a central question.

Anaximander also speculated about the structure of the world, including its dimensions, and again he avoided mixing his cosmology with mythology. He assumed that the shape of the Earth was cylindrical ('like a stone column'), with the height of the cylinder being one-third of its breadth. Humans and other inhabitants of the Earth would occupy one of the plane surfaces. As to the size of the Sun and its distance from the Earth, 'Anaximander says that the Sun is equal to the Earth, and the circle . . . on which it is carried is 27 times the size of the Earth.' He further held that the Earth is at the centre of the universe, and gave a kind of symmetry argument to the effect that the Earth therefore had to be immobile (for why should a central body move in one direction rather than any other?). It is not clear if Anaximander, in saying that 'the Sun is equal to the Earth', also implied that the two celestial bodies had the same physical composition. But Anaxagoras, a later philosopher in the Ionian tradition, did believe as much, since he claimed that the Sun, far from being divine, was just a hot stone. He likewise surmised that the Moon was Earth-like, with mountains, plains, and ravines. Because of his heretical view, he was prosecuted and exiled from Athens, where he lived. Anaxagoras adopted the flat Earth, but his explanation of why the Earth stays aloft in the middle of the universe (rather than falling down) differed from that of Anaximander. According to Anaxagoras, the Earth was supported by air, which he described as an ocean upon which the Earth rested.

Among the Presocratic philosophers should also be mentioned Empedocles, born around 490 BC, who was the first to suggest that all matter consisted of four basic and unchanging elements, namely earth, water, air, and fire. Because Aristotle adopted his view, it came to serve as the foundation of matter theory, alchemy, and much else for a period of nearly two thousand years. Empedocles stated that originally the elements were mixed, but eventually some vortex mechanism caused a separation of them, first separating off the air and next the fire. 'He declares that the Moon was formed separately out of the air that was cut off by the fire.' As to the Sun, he seems to have believed that it was either a vast aggregation of fire or a reflection of fire. Empedocles realized that the Moon was not a luminous body, but that it reflected the light from the Sun, and also that a solar eclipse occurred when the Moon stood between the Earth and the Sun. Like other natural philosophers, he came up with an explanation for the immobility of the Earth. According to Aristotle, Empedocles explained the stable, circular motion of the stars and planets by their great velocity. 'For when the cup [filled with water] is whirled in a circle, the water, whose natural movement is downward, does not fall down, even though it is often underneath the bronze.' Empedocles seems to have believed that the swift rotation of the heavens prevented the Earth from moving.

Empedocles' cosmos, materially consisting of the four elements, was governed by two gods or motive forces called 'Love' and 'Strife'. Since the elements had always existed, there was no need to explain how they originally came into existence. Depending on the influence of either of the two cosmic forces, the universe alternated in a cyclical pattern.[13] Thus, when Love dominated, the elements were mixed up into a uniform mass; and at the time of Strife's complete dominance, they were fully separated from one another and arranged in concentric spheres. Only in between the two extremes was the universe hospitable to life-generating processes, as we experience them. The changes between dominance by Love and Strife proceeded eternally, corresponding to continual creations and destructions of the world. However, the two forces were not simply creative and destructive, for the conditions of life demanded a certain balance between them. The cycles were symmetric, so that the events in one phase were repeated in the opposite phase, but in reverse time order (a process from birth to death will be followed by one from death to birth). The periods of the cycles would be very long, but Empedocles did not specify their length.

During the Presocratic period, the emphasis was upon explaining what was known, whereas there was little interest in extending the empirical basis by means of new observations. Moreover, the explanations that the Presocratic philosophers came up with were crude analogies of a purely qualitative nature. Indeed, from a later perspective the explanations of Anaximander, Anaxagoras, Empedocles, and their kindred spirits appear primitive and speculative. But what matters is not their answers, but their questions and the conditions they posited for acceptable explanations.

1.1.4 *Pythagoreans and atomists*

Whereas Pythagoras is a somewhat shadowy figure who left nothing in writing to posterity, the philosophical school he founded in southern Italy was influential throughout antiquity.[14] The early Pythagoreans formed a secret religious fraternity and they continued to emphasize religious and mystical aspects of their philosophy rather than scientific aspects. Nonetheless, their thoughts came to exert a strong influence on early Greek science. In the present context, we only need to draw attention to their original idea of associating numbers with material substances, an idea which pointed the way to a mathematization of physics and cosmology. It is not very clear what the Pythagoreans meant by relating numbers to things (but then, presumably, it was not meant to be clear). Some of them apparently claimed that things *are* numbers, which clearly is an implausible claim; others may have meant, less implausibly, that material objects resemble numbers and that physical phenomena can be explained by numbers.

The Pythagoreans were aware of the five regular polyhedra, also known as the Platonic bodies, and claimed that the element earth was made from the cube, fire from the tetrahedron, air from the octahedron, and water from the icosahedron; the fifth of the regular bodies, the dodecahedron, they associated with the whole of the cosmos, which they believed was spherical and limited in extent. They were among the first to adopt a spherical Earth, a conceptual innovation which dates from around 430 BC. Even more remarkably, some of the Pythagorean thinkers removed the Earth from its privileged position in the centre of the universe. According to Philolaus, one of Pythagoras' successors in Italy, the central place was occupied by a fire—'the guard of Zeus'—around which rotated the planets and the stars. It is to be noted that Philolaus' cosmos was not heliocentric, as he did not identify the central fire with the Sun, which he took to revolve around the centre. Furthermore, he postulated a dark 'counter-Earth' which moved opposite to the real Earth and with the same period of revolution. The Earth described a circle around the central fire, which was, however, invisible to us because humans only lived on the side of the Earth that was turned away from the centre of revolution. Another Pythagorean, Ecphantus (who may or may not have been a real person), was said to have maintained that the Earth performed a daily rotation around its axis from west to east.

The reason for the introduction of the counter-Earth was numerological, not astronomical. According to Aristotle, the Pythagoreans held that the number 10 was perfect, and for this reason they maintained that there must be 10 celestial bodies. Taking the Earth, the Moon, the Sun, the planets, and the sphere of the fixed stars they counted nine, and by including the counter-Earth they got the right number. The order of the bodies was as mentioned, with the counter-Earth innermost and followed by the Earth, the Moon, etc. Aristotle was not impressed by Pythagorean cosmology, which he found to be speculative and unrelated to observations. As he wrote in *De caelo*, 'They are not inquiring for theories and causes with a view to the phenomena, but are forcing the phenomena to fit certain theories and opinions of their own, and trying to bring them into line.'[15] All the same, some 2000 years later Copernicus would refer to Philolaus' pyrocentric world model for support of the idea that the Earth is a circularly moving planet.

According to Aristotle, the atomistic school of natural philosophy was founded by Leucippus, a philosopher possibly from Miletus. However, atomism is usually associated with the better-known Democritus from Abdera in Thrace, a contemporary of Socrates and with the reputation of being a prolific author. Leucippus may have been a pupil of Zeno and somewhat older than Democritus. Both of the founders of atomistic natural philosophy are rather shadowy figures, known only through the works of later authors.

The basic idea of ancient atomism was the postulate that all that truly exists in the world is atoms, indivisible and invisible particles which move incessantly in an unlimited void, a cosmic vacuum. Whereas the atoms are being, the void is non-being. Although Democritean atomism is often represented as a monistic theory, it operated with non-being as well as being, and the non-being void was ascribed an ontological status somewhat similar to that of the being atoms. This is what lies behind Democritus' paradoxical statement that 'nothingness exists'. The atoms were uniform in substance and differed only in size and shape. Because there were an infinite number of shapes, there were an infinite number of different atoms too. Material objects were formed by chance congregations of atoms, which first resulted in compounds or, anachronistically, 'molecules'. The process might also give rise to a vortical motion with larger and slower objects tending toward the middle, whereas smaller and faster objects tended toward the periphery. Out of such vortices entire worlds might originate. The general idea of ancient atomism was to explain the complexity of the phenomenal world solely in terms of atoms moving in a void, to reduce observed qualities and changes to changes in the relative position of atoms which were themselves qualityless and eternal.

Atomistic philosophy included a particular cosmological view in which a distinction was made between the infinite world at large and world systems within it, sub-universes, which were limited in space and time. Our cosmos was just one out of an infinite number of roughly similar systems, some larger and some smaller; like the other world systems, ours had come into being and would one day perish. 'There are an infinite number of universes [*kosmoi*] of different sizes. In some there is no Sun and Moon. In some the Sun and Moon are larger than ours and in others there are more. . . . Some are growing, some are at their

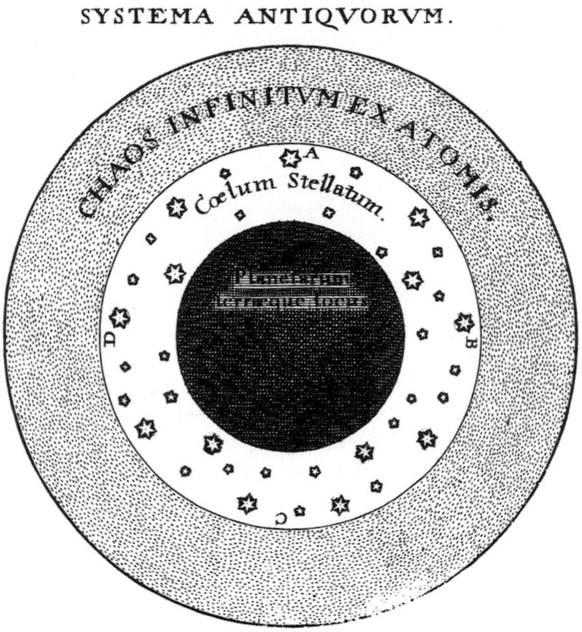

Fig. 1.4 Democritus' atomistic universe as depicted in a book published in England in 1675. The dark central area is made up of the Earth and the planets, surrounded by a thick stellar sphere. Outside the stars is the infinite chaos of randomly moving atoms. Although pictured as a shell, it is supposed to have no outer limit. From Heninger 1977, p. 193.

peak, and some are declining, and here one is coming into being, there one is ceasing to be. They are destroyed when they collide with one another.'[16]

As to the arrangement of the celestial bodies, Democritus placed the Moon nearest the Earth, then the Sun followed, and outside it the fixed stars; the planets were said to 'have different heights'. Leucippus believed that the Sun was farthest away. The two philosophers agreed that the Earth was at the centre of our universe, while for the world at large there was not, of course, any central place. Contrary to the Pythagoreans, Democritus did not accept a spherical Earth, but suggested that it had an oval shape with a length one and one-half times its width.

In the atomists' conception of the universe, there was no room for design, purpose, or divine agency. All that existed were material atoms moving randomly in a void. This does not mean that the atomists denied the existence of the gods, but they did deny that the gods had anything to do with natural processes. Some four hundred years after Democritus, the Roman poet Titus Lucretius Carus wrote his famous text *De rerum natura*, in which he presented his own version of atomism. Although this version derived more from Epicurus than from Democritus, in general it agreed with ancient atomist cosmology. Here is Lucretius' description of the cosmos:

All that exists, therefore, I affirm, is bounded in no direction; for, if it were bounded, it must have some extremity; but it appears that there cannot be an extremity of any thing, unless there be something beyond, which may limit it . . . Now, since it must be confessed that there is nothing beyond the WHOLE, the whole has no extremity; nor does it matter at what part of it you stand, with a view to being distant from its boundary; inasmuch as, whatever place any one occupies, he leaves the WHOLE just as much boundless in every direction.

Having argued in this way for an infinite universe, Lucretius proceeded with arguing for an infinity of inhabited worlds:

Further, when abundance of matter is ready, and space is at hand, and when no object or cause hinders or delays, things must necessarily be generated and brought into being. And now, if there is such a vast multitude of seminal-atoms as the whole age of all living creatures would not suffice to number, and if there remains the same force and nature,. . . you must necessarily suppose that there are other orbs of earth in other regions of space, and various races of men and generations of beasts.

Lucretius further explained that although the cosmos is infinite in space, it is of finite age and 'there will be an end to the heaven and the Earth'. He based his argument on the shortness of human history, which he found to be inexplicable if the world had always existed:

If there was no origin of the heavens and Earth from generation, and if they existed from all eternity, how is it that other poets, before the time of the Theban war, and the destruction of Troy, have not also sung of other exploits of the inhabitants of Earth? How have the actions of so many men thus from time to time fallen into oblivion? . . . But, as I am of opinion, the whole of the world is of comparatively modern date, and recent in its origin; and had its beginning but a short time ago.

Not only did the universe have a beginning, it was also decaying, on its way to an end. Lucretius spoke of a cosmic deterioration, a theme which can be followed throughout the history of cosmological thought. 'The walls of the great world, being assailed around, shall suffer decay, and fall into mouldering ruins. . . . It is vain to believe that this frame of the world will last for ever.'[17] As has become clear, the atomist cosmology followed the trend in Presocratic natural philosophy in being grand and speculative. It included many visions, including the bold proposal of many worlds, that are still considered interesting by modern cosmologists.

1.2 The Greek cosmos

During the centuries after 400 BC, natural philosophy partly transformed into science. For the first time Greek thinkers focused on observations of nature and attempted to construct explanations or models that agreed quantitatively with the observations. In no area was the new kind of science pursued with more vigour and success than in astronomy. Yet, as the science of the heavens became more mathematical and better founded in observational data—in short, more scientific—the more narrow did it become. Whereas interest in cosmology and cosmogony had flourished among the Presocratic philosophers, such speculations declined drastically in the long period between Plato and Ptolemy.

Two points are worth emphasizing. First, cosmogony, in the strict meaning of the term, practically came to a halt. Scientists and natural philosophers rarely addressed questions concerning the origin of the universe or how it had developed into its present state. From Aristotle onwards, most astronomers tacitly assumed that the world had always existed and that it would continue to do so into an indefinite future. Of course, granted this assumption, there was no room for cosmogony. The second point I want to mention is that the meaning of 'the universe' (or 'cosmos') changed. It was still everything physical in the world, but in astronomical practice the universe tended to be identified with the seven planets encircling the Earth. Although the fixed stars belonged to the universe too, there was little that astronomers could do about them except to count and classify them. (The first classification into magnitudes was due to Hipparchus, who divided the stars into six classes with the most luminous belonging to magnitude 1, and the least luminous belonging to magnitude 6.) The narrower view and the emphasis on mathematical models meant that cosmology became peripheral to the astronomers' research programme, a state of affair that was to continue throughout the Middle Ages and the Renaissance.

This is not to say that cosmology vanished from the scene of Greek science, only that it was given little priority and, when it was cultivated, appeared in different forms than previously. Among the more interesting cosmological theories in the period were those of Aristotle, Aristarchus, and Ptolemy. Most astronomers preferred to leave cosmology to the philosophers, and here we do find an interest in the subject along lines similar to those of the Presocratics. The Stoics, for example, were much interested in cosmological questions, but did not combine them to any extent with astronomical knowledge. To mention but one aspect of Stoic cosmology, they held a cyclical world view in which the formation and destruction of the cosmos was associated with thermal phenomena. The world was a gigantic sphere oscillating through cycles of expansion and contraction in the void surrounding it. Chrysippus, a leader of the Stoic school in Athens in the third century BC, is said to have believed that 'after the conflagration of the cosmos everything will again come to be in numerical order, until every specific quality too will return to its original state, just as it was before and came to be in that cosmos.'[18]

1.2.1 *Aristotle's world picture*

Although Plato discussed astronomical issues in several of his writings, his attitude was idealistic in the sense that he denied the epistemic value of observations. The cosmos could be comprehended mathematically, by pure thought, whereas empirical investigations would only obscure the truth; they would at most lead to a 'likely story' of the real world. In the *Republic*, he insisted that astronomy should be pursued as if it was geometry. 'We shall dispense with the starry heavens, if we propose to obtain a real knowledge of astronomy,' he wrote.

All the same, according to tradition Plato was the first to state what soon became the basic problem of astronomy and an approach to this science of huge importance. According to Simplicius' *Commentary on Aristotle's De Caelo*, a work written in the early part of the sixth century AD, Plato suggested that the business of the astronomers was to reduce the apparent motions of the planets (including the Sun and the Moon) to uniform, circular motions—to 'save the phenomena'. It is now believed that the demand for uniformity and circularity of celestial motions was a later innovation, which cannot be found in Plato and to which he did not subscribe.[19] The principle was to shape the paradigm that would dominate astronomy and cosmology until the time of Kepler, over a period of two thousand years. Whatever Plato's priority, it was a pupil of his who first answered the challenge, that is, who first proposed a single system which accounted for the observed motions of the planets in terms of circular orbits.

Eudoxus of Cnidos had for a short period stayed with Plato at his Academy in Athens, and later in life he constructed a system of revolving concentric spheres which accounted for many of the observed features of the heavens.[20] None of Eudoxus' writings have survived, but the basic content of his world model is known from later writers, Aristotle and Simplicius in particular.[21] Eudoxus considered each of the heavenly bodies as a point on the surface of one of several interconnected spheres, which were all concentric—or 'homocentric'—with the Earth at the centre. He imagined the spheres to turn around different axes and with different speeds, but in accordance with Plato's paradigm he only allowed uniform revolutions. In the case of the five planets, he made use of four spheres, the outer one of which represented a motion around the Earth with a period of 24 hours. For the Sun and the Moon, he postulated three spheres.

Among the irregular motions that had to be explained was the fact that some of the planets appeared to reverse their motion and then, after some time, continue their regular course towards the east. Such retrograde motion was considered most undignified for a heavenly, divine body, and hence something that had to be explained as apparent only. This Eudoxus' model succeeded in doing, if only in a qualitative and incomplete way, and it also largely accounted for another disturbing irregularity, the planets' variation in latitude. Because the model had only two parameters that could be varied, one corresponding to the speeds of revolution and the other to the inclination of the spheres, it was, however, unable to give the right motions of the planets.

In his *Introduction to Astronomy*, a work from around 70 BC, the Stoic philosopher Geminus gave an excellent exposition of the research programme adopted by Eudoxus and his followers. 'Their view was that, in regard of divine and eternal beings, a supposition of such disorder as that these bodies should move now more quickly and now more slowly, or should even stop, as in what are called the stations of the planets, is inadmissible.' Interestingly, Geminus drew an analogy to the social norms of his time:

Even in the human sphere such irregularity is incompatible with the orderly procedure of a gentleman. And even if the crude necessities of life often impose upon men occasions of haste and loitering, it is not to be

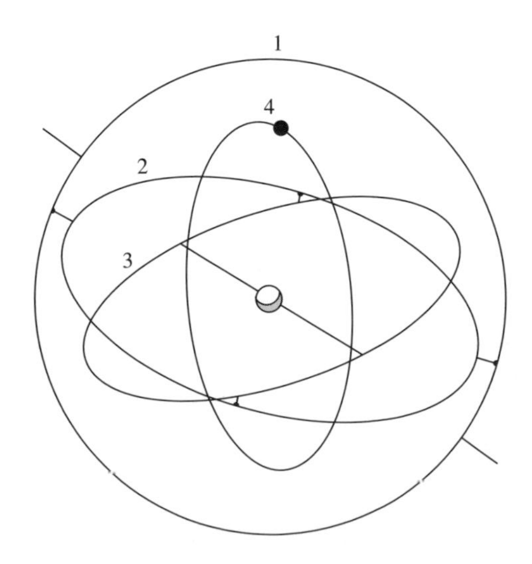

Fig. 1.5 Planetary mechanism based on Eudoxus' model with four concentric spheres.

supposed that such occasions inhere in the incorruptible nature of the stars [planets]. For this reason they defined their problem as the explanation of the phenomena on the hypothesis of circular and uniform motion.[22]

In a commentary on Aristotle, Geminus, as quoted by Simplicius, further spelled out the difference between physics and astronomy, as these disciplines were conceived in Greek antiquity:

It is the business of physical inquiry to consider the substance of the heaven and the stars, their force and quality, their coming into being and their destruction, nay, it is in a position even to prove the facts about their size, shape, and arrangement; astronomy, on the other hand, does not attempt to speak of anything of this kind, but proves the arrangement of the heavenly bodies by considerations based on the view that the heaven is a real κόσμος [kosmos], and further, it tells us of the shapes and sizes and distances of the Earth, Sun, and Moon, and of eclipses and conjunctions of the stars, as well as of the quality and extent of their movements.[23]

This was a distinction that would last for about eighteen centuries and have a crucial impact on the histories of astronomy and cosmology. Whereas Eudoxus needed 26 spheres to account for the workings of the heavens, Callipus of Cyzicus, a near-contemporary of Aristotle, refined the model by adding seven more spheres (one each for Venus, Mars, and Mercury, and two each for the Sun and the Moon). Eudoxus and Callipus seem to have conceived their world models purely geometrically and the celestial spheres to be merely theoretical entities.

The homocentric model adopted by Aristotle was a modification of the models of Eudoxus and Callipus, but at the same time it marked an important change in the research programme in that Aristotle introduced a physical perspective. His spheres were corporeal, not mathematical constructs, and his planets and stars were physical bodies attached to a series of interconnected rotating shells. This made him propose a mechanism to explain *why* the bodies moved as they did. According to Aristotle, the spheres of an outer planet were physically connected with those of an inner planet, a model which forced him to introduce some countermeasures in order to reproduce the observed motions. In his *Metaphysics*, he wrote: 'If all the spheres combined are to give an account of the phenomena, then for each planet there must be other spheres . . . which counteract and restore to the same position the first sphere of the innerlying planet, for only in this way will the whole system produce the required motion of the planets.'[24] There clearly was a cost to Aristotle's physicalization of the cosmos, namely a drastic increase in complexity. No fewer than 55 spheres were now needed, 22 of them introduced to restore the independence of the seven planetary systems.

Aristotle's great innovation was to provide a physical model of the actual heavens in agreement not only with the postulate of uniform circular motion but also with the general principles of his natural philosophy. This connection was a leading theme in his famous treatise on the heavens, known by its Latin title *De caelo*. Perhaps the most important feature in Aristotle's cosmos was that it was a two-region universe, as he drew a sharp distinction between the sublunar and the superlunar world. The first region, covering the Earth and the air up to the Moon, was composed of bodies made up of the four Empedoclean elements with their natural motions, which were rectilinear, either towards the centre of the Earth (earth and water) or away from it (air and fire). Beyond the Moon, the celestial bodies moved naturally in eternal, uniform circular motions, without being subject to the terrestrial laws of physics. The stars, planets, and celestial spheres were composed of an entirely different kind of matter, an ethereal, divine substance or fifth element, *quinta essentia* in Latin. Unlike the matter of the sublunar world, the heavenly ether was pure and incorruptible. Whether in the sublunar or superlunar region, a void could not possibly exist, and hence the universe was a plenum.

Aristotle's cosmos enjoyed general respect in the ancient world, but it was not beyond criticism. Xenarchus of Seleuchia, who was a contemporary of Cicero, wrote a treatise entitled *Against the Fifth Substance*, in which he challenged two of Aristotle's basic notions, the existence of a fifth element and the circular motion of the celestial bodies. Among his arguments against the heavenly ether was that the hypothetical substance was superfluous. He denied that a simple or perfect body by its nature would follow a circular path, as claimed by Aristotle and most other astronomers. For, as Xenarchus argued, in circular motion those parts nearer to the centre move with a smaller linear velocity than those nearer to the periphery, whereas a simple body must necessarily have the property that all its parts move with the same velocity.

Although Aristotle held that the Earth was located at the centre of the universe, this was in a geometrical sense only. Contrary to the Pythagoreans, he saw no reason to identify the geometric centre with the true or 'natural' centre of the universe, understood in a physical and ontological sense. On the contrary, in *De caelo* he suggested that this more elevated status belonged to the sphere of the fixed stars, from where motion was transmitted to the interior parts of the world. That which contains is more precious than that which is contained, he wrote. Thus, one may say that Aristotle operated with two centres of heavenly motion, an idea which was taken over into the medieval conception of the universe. Not only was the stellar sphere of a nobler nature than the corruptible Earth, it was also the origin of universal time and closer to the unmoving prime mover (corresponding to God).

Based as it was on Eudoxus' homocentric model, Aristotle's system shared most of its weaknesses, the most serious of which was its inability to account for the variations in brightness shown by some of the planets. It was well known that the brightness of Venus and Mars varied considerably during their course, which is easily explained if their distances from the Earth change. However, it followed from the premises of the homocentric system that the planets must always be at a constant distance from the Earth. This and other problems were pointed out by Autolychos only a generation after Eudoxus and later also by Simplicius, who quoted Sosigenes, a contemporary of Julius Caesar. 'Nevertheless the theories of Eudoxus and his followers fail to save the phenomena', Sosigenes is said to have said. The inability to explain the variable brightness was the main reason why the homocentric model, whether in the version of Eudoxus or of Aristotle, did not survive for long.

Aristotle did not only establish a kind of physical astronomy, he was also much concerned with the greater questions of cosmology. One of these questions related to the temporal aspect of the world. Had it once come into existence? Would it come to an end? In his famous dialogue *Timaeus*, Plato discussed these questions, although in a form far away from a scientific discourse. According to Plato, the world had come to be, it was created. He pictured the creation as made by a 'demiurge', a divine craftsman who first made the soul of the cosmos and subsequently its body, the two fitting perfectly. Moreover, Plato made clear that there could be only one world, not many images of the ideal world. He stated that the stars and planets were divine and in perpetual motion, and 'whatever is in perpetual motion is immortal'. The creation of the world was not *ex nihilo*, for the demiurge made the cosmos as a copy of an eternal and divine original, a kind of pre-existing universe-idea.[25] Since Plato formulated his creation story as a myth, one should be careful not to read into it later ideas of cosmic creation, whether in a theological or a scientific sense. Most modern interpreters warn that *Timaeus* should be read metaphorically rather than literally.

At any rate, Aristotle disagreed with his former teacher and vehemently denied that the universe was created and also that it was spatially infinite. On the contrary, he argued that the universe as a whole was ungenerated as well as indestructible, in short eternal. A spatially infinite world was impossible, for by its very nature the world revolved in a circle, and Aristotle argued that such motion was impossible for an infinite body as it would lead to an infinite velocity. This conclusion would not hold true in a universe consisting of a finite material cosmos surrounded by an infinite void; but such a picture (which was adopted by some Stoic philosophers) ran counter to Aristotle's notion of space as volume filled with matter. According to Aristotelian natural philosophy, a large empty space was ruled out by definition. What was enclosed by the outermost sphere included everything. In opposition to some earlier philosophers, Aristotle maintained that the universe was unique, eternal, and all-inclusive:

> The world in its entirety is made up of the whole sum of available matter . . . and we may conclude that there is not now a plurality of worlds, nor has there been, nor could there be. This world is one, solitary and complete. It is clear in addition that there is neither place nor void nor time beyond the heaven; for (*a*) in all place there is a possibility of the presence of body, (*b*) void is defined as that which, although at present not containing body, can contain it, (*c*) time is the number of motion, and without natural body there cannot be motion.[26]

As to the central body, the Earth, Aristotle argued that it was spherical and immobile, neither of which claims was controversial. Although the celestial spheres would move naturally, Aristotle introduced in his *Physics* an 'unmoved mover', a spiritual something at the outermost part of the universe which he conceived as the ultimate source of all celestial movement. However, he did not develop the topic, nor did he provide any explanation of how the transmission of movement took place. In his *De caelo*, Aristotle referred briefly and somewhat cryptically to the question of the Earth's axial rotation 'as is stated in the *Timaeus*'. This passage has been discussed endlessly, from Plutarch in antiquity, through Thomas Aquinas in the Middle Ages, to scholars in the twentieth century. Did Plato really assume a rotating Earth? It is pretty certain that he did not, for other reasons, because such a notion would have been wholly inconsistent with his astronomical system. Plato shared the standard view of the Earth sitting motionless in the centre of the universe.

Aristotle's assumptions about a finite and eternal cosmos, and his denial of a vacuum, were not generally accepted in ancient Greece and Rome. For example, they were opposed by the Stoics and Epicureans, who not only returned to Presocratic ideas of cosmic evolution but also operated with versions of an infinite universe. As we have seen, Lucretius' exposition of cosmology in *De rerum natura* was most un-Aristotelian.

The Stoic school, which included Chrysippus and later Poseidonius as prominent members, developed a cosmology where the element fire was essential and was seen as the source of the other three elements. They agreed with Aristotle that there could be no void within the material world, but not that an extra-cosmic void was impossible. On the contrary, they supposed that 'beyond the cosmos there stretches an infinite, non-physical world'. Stoic philosophers pictured the universe as slowly pulsating, performing cycles of condensation and rarefaction. An extra-cosmic void would not cause matter to dissipate into the void, as Aristotelians argued, for 'the material world preserves itself by an immense force, alternately contracting and expanding into the void following its physical transformations, at one time consumed by fire, at another beginning again the creation of the cosmos'.[27]

The problem of the eternity of the world (or the Earth) remained a matter of dispute, especially among Stoic philosophers, who objected to Aristotle's thesis with empirical arguments based on the observed surface of the Earth. They reasoned that erosion is a unidirectional process and if it had been at work for an infinite time, all mountains and valleys would by now have been planed down; they clearly are not, and hence the Earth must have existed only over a limited span of time. This argument against the eternity of the world was developed by the Stoic philosopher Zeno of Citium around 300 BC and reported by Theophrastus as follows:

> If the Earth had no beginning in which it came into being, no part of it would still be seen to be elevated above the rest. The mountains would now all be quite low, the hills all on a level with the plain . . . As it is, the constant unevenness and the great multitude of mountains with their vast heights soaring to heaven are indications that the Earth is not from everlasting.[28]

This is the first time we meet a theme that would come to occupy a prominent position in cosmological thinking more than two thousand years later: there exist in nature unidirectional processes—whether given by erosion, radioactivity or entropy increase—that speak against an eternal world (see Section 2.4). Faced with the Stoics' argument, proponents of Aristotelian physics postulated that corruptive geological processes were counteracted by generative processes, but they were unable to provide a satisfactory account, based on Aristotle's matter theory, of how these compensating processes operated.

1.2.2 *Aristarchus and the dimensions of the universe*

It has always been an important task of astronomers and cosmologists to determine distances in the universe, from the surface of the Earth to objects as far away as possible. It is also one of the most difficult tasks.[29] How big was the universe of the ancient Greeks? Nobody knew, for there were no ways in which the distances to the stars and the planets (except the Sun and the Moon) could be measured. In fact, not even the order of the planets could be unambiguously determined, except that the sphere of the fixed stars was obviously the farthest away from the Earth, and the Moon was the closest. Yet the Greeks were not totally at a loss and they did make some progress in determining cosmic distances, if only in the neighbourhood of the Earth.[30]

Alexandria and Syene (now Aswan) in southern Egypt are located roughly on the same meridian. In the third century BC, Eratosthenes, director of the famous library in Alexandria, estimated the distance between the two cities to be 5000 stades. Assuming that the Sun was sufficiently distant that its rays could be treated as if they were parallel, he concluded from a simple measurement that the circumference of the Earth was close to 250 000 stades. We do not know the value of the stade he used, but if one stade equals 157.7 m, as often assumed, the result corresponds to 39 370 km, in excellent agreement with later determinations. However, the numerical agreement may to some extent have been fortuitous and should not be given much weight. What matters is that from the time of Eratosthenes the order of magnitude of the size of the Earth was known and generally accepted.

Aristarchus of Samos, Eratosthenes' senior by some 40 years, was an accomplished mathematician and astronomer. In his only extant writing, *On the Sizes and Distances of the Sun and Moon*, he undertook to establish the relative distances of the Sun and Moon from the Earth and also to determine the sizes of the Sun and Moon.[31] His main method was to measure the angle between the directions from the Earth pointing towards the Moon and the Sun at the moment when the Moon was observed to be exactly half illuminated (Fig. 1.6). He found the value 87° and, from lunar-eclipse observations, which he used to determine the sizes of the Sun and Moon, he found that the Moon's apparent diameter was 2°. Here, in the words of Aristarchus, is what he concluded:

1. The distance of the Sun from the Earth is greater than eighteen times, but less than twenty times, the distance of the Moon [from the Earth].
2. The diameter of the Sun has the same ratio [as aforesaid] to the diameter of the Moon.
3. The diameter of the Sun has to the diameter of the Earth a ratio greater than that which 19 has to 3, but less than that which 43 has to 6.[32]

Aristarchus' conclusions were wide of the mark. The reason was errors in his two basic data values, which should have been 89°50′ and ½° rather than 87° and 2°. His method was clever and correct, but his results wrong; or, as a historian has expressed it, it was 'a geometric success but a scientific failure'.[33]

As a result of his wrong data, Aristarchus obtained values that were much too small, especially for the Earth–Sun distance, where his result was wrong by a factor of no less than 65 (Table 1.1). Nonetheless, his methods were sound, and a refined use of them later led Hipparchus to a much better value of the distance between the Earth and the Moon (the distance to the Sun was also much improved, if still off the mark by a factor of 9.5).

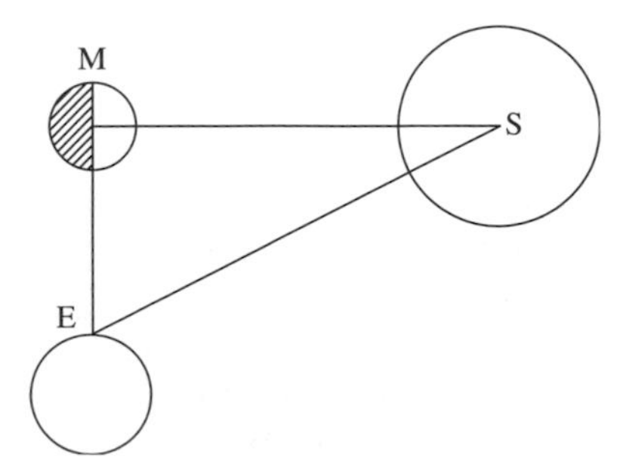

Fig. 1.6 Aristarchus' method for determining the relative distances of the Sun and the Moon. An observer on the Earth E sees the Moon M when it is in its first quarter and measures the angle MES, from which the ratio EM : ES follows.

Table 1.1 Ancient values of mean distances and sizes of the Moon and the Sun expressed in diameters of the Earth. Adapted from Heath 1959, p. 350.

	Moon, distance from Earth	Moon, diameter	Sun, distance from Earth	Sun, diameter
Aristarchus	9.5	0.36	180	6.8
Hipparchus	33.7	0.33	1 245	12.3
Poseidonius	26.2	0.16	6 545	39.3
Ptolemy	29.5	0.29	605	5.5
Modern values	30.1	0.27	11 728	109.1

Aristarchus is today best known for having proposed a heliocentric system, for which reason he is sometims referred to as 'the Copernicus of Antiquity'. Although Copernicus knew about Aristarchus' world system, he did not refer to it in *De revolutionibus*. Apparently the Polish reformer of astronomy did not think highly of his Greek predecessor, whose ideas did not influence him to any extent.[34] Aristarchus' original text no longer exists, but Archimedes gave a brief account of it in a fascinating work known as *The Sandreckoner*:

Now you are aware that 'universe' is the name given by most astronomers to the sphere whose centre is the centre of the Earth and whose radius is equal to the straight line between the centre of the Sun and the centre of the Earth. This is the common account, as you have heard from astronomers. But Aristarchus of Samos brought out a book consisting of some hypotheses, in which the premisses lead to the result that the universe is many times greater than that now so called. His hypotheses are that the fixed stars and the Sun remain unmoved, that the Earth revolves around the Sun in the circumference of a circle, the Sun lying in the middle of the orbit, and that the sphere of the fixed stars, situated about the same centre as the Sun, is so great that the circle in which he supposes the Earth to revolve bears such a proportion to the distance of the fixed stars as the centre of the sphere bears to its surface.[35]

Aristarchus' reform of the world picture was presumably rooted in his determinations of the relative sizes of the Moon and the Sun. The Moon revolved around the more bulky Earth, which had a volume about thirty times as large as its satellite. If the Sun was some 300 times larger than the Earth in volume, as he had found, it was natural to think of the Sun as the central body instead of the Earth.

Archimedes' interest in the matter was mathematical, not astronomical. That his work was not an attempt to obtain a correct figure for the size of the universe is illustrated by the fact that he intentionally overestimated the cosmic dimensions. For example, he took the Sun to be 30 times as large as the Moon, where Aristarchus had a value of 18–20; and for the circumference of the Earth he used the value 3 million stades, which he knew was much too large. What appealed to Archimedes was the enormous size that must be ascribed to Aristarchus' universe in order to account for the absence of an observed stellar parallax.

Was it possible to express a number greater than the number of sand grains needed to fill up the entire heliocentric universe? In order to solve this problem—clearly of mathematical interest only—Archimedes developed a number system which allowed him to express numbers of gigantic magnitude. His result was that 'a sphere of the size attributed by Aristarchus to the sphere of the fixed stars would contain a number of grains of sand less than 10,000,000 units of the eighth order of numbers'. The number referred to by Archimedes can be written in modern notation as 10^{63}, the first 'very large number' that appears in the history of science. Much later, such dimensionless numbers would become important in cosmology. There is a similarity, if more in spirit than in substance, between Archimedes' number and Eddington's cosmical number 10^{79}, which denotes the number of fundamental particles in the observable universe.[36]

Aristarchus' heliocentric system was not considered a serious rival to the geocentric models and soon went into oblivion. The only astronomer in antiquity who is known to have supported the idea was Seleucus, who lived about 150 BC. It was hard to see the advantages of a system that contradicted common sense and could only account for the absence of a parallax by placing the stellar sphere at a ridiculously far distance from the Earth. There is no indication that Aristarchus worked out the details of his hypothesis, for example that he developed a planetary theory on the basis of a moving Earth, such as Copernicus would do some eighteen centuries later. In addition, it may have added to the theory's lack of acceptability that it was accused of being impious 'for putting in motion the hearth of the universe'. This we know from Plutarch's *On the Face in the Orb of the Moon*, where there is a reference to charges raised against Aristarchus by Cleanthes, a Stoic philosopher. Yet, he also says that Aristarchus was a mathematician, not a physicist (or philosopher), and for this reason his hypothesis should not be taken too seriously. The distinction between the physicist's and the mathematician's view of the universe would later reappear in connection with Copernicus' world system and would in general constitute an important theme in the history of scientific cosmology.

We may get an impression of the cosmological views of the early Roman empire from Pliny the Elder's voluminous compilation *Historia naturalis*, a work consisting of 37 'books' and which exerted a great influence on late antiquity and the Middle Ages.[37] Astronomy, presented in a qualitative way in Book II, was but a small part of the erudite Roman's work, but it may have been representative of what non-astronomers knew and thought about cosmology at the time. Pliny rejected astrology and conceived the world (*mundus*) as 'sacred, eternal, immeasurable, wholly within the whole'. What may be outside it 'is not within the grasp of the human mind to guess'. Pliny was aware that some philosophers had made suggestions about the dimensions of the universe, but these he discarded as 'mere madness', a phrase he also used for attempts to investigate what lies outside the world. The general features of Pliny's universe were in agreement with Hellenistic cosmology in so far that he adopted a spherical, Earth-centred world with the fixed stars at its outer boundary. He had no doubt that the Earth was the central body of the universe, which he substantiated with 'irrefragable arguments' of which the most important was the equal hours of day and night. On the other hand, the Sun was not merely one planets among others, for,

> In the midst of these [planets] moves the Sun, whose magnitude and power are the greatest and who is the ruler not only of the seasons of the lands, but even of the stars themselves and of the heaven. . . .[The Sun is] the soul, and more precisely the mind, of the whole world, the supreme ruling principle and divinity of nature. He . . . lends his light to the rest of the stars also; he is glorious and pre-eminent, all-seeing and even all-hearing.

Pliny further accepted the doctrine of the four elements, arranged in such a way that the element fire was nearest the stars, followed by air, which was thought to exist throughout the universe. Of course, in between the immobile Earth and the revolving stellar sphere he placed the seven planets, taking their order to be the Moon, Mercury, Venus, the Sun, Mars, Jupiter, and Saturn. The Earth was spherical and kept in place by the force of air. It is uncertain if Pliny accepted the Aristotelian distinction between a sublunar, elementary world and a supralunar, ethereal region, as he wrote somewhat ambiguously on the matter. He did, however, agree with Aristotle that the universe was uncreated and eternal. Although he knew of the idea of a cyclical universe repeating itself eternally, he seems to have found the notion unattractive.

1.2.3 *Ptolemaic planetary astronomy*

The troubles that faced the homocentric models of Eudoxus and Aristotle were largely solved with the introduction of an alternative planetary model in the second century BC. It is believed that this alternative was first proposed by the Alexandrian mathematician Apollonius, who is especially known for his unified theory of conic sections, including the circle, parabola, ellipse, and hyperbola. As an astronomer, Apollonius investigated the motion of a planet revolving around a point displaced from the fixed Earth. This *eccentric* model is equivalent to a model in which the planet moves uniformly in a small circle (the *epicycle*), whose center revolves in a larger circle (the *deferent*) with the Earth at its centre. This combination of two circular motions could reproduce the observations of apparently non-circular and non-uniform celestial phenomena.

Apollonius' writings on astronomy have not survived, but his idea was developed by Hipparchus, who was the first to supply it with numerical parameters based on observations. With Hipparchus, the idea was turned into a geometrical model of epicycles and deferents that initiated a new chapter in the history of theoretical astronomy. What was most important was that Hipparchus' solar theory led him to conclude that all the fixed stars had small motions parallel to the ecliptic, a phenomenon known as the precession of the equinoxes. The value he gave for the precession was one degree per century or 36″ per year, which is in reasonable agreement with the true value of 50″ per year. The discovery of the precession turned out to be cosmologically important, as it led Ptolemy to conclude that the stellar sphere needed to be extended with yet another sphere. According to Ptolemy, the precession was due to the stellar sphere, but outside it there was a ninth sphere which caused the daily revolution. The ninth sphere was empty, yet it was the prime mover of the celestial revolutions. He described the two movements in the heavens as follows: 'One of them is that which carries everything from east to west: it rotates them with an unchanging and uniform motion . . . The other movement is that by which the spheres of the stars perform movements in the opposite sense to the first motion, about another pair of poles, which are different from those of the first rotation.'[38]

The zenith of ancient astronomy was reached in the second century AD with the famous *Almagest* by Claudius Ptolemy, an Alexandrian mathematician and astronomer who also wrote important texts on optics, astrology, and geography. The original title was *Megale syntaxis* ('Mathematical Compilation'), and in the Arabic world it became *al-majisti*, meaning 'the greatest', which in medieval Latin was rendered as *almagestum*. In his introduction to the *Almagest*, Ptolemy praised mathematical astronomy as the only science that could provide unshakeable knowledge and, at the same time, was morally uplifting: 'From the constancy, order, symmetry and calm which are associated with the divine, it makes its followers lovers of this divine beauty, accustoming them and reforming their natures, as it were, to a similar spiritual state.'[39] This theme would later play an important role in the Christian world, both in the Middle Ages and during the scientific revolution, but Ptolemy did not elaborate. The *Almagest*, structured in thirteen books, was a mathematically demanding, highly technical work, not a discourse on natural philosophy or cosmic theology.

Whereas Ptolemy adopted Hipparchus' solar theory, he offered a new and much improved theory of the five planets that agreed excellently with observations. His planetary theory was based on a sophisticated use of eccentrics, epicycles, and deferents that allowed him to explain, for example, retrograde motions and the limited elongations of Mercury and Venus (which never deviate from the Sun by more than 23° and 44°, respectively). In Ptolemy's theory, the centre of the epicycle did not move uniformly with respect to either the Earth or the centre of the deferent, but with respect to a point located at the opposite side of the centre and at an equal distance from it. This point is called the *equant*. With the use of the equant, Ptolemy was able to compute planetary positions accurately. On the other hand, it was a technical device that violated the philosophical doctrine of uniform motion and for this reason it later became controversial, first among Islamic astronomers and later in the medieval West. Ptolemy's world system differed technically from Aristotle's, yet it also had much in common with it. Thus, in the beginning of the *Almagest*, Ptolemy stated the physical premises of his theory in terms that Aristotle would have fully agreed with:

> The heaven is spherical in shape, and moves as a sphere; the Earth too is sensibly spherical in shape, when taken as a whole; in position it lies in the middle of the heavens very much like its centre; in size and distance it has the ratio of a point to the sphere of the fixed stars; and it has no motion from place to place.[40]

Not only did the Earth not move from place to place, it also did not rotate around its axis. Ptolemy was aware that the possibility had been discussed by 'certain people'—he most likely thought of Heracleides of Pontus—but he dismissed it as 'ridiculous' and 'unnatural' because it was contrary to experience. Although he recognized that an axial rotation might account for the celestial motions, he argued that it led to consequences incompatible with observations, such as clouds being left behind in a westward direction. Ptolemy's arguments against a daily rotation would later be reconsidered by philosophers in the Middle Ages and the Renaissance.

The *Almagest* marked the culmination of Greek astronomy, just as Euclid's *Elements* marked the culmination of geometry. However, it was essentially a mathematical theory of the planets revolving around the Earth, and for this reason the *Almagest* is of no particular cosmological significance. As far as cosmology is concerned, another and later of Ptolemy's works is of far greater interest, the *Planetary Hypotheses*.[41]

Ptolemy's physical cosmology was based on Aristotelian natural philosophy, including the doctrines of the five elements and their natural motions. He believed that the ether consisted of tiny spherical particles and that this was a physical argument in support of the sphericity and circular motion of the celestial bodies. Ptolemy agreed that there could be no void in the universe, which became the foundation of his cosmological theory as described in *Planetary Hypotheses*. He found the arrangement of nested planetary spheres he arrived at to be 'most plausible, for it is not conceivable that there be in nature a vacuum, or any meaningless and useless things'.[42] The basic principle of Ptolemy's theory was to arrange the shells of the celestial bodies one within another, with the thickness of each shell being determined by the eccentricity of the planet's deferent circle and the radius of its epicycle. The whole system was arranged in such a way that no empty space appeared between the

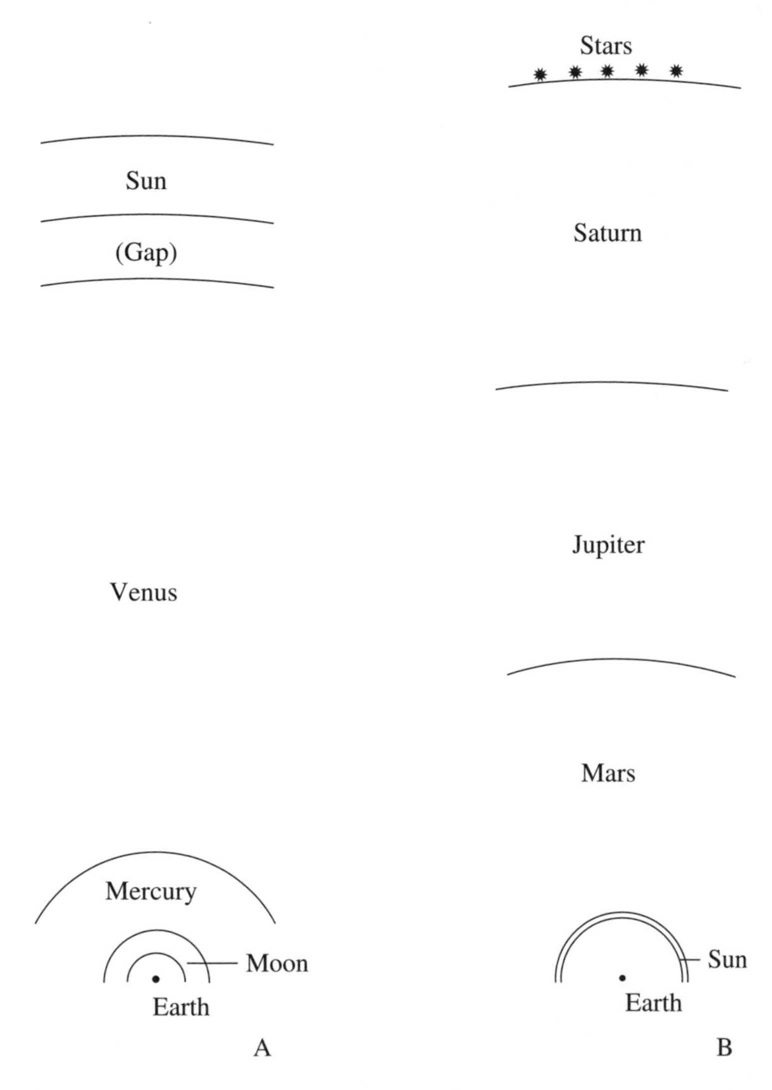

Fig. 1.7 Distances in Ptolemy's cosmology in *Planetary Hypotheses*, drawn to scale. The scale of the left part is 15 times as small as in the right part. Copyright © 1993, from *Encyclopedia of Cosmology* (Hetherington 1993). Reproduced by permission of Routledge/Taylor & Francis Group, LLC.

shells, meaning that the greatest distance of one planet was equal to the least distance of the planet outside it.

The general idea of incorporating epicycles and deferents into the Aristotelian model of nested spheres was anticipated by Theon of Smyrna, a philosopher who lived in the early part of the second century. Contrary to most other philosophers and astronomers, Theon was careful to distinguish apparent from real motions, and he emphasized the need to understand the heavens in physical terms. His epicycles and deferents were not mere mathematical tools, but had a real existence.

The space between the Earth and the Moon was filled with air and fire, and in the *Almagest* Ptolemy determined the Moon's distance from the Earth to vary between 33 and 64 Earth radii. The variation was much too great to fit with observations, which he must have known. However, it was not possible to determine the order of the other planets by means of astronomical data, and Ptolemy therefore had to rely on physical arguments of a somewhat arbitrary nature. Whatever the soundness of these arguments, he concluded in the *Planetary Hypotheses* that following the Moon's sphere there came, in this order, the spheres of Mercury, Venus, the Sun, Mars, Jupiter, and Saturn. The sphere of the fixed stars completed the system. Since the greatest distance of the Moon was 64 Earth radii, this must also be the least distance of Mercury. By means of the theory of epicycles and deferents, as developed in the *Almagest*, he found that the ratio of the least to the greatest distance for Mercury was 34:88, which implied that Mercury's greatest distance was $64 \times (88/34) = 166$ earth radii. The construction of the thicknesses of the remaining planetary shells followed the same procedure and resulted in a cosmological distance scale (Table 2), which Ptolemy summarized as follows:

> In short, taking the radius of the spherical surface of the Earth and the water as the unit, the radius of the spherical surface which surrounds the air and the fire is 33, the radius of the lunar sphere is 64, the radius of Mercury's sphere is 166, the radius of Venus' sphere is 1,079, the radius of the solar sphere is 1,260, the radius of Mars' sphere is 8,820, the radius of Jupiter's sphere is 14,187, and the radius of Saturn's sphere is 19,865.

It is to be noted that there is a gap of 81 Earth radii between the maximum distance of Venus and the minimum distance of the Sun. The gap was embarrassing, as it could not consist of void space. Ptolemy argued that it might be reduced by increasing the distance to the Moon slighly, but he nonetheless kept to his numbers, which, he stated, were inescapable.

Table 1.2 Ptolemy's cosmological distance scale. All numbers are in Earth radii

	Least distance	Greatest distance	Mean distance
Moon	33	64	48
Mercury	64	166	115
Venus	166	1 079	622.5
Sun	1 160	1 260	1 210
Mars	1 260	8 820	5 040
Jupiter	8 820	14 187	11 504
Saturn	14 187	19 865	17 026

Having found the cosmic distances expressed in Earth radii, Ptolemy converted them to stades: 'The boundary that separates the sphere of Saturn from the sphere of the fixed stars lies at a distance of 5 myriad myriad and 6,946 myriad stades and a third of a myriad stade.' Expressed in more familiar terms, the radius of Ptolemy's universe was about 570 million stades, or roughly 85 million kilometres. Like other Greek astronomers, he had nothing to say about the thickness of the sphere of the fixed stars. The stars were usually conceived to be at the same distance from the Earth, but it was realized that this was just an assumption with no justification in either theory or observation. For example, in his *Elements of Astronomy*, Geminus wrote that 'we must not suppose that all the stars lie on one surface, but rather that some of them are higher [i.e. more distant] and some lower [less distant]; it is only because our sight can only reach out to a certain equal distance that the difference in height is imperceptible to us.'[43]

Ptolemy went on to determine the sizes of the celestial bodies, which he did from estimates of their apparent diameters. He found that the Sun was the largest of the planets, with a diameter 5.5 times that of the Earth, followed by Jupiter (4.4) and Saturn (4.3). With a diameter of only 0.04 times that of the Earth, Mercury was the smallest planet.

Unlike the *Almagest*, Ptolemy's *Planetary Hypotheses* did not circulate widely. Its content was mostly known from other works, especially by Islamic astronomers. Thâbit ibn Quarra wrote in the ninth century a work that surveyed Ptolemy's cosmology and was partly based on the *Planetary Hypotheses*. Thâbit used Ptolemy's numbers, except that he changed the Sun's least distance to 1079 Earth radii in order to get rid of the gap between the spheres of Venus and the Sun. He kept the Sun's greatest distance (1260 Earth radii) and thus increased the thickness of the Sun's sphere. Such a change had astronomical consequences—it resulted in a solar eccentricity much larger than allowed by observations—but Thâbit chose to ignore these. The important thing was to fill the gap and thus avoid an embarrassing cosmic void.

1.3 Medieval cosmology

The highly developed Hellenistic science, such as that represented by Ptolemy, came to a halt in the late phase of the Roman empire. Since its language was Greek, it remained unknown to most learned people in the early Middle Ages, and it was only after the Greek literature was translated into Arabic that it eventually found its way to Latin-using medieval Europe.

For a long time the best known of the ancient cosmological works was Plato's *Timaeus*, most of which was translated into Latin by Chalcidius, who worked in either the fourth or the fifth century. With the translations in the twelfth century of Aristotle and Ptolemy, the European scene was ready for a change. For nearly four centuries, Aristotle's natural philosophy served as the basis of a stable and harmonious world picture which was strongly influenced by Christian thought. A form of Christianized Aristotelianism became the foundation of a cosmology that gained a paradigmatic status. The medieval cosmos was finite and geocentric, with the seven planets and the stellar sphere revolving around the immobile Earth; the celestial bodies moved with uniform speed in circles or spheres; whereas the terrestrial region was corruptible and made up of the four elements, the heavens constituted a changeless world made of a fifth element unknown on Earth; and, finally, the spheres surrounded one another contiguously, excluding all void or empty space.

The thirteenth and fourteenth centuries witnessed lively discussions of cosmological issues, many of them focusing on possible universes rather than the one actually existing. Could God have created a different universe, say one that violated the doctrines of Aristotelian physics? Could he have created many universes? Because God was omnipotent, his creative power was limited only by what is logically impossible. This kind of scholastic exercise, led to debates of great ingenuity and several remarkable ideas, but when it came to the real universe imagination was much more restrained and only very few scholars dared to question the standard cosmology.

One of the most notable features of the high Middle Ages was that the temporal dimension, which had been largely ignored in Greek cosmology, was brought back into focus. The Christian universe was created by God, which was generally taken to mean that the universe had only existed for a finite period of time. However, although cosmology was thus provided with a temporal marker, that was restricted to the act of creation; there was still no perspective of development. The absolute age of the universe—or of the Earth, a distinction was rarely made—was not an issue of great importance in the Middle Ages, but it was generally conceded that a reliable figure could be derived from Biblical chronology.[44] As early as the the late second century, Theophilus of Antioch concluded that creation had taken place in 5529 BC, and Augustine affirmed that this was of the right order of magnitude. During most of the medieval era it was accepted that the world had come into existence by a supernatural act about 6000 years ago, a belief that would persist until well into the eighteenth century.

1.3.1 *Athens or Jerusalem?*

What little was known about the universe in the early Middle Ages included the idea that it was created *in toto* in a supernatural act rather than shaped out of some pre-existing state of matter. It was a true *creatio ex nihilo*. Given that this is a fundamental doctrine of Christianity, and in view of the overwhelming impact of Christian thought on cosmology through a large part of history, it is not irrelevant to repeat that *creatio ex nihilo* is nowhere explicitly stated in the Bible, neither in the Old nor in the New Testament. It is a doctrine not to be found in the earliest form of Christianity, when the form of creation was rarely a matter of discussion. Only in the second half of the second century can the doctrine be found in its strict sense, as an ontological and theological statement that expresses the contingence of the creation and the omnipotence and absolute freedom of God.[45]

St Augustine went a step further by arguing that cosmic creation did not only mean that God caused the universe to exist, but also that creation was timeless and implied a continual existence of the world. He may have been the first to state that, paradoxically, the created universe has always existed. When the doctrine of creation out of nothing was first formulated, it quickly became accepted as almost self-evident. Church fathers of the third century, such as Tertullian, Hippolytes, and Origen, all found *creatio ex nihilo* to be a fundamental doctrine that must necessarily be true. When it was officially accepted by the fourth Lateran Council in 1215, it had been widely adopted for a millennium.

The early Middle Ages—roughly the period from 400 to 800—witnessed a drastic decline in science, including astronomy and cosmology. The new spiritual power, the Christian church, had no unified view of what little was still known about Greek science, but for a time it expressed strong hostility towards any form of natural philosophy which could not be derived from the Bible or otherwise be justified theologically. The astronomical knowledge of even the most learned of the church fathers was pitifully small. At least some of the Christian leaders flatly rejected the Greek conception of the world and supported a Biblical fundamentalism. 'What indeed has Athens to do with Jerusalem?' asked Tertullian. Not much, he thought: 'We want no curious disputation after possessing Christ Jesus, no inquisition after enjoying the gospel! With our faith, we desire no further belief.'[46]

The changed intellectual atmosphere in early Christianity is illustrated by the remarkable if short-lived return of a non-spherical Earth. According to Lactantius, a bishop who lived in the first half of the fourth century (and whose real name was Lucius Caecilius Firmianus), the sphericity of the Earth was a ridiculous as well as heretical belief. In his *Divinae institutiones*, he asked: 'Is there anyone as stupid as to believe that there are men whose footprints are higher than their heads? Or that things which lie straight out with us hang upside down there; that grains and trees grow downwards; that rain and snow and hail fall upwards upon the Earth?'[47]

Lactantius—'a poor mathematician' according to Copernicus—was not the only Christian who believed in a literal interpretation of Scripture. Some of the church leaders were flat-earthers, accepted the supracelestial waters, and denied the spherical shape of heaven.[48] They suggested that heaven was rather like a tent or the Tabernacle, a view they could easily find evidence for in the Holy Book, such as in Isaiah 40:22: 'It is he who sits above the circle of the Earth, and its inhabitants are like grasshoppers; who stretches out the heavens like a curtain, and spreads them like a tent to dwell in.' This was the opinion of Diodorus, a bishop of Tarsus in the fourth century. Most of the patristic writers were hostile to Hellenistic cosmology but did not attempt to replace it with a detailed cosmological system based on the Bible.

Such a system was what Cosmas Indicopleustes, a widely travelled Byzantine or Egyptian merchant of the sixth century, provided in his *Christian Topography*. Cosmas argued against

Fig. 1.8 Cosmas' universe. The left figure is a cross-section of the vaulted box containing the Earth or universe. The Sun moves round the large conical mountain. The figure to the right is Cosmas' diagram of the box with the mountain and four gulfs (from the left: Mediterranean, Red Sea, Persian Gulf, and Caspian Sea). From Cosmas 1897.

the spherical shape of the Earth, summarily rejected the epicyclic theory, and also declared it ridiculous to believe that the Earth was at the centre of the universe. Such views were 'absurdities contrary to nature, in opposition to scripture'. The Earth, an incredibly heavy body, must surely be at the bottom of the universe, he argued. It will come as no surprise that Cosmas believed that heaven was designed like the Tabernacle and that the only way to understand its construction was to pay close attention to the Mosaic writings.

Cosmas included in his Christian topography a figure of the civilized world, which he pictured as a vaulted box. Above was the vault of the sky, with the firmament between it and the ground. 'There is also the firmament which, in the middle, is bound together with the first heaven, and which, on its upper side, has the waters according to divine scripture itself.' The heavenly bodies did not revolve around the Earth, but were placed below the firmament and moved by angels. The Sun and the Moon disappeared each day behind a huge mountain, which to Cosmas explained the difference between night and day. The stars were not at immense distances, as the pagans held, but belonged to the aerial spaces together with the planets. For, 'How is it that many of the fixed stars are equal and like to the planet we call Mars, to which a lower sphere has been assigned, and how do we in like manner see not a few of them to be like the planet Jupiter?'[49]

It would be wrong, though, to believe that all early Christians were enemies of secular philosophy or fundamentalists of the same breed as Lactantius and Cosmas. In fact, the two were exceptions when it came to the non-spherical shape of the Earth. By far the most influential of the church fathers, St Augustine, was a learned man and much more moderate in his views. Augustine sometimes warned against natural philosophy, but in so far as it did not conflict with Scripture he was willing to take it seriously, if for no other reason because it might in some cases help in Biblical exegesis. As far as astronomy was concerned, he did not reject the spherical Earth, although he did not endorse it either. He had no doubt about the water above the firmament—after all, there was solid Scriptural evidence for it. As to the Aristotelian idea of an element particular to the heavens, the ether, he rejected it. That Augustine was not simply antiscientific may also be judged from Galileo's *Letter to Grand Duchess Christina*, where Galileo quoted Augustine extensively in support of the view concerning science and faith favoured by himself. One of the quotations reads:

> What is it to me [Augustine] whether heaven, like a sphere, surrounds the Earth on all sides as a mass balanced in the centre of the universe, or whether like a dish it merely covers and overcasts the Earth?. . . Hence, let it be said briefly, touching the form of the heaven, that our authors [of the Bible] knew the truth but the Holy Spirit did not desire that men should learn things that are useful to no one for salvation.[50]

It was only in the seventh century that a new scientific literature began to appear, and even then it relied heavily on earlier, mostly Roman authors. Writers such as Martianus Capella and Ambrosius Macrobius, who lived in the early fifth century, preserved the rudiments of Greek astronomy, such as the distinction between the planets and the fixed stars, and the spherical, Earth-centred universe. But this was about all that was left from the glorious past. As two historians of science have expressed it, 'Compared to the sophistication of the *Almagest*, knowledge of astronomy among the Latins in the second half of the first millennium was primitive in the extreme.'[51]

The ideas of John Philoponus, a philosopher from Alexandria who lived in the sixth century, were, however, far from primitive. A Christian strongly influenced by Neoplatonism, Philoponus criticized Aristotle's natural philosophy and sought to replace it with a system in harmony with monotheism. Thus he attacked the traditional doctrines of the eternity of the world and the essential difference between the terrestrial and heavenly parts of the world. According to Philoponus, heaven and Earth were made of the same elements, created by God but with no divine qualities. The light from the stars did not differ from light from terrestrial sources, a most un-Aristotelian view: 'There is much difference among the stars in magnitude, colour, and brightness; and I think the reason for this is to be found in nothing else than the composition of the matter of which the stars are constructed. . . . Terrestrial fires lit for human purposes also differ according to the fuel, be it oil or pitch, reed, papyrus, or different kinds of wood, either humid or in a dry state.'[52]

Since God had created the world out of nothing, it must have a finite age, contrary to what Aristotle had taught. Philoponus did not rest content with basing his conclusion on the authority of the Bible, but proved, to his own satisfaction, by means of *reductio ad absurdum* arguments, that an eternal universe would lead to absurdities. For example, the celestial bodies move with different periods, Saturn more slowly than Jupiter and much more slowly than the fixed stars. Now, if Saturn had revolved an infinity of times, Jupiter would have performed three times as many revolutions and the stars more than 10 000 times an infinite number of revolutions! This Philoponus thought was an impossible notion, and 'Thus necessarily the revolution of the heavenly bodies must have a beginning.'

Bishop Isidore of Seville, who lived around 600, was the author of a large encyclopedia in twenty books, *Libri etymologiarum*, which included many references to scientific subjects. Contrary to most other authors, he drew a sharp distinction between astronomy and astrology, rejecting prognostic astrology as superstition. In a smaller work, *De natura rerum*, Isidore compiled contemporary knowledge of the Earth and heaven. His Earth was a flat disc, and outside the firmament he assumed a watery heaven in accordance with Genesis. 'The sphere of heaven is a certain form, spherical in shape,' he wrote:

> Its center is the Earth and it is shut in equally on all sides. They say that the sphere has neither beginning nor end; since it is round like a circle its beginning and end cannot readily be seen. . . . Heaven has two gates, east and west, for the Sun issues from one and retires into the other. . . . The rising Sun follows a southerly path, and after it comes to the west and has dipped into the ocean it passes by unknown ways beneath the Earth and again returns to the east.[53]

The Venerable Bede, an English monk living a generation after Isidore, had an impressive mastery of conventional learning. He wrote a work on calendars which enjoyed a high reputation throughout the Middle Ages, and he was also the author of a cosmological treatise, again titled *De natura rerum*, which to a large degree relied on Pliny. Contrary to some of his predecessors, Bede had no problem with the spherical Earth, and he stated that the Sun was much larger than the Earth (he still stuck to the idea of water above the heaven). Bede was neither a scientist nor an innovative thinker, but he did provide some continuity through a difficult period. In a commentary on *De natura rerum* from the ninth century, the anonymous commentator made the interesting suggestion that whereas Mars, Jupiter, and Saturn revolved around the Earth, Venus and Mercury were satellites to the Sun. This kind of geo-heliocentric system was known in ancient Greece and was often ascribed to Heracleides of Pontus, a pupil of Plato. It bears some similarity to the world system devised by Tycho Brahe in the late sixteenth century.

By 900, astronomy and cosmology in the Christian West were still at a low ebb. The problem was not so much Scriptural fundamentalism, or the tension between Jerusalem or Athens, but rather that most of the products of Athens (and Alexandria) were unknown or only known in highly diluted versions from secondary sources. Only when the masterpieces of Greek philosophy and science became available in Latin versions could progress start anew.

1.3.2 *Aristotelianism revived*

The revival of learning in Western Europe relied crucially on translations of Greek scientific texts. As far as cosmology was concerned, Latin translations began to appear around 1150, and after a century or so almost the entire corpus of Greek astronomy and cosmology was available to European natural philosophers. Some of the works were translated directly from Greek to Latin, but most were based on Arabic books and commentaries. Spain, where Arabic and Islamic culture flourished, became the centre of the new translation movement. For example, this is where the best known of the translators, Gerard of Cremona, worked. The industrious Gerard produced translations from Arabic to Latin of Euclid's *Elements* and Aristotle's treatises on natural philosophy, including *De caelo*; but his greatest service to the revival of Greek science was probably his direct translation from the Greek of Ptolemy's *Almagest*, which he completed in 1175 (the first Arabic translation had appeared more than three hundred years earlier). Although the main texts of Greek astronomy and cosmology had been translated by the closing years of the twelfth century, it took another half-century until Aristotelian and Ptolemaic cosmology was generally known and made its impact on the teaching in the newly founded universities.

Even before the results of the translation movement became apparent, scholars produced texts of a cosmological orientation. These were influenced by Platonic and Neoplatonic thoughts, and of course also by Christian theology, whereas Aristotelian philosophy was of limited significance only. Scholars such as Thierry of Chartres, William of Conches, and Adelard of Bath, who all were active in the first part of the twelfth century, advocated a naturalistic approach to the study of nature. They conceived nature as an autonomous entity which proceeded in accordance with its own laws or inherent order. God had of course created the universe, but all of what happened after the creation was a result of natural causation. This view implied that it was the task of philosophers to find natural explanations and to have recourse to divine intervention only if such explanations should utterly fail. The message in William of Conches' *Philosophia mundi* was that the cosmos could be studied scientifically and that such a study of secondary causes would only affirm the glory of the undisputed primary cause, God.

Bernard Sylvester, who lived in the mid twelfth century, wrote a large treatise, *Cosmographia*, structured in two books, which included a good deal of natural philosophy. The first book (*Macrocosmus*) dealt with the creation of the world in a way that differed considerably from the account in Genesis. Bernard started 'before the beginning' with *Hyle*, a primeval and formless substance, the origin of which he did not explain and which he may have thought of as unoriginated. Out of the chaotic *Hyle*, the elements were shaped and order was introduced in the universe in a process closer to re-creation than to creation in the traditional Christian sense. In Bernard's poetic creation account, which shared some of the features of pre-Christian cosmogonies, matter was seen as an active power.

Robert Grosseteste, the first chancellor of the University of Oxford and known in particular for his works on optics, wrote in the 1220s two cosmological treatises, *De luce* and *De motu corporali et luce*, in which he constructed a cosmology of light. The universe, he said, was originally created by God in the form of a point of light in a primeval, transparent, dimensionless form of matter; the light instantaneously propagated itself into an expanding sphere, thereby giving rise to spatial dimensions and eventually, by means of light emanating inwards from the expanding light sphere, to the celestial spheres of Aristotelian cosmology. Grosseteste described the essence of his cosmogony as follows:

> I hold that the first form of a body is . . . light (*lux*), which as it multiplies itself and expands without the body of matter moving with it, makes its passage instantaneously through the transparent medium and is not motion but a state of change. But, indeed, when light is expanding itself in different directions it is incorporated with matter, if the body of matter extends with it, and it makes a rarefaction or augmentation of matter . . . From this it is clear that corporeal motion is a multiplicative power of light, and this is a corporeal and natural appetite. [54]

Grosseteste's light-cosmogony was of course speculative, but it was a naturalistic explanation of the origin of the universe in so far as it did not rely on miracles or other divine intervention. And then the scenario has a curious, if of course superficial, similarity to modern accounts of the radiation-dominated expanding universe—inflation included!

During the first half of the thirteenth century, scholars became increasingly aware of the power of the Aristotelian thought system, with the result that Aristotle gradually replaced Plato as the authority in natural philosophy. The consequence was a world picture which was basically Aristotelian, but which included elements of the Ptolemaic system in the form of eccentrics, deferents, and epicycles.[55]

Everybody agreed that the spherical Earth was at the centre of the universe and that it was surrounded by seven planetary spheres in perfect contact. Outside the sphere of Saturn was the *primum mobile*, with the stars. However, to these eight spheres two or three more were usually added, mostly for theological reasons. The Bible speaks of the waters above the firmament, which had to be taken seriously; the general interpretation was that it referred to a 'crystalline' sphere above the stars consisting of water in either fluid or hard form. This ninth sphere—but it could also be two spheres, a ninth and a tenth—was starless and perfectly transparent. Some scholars added yet another sphere, an immobile 'empyrean heaven', the ultimate container of the universe and the abode of the angels. There was some discussion of whether the celestial spheres were fluid or solid, but from around 1300 a solid or crystalline theory was commonly adopted. As to the celestial spheres and the bodies revolving along with them, it was generally assumed that they were made of some incorruptible, perfect, unalterable substance, which in most cases was identified with Aristotle's ether or quintessential element. The stars and planets, assumed to be spherical like the Earth, did not differ physically from the orbs, as they were thought to consist of the same ethereal element, only in a much denser form. Most scholars believed that the stars and planets received their light from the Sun, but a few argued that they were self-luminous bodies.

Whatever the opinions on these questions, it was agreed that the celestial spheres were three-dimensional. They were endowed with thickness and arranged in such a way that the convex surface of one sphere was equal to the concave surface of the sphere following it. In this way, gaps in the heavens and problems with celestial voids were avoided. The model also made it possible to calculate the dimensions of the cosmos, very much along the lines that Ptolemy had used in his *Planetary Hypotheses*. Campanus of Novara, who flourished around 1260, may not have known about Ptolemy's work but his calculations nonetheless led to a universe strikingly similar to that of the Alexandrian mathematician (Table 1.3). According to Campanus' *Theorica planetarum*, the inner surface of the Moon's sphere was about 108 thousand miles, and the outer surface about 209 thousand miles. At the farthest end of the universe, Saturn was located between 52 and 73 million miles away from the centre of the Earth. Since the sphere of the fixed stars was assigned no thickness, Campanus' universe was a huge sphere of radius 73 million miles, of the same magnitude as Ptolemy's. Also like Ptolemy, Campanus believed he could calculate the sizes of the planets.

The picture of the medieval universe as outlined here, was basically qualitative and of more interest to the philosophers than to the astronomers. Astronomy was predominantly a mathematical science aimed at calculating the positions of planets and stars, and for this purpose cosmological problems such as the nature of the celestial substance were not of great relevance. The attitude of many medieval astronomers, if by no means all, was instrumentalistic. Was astronomy to provide a true representation of celestial phenomena or merely mathematical models that saved the phenomena? There was no unified position on this point during the Middle Ages. Moses Maimonides, the Jewish–Spanish philosopher of the late twelfth century, was in favour of an instrumentalist position. Concerning astronomy, he wrote:

> The object of that science is to suppose as a hypothesis an arrangement that renders it possible for the motion of the star [planet] to be uniform and circular . . . and to have the inferences necessarily following from the assumption of that motion agree with what is observed. At the same time the astronomer seeks, as much as possible, to diminish motions and the number of spheres.[56]

According to Maimonides, it was only God who knew the true reality of the heavens. Man could not possibly know this truth, and could only devise models that accounted as well as

Table 1.3 Cosmic dimensions according to Campanus of Novara. All figures are in miles.

	Least distance	Greatest distance	Thickness of sphere	Diameter of planet
Moon	107 936	209 198	101 261	1 896
Mercury	209 198	579 321	370 122	230
Venus	579 321	3 892 867	3 313 546	2 885
Sun	3 892 867	4 268 629	375 762	35 700
Mars	4 268 629	32 352 075	28 083 446	7 573
Jupiter	32 352 075	52 544 702	20 192 626	29 642
Saturn	52 544 702	73 387 747	20 843 044	29 209

possible for observed phenomena. Maimonides' position was not generally accepted, though, and most medieval natural philosophers denied that astronomy was merely model-making. In spite of different attitudes, astronomers realized that they were dealing with the same universe as the cosmologists and natural philosophers. As David Lindberg, a leading scholar of medieval science, has expressed it, 'astronomy and cosmology were not glaring at each other across a methodological chasm, but rubbing shoulders along a methodological continuum'.[57]

Islamic astronomers saw Ptolemy's *Almagest* in a different and more critical light than did their European colleagues. Ibn al-Haytam, who in Christian Europe was known as Alhazen, criticized the Ptolemaic system in about 1000 for being abstract geometry with no physical reality behind it. As Copernicus would do 500 years later, he objected to Ptolemy's use of the equant. The influential philosopher Averroes, or Muhammad ibn Rushd, later argued that although the deferent–epicycle theory might save the phenomena it was unsatisfactory. He, too, wanted a world system that made physical and not only mathematical sense. In a commentary on Aristotle, he wrote: 'The astronomer must, therefore, construct an astronomical system such that the celestial motions are yielded by it and that nothing that is from the standpoint of physics impossible is implied. . . . Ptolemy was unable to see astronomy on its true foundations. . . . The epicycle and the eccentric are impossible.'[58]

From the middle of the thirteenth century there appeared several books, usually with the title *Theoria planetarum*, which focused on planetary theory in the Ptolemaic tradition. They were mathematical in orientation and aimed at producing astronomical tables and calculating positions of the planets. *Tractatus de sphaera*, written by Johannes de Sacrobosco (John of Holywood), was an elementary and highly successful textbook which outlined the Aristotelian world picture but included only the most rudimentary planetary theory. As to the nature of the heavens, Sacrobosco wrote:

> Around the elementary region there is the ethereal, which is lucid and immune from all variation in its unchanging essence, and which turns in a circular sense with a continuous motion. It is called the 'fifth essence' by philosophers. Of this there are nine spheres . . . namely of the Moon, Mercury, Venus, the Sun, Mars, Jupiter, Saturn, the fixed stars, and the final heaven. Each of these spheres encloses the one below spherically.[59]

Sacrobosco's *De sphaera* was used as a textbook for nearly three hundred years.

We meet a different kind of cosmology in some of the literary masterpieces of the medieval world, such as Dante Alighieri's *Divina commedia* and Geoffrey Chaucer's *Canterbury Tales*.[60] In Dante's *Divina commedia*, written between 1306 and 1321, the reader is presented with a simplified Aristotelian cosmos consisting of the seven planetary spheres, an immense sphere of the fixed stars (the *stellatum*), and a starless *primum mobile*. When Dante and his beloved Beatrice enter this outermost sphere he notes with surprise that it is so uniform that he cannot say where he entered it. Dante believed in the actual existence of the crystalline spheres made up of 'rounded ether', but he had ten spheres rather than Aristotle's nine. The tenth was, however, non-physical, endowed with neither dimensions nor extension. It was the empyrean heaven, the mind of God himself and a kind of paradise where the souls of the blessed were found. Dante described the speed of revolution of the *primum mobile* as incomprehensible, a result of the desire of each part of this

Fig. 1.9 The medieval Christian universe in a folk version, with the empyreum surrounding ten heavenly spheres. Illustration from Petrus Apianus's *Cosmographicum liber* of 1533.

sphere to conjoin with the divine empyreum. This 'heaven of pure light' had no limits and was not located in space. In the later *Il convivio*, he described the empyreum as 'the sovereign edifice of the world, in which all the world is enclosed, and beyond which is naught; and it exists not in space, but received form only in the Primal Mind, which the Greeks call Protonoe'.[61]

1.3.3 *Scholastic controversies*

Many theologians welcomed Aristotelianism if only it could be presented in a decent, Christianized version; but they were also aware of its dangers and the incompatibility of Aristotelian philosophy and certain Christian doctrines such as God's creation of the world. Around 1270, the faculty of arts in Paris housed a group of radical thinkers who were willing to carry Aristotle's rationalism and naturalism as far as possible, even to the point where it conflicted with religious dogma. Siger of Brabant and Boethius of Dacia were the most prominent of the group. Inspired by Averroes, they argued that it is the task of the philosopher to investigate every question that can be disputed on rational grounds; the arguments should be followed to their logical conclusion, without regard for the true faith. From the church's point of view, this was a deeply troubling position that had to be opposed. Action came in 1270, when the Bishop of Paris, Etienne Tempier, issued a list of 13 propositions which were declared false and heretical. Apparently this was not enough, for seven years later the list was greatly expanded, now covering 219 articles. To defend any of these propositions, many of which related to the opinions of Siger and other radical Aristotelians, could lead to excommunication.[62] The views of the radical Aristotelians (or Averroists) were condemned not only in Paris, but also in England, where the Archbishop of Canterbury issued condemnations in 1284 and again in 1286.

More than 20 of the propositions condemned by Tempier referred to cosmology; for example, it was an error to claim the following:

6. That when all celestial bodies have returned to the same point—which will happen in 36,000 years—the same effects now in operation will be repeated.
34. That the first cause [God] could not make several worlds.
49. That God could not move the heavens [the world] with rectilinear motion; and the reason is that a vacuum would remain.
87. That the world is eternal as to all species contained in it; and that time is eternal, as are motion, matter, agent, and recipient . . .
185. That it is not true that something could be made from nothing, and also not true that it was made in the first creation.
201. That He who generates the whole world assumes a vacuum because place necessarily precedes what is generated in that place; therefore, before the generation of the world there was a located place which is a vacuum.

Let us now consider some of the cosmological questions that were discussed in the Middle Ages, irrespective of whether they were mentioned specifically in the condemnations. First, it was generally agreed that the world was spatially finite. The possibility of an infinite world was sometimes discussed, but only to reject it as absurd and incompatible with Aristotelian physics. For example, Jean Buridan, an important Parisian scholar of the middle of the fourteenth century, argued that an infinite body cannot possibly move with a circular motion; for to do so there must be a centre, and an infinite body cannot have a centre. In spite of consensus on this point, there remained the possibility of an infinite, non-material universe, a possibility that was often discussed (see below).

Much more difficult was the question of temporal finitude, where Aristotle's insistence on an eternal world clashed head-on with the fundamental dogma of a world created in time. No wonder that article 87 specifically condemned the eternity of time, motion, and matter. Siger of Brabant was convinced of the truth of Aristotle's arguments and was consequently led to conclude that the world was not created. This was of course a decidedly heretical conclusion, and Siger was careful to point out that it rested wholly on reason; since it conflicted with faith, in this case reason could not be relied on. Other great medieval scholars, such as Buridan and Nicole Oresme, expressed a similar opinion. Logically and naturally, heaven could not have come into being, nor could it be annihilated. Nonetheless, it *was* created a finite time ago, and only in a supernatural act, by the will of God.

In his *De aeternitatis mundi* from about 1270, Thomas Aquinas discussed whether something that had always existed could be made; only if this was logically impossible would he concede that God could not have created an eternal universe. He argued that creation, in its theological meaning, differs from the generation of change or processes such as that studied by the natural philosophers. *Creatio non est mutatio*. Creation is to give existence to things, to cause them. God does not take 'nothing' and transform it into something, he causes things to exist continually in the sense that 'if the created thing is left to itself, it would not exist, because it only has a being from the causality of the higher cause'.[63]

Thomas distinguished between a temporal beginning of the universe and its creation, where the latter concept refers to the existence of the universe as such. Even if the universe had always existed, it would still depend on God for its very being; it would be created. As a Christian, Thomas believed that Aristotle was wrong and that the universe was of finite age; as a philosopher, he was willing to concede that the universe was eternal. At any rate, the question could not be answered on the basis of reason alone. What mattered was that God had caused the universe to exist, and this involved no contradiction with either reason or faith. Another line of reasoning, adopted by Thomas and his contemporaries, was that Aristotle's argument for the eternity of the world was not a formal proof and was therefore not in need of formal rejection; it could be dismissed on the sole ground that it was contrary to faith.

The possibility of other worlds was eagerly discussed during the Middle Ages.[64] Aristotle had emphatically rejected the possibility, but almost all medieval philosophers agreed that God could have created other worlds, had he so wished. Yet they also agreed that in fact God had chosen to create only one world. Article 34 of the condemnation of 1277 demanded that the faithful had to concede that God could create other worlds, but not that such worlds actually existed. Nicole Oresme, a Parisian philosopher and mathematician, was one of several scholars who examined the question and tried to find weaknesses in Aristotle's conclusion that only one world was possible. Oresme, who had translated Aristotle's *De caelo* into French, distinguished between three different ways in which the plurality of worlds could be conceived:

> One way is that one world would follow another in succession of time, as certain ancient thinkers held . . . Another speculation can be offered which I should like to toy with as a mental exercise. This is the assumption that at one and the same time one world is inside another so that inside and beneath the circumference of this world there was another world similar but smaller. . . . The third manner of speculating about the possibility of several worlds is that one world could be entirely outside the other in an imagined space, as Anaxagoras held.[65]

After a lengthy analysis of the three possibilities, Oresme concluded that God in his omnipotence could make more worlds. 'But, of course, there has never been nor will there be more than one corporeal world.'

Related to both the question of the finitude of the universe and the question of other worlds, there was the question of whether or not the corporeal world was surrounded by an infinite void space, an idea with roots in ancient Greece (see Section 1.2). By and large, the favoured answer was—once again—that God could have created such a space, but that there was no reason to believe that he did. Buridan's conclusion represented the majority view: 'An infinite space existing supernaturally beyond the heavens or outside this world ought not to be assumed . . . Nevertheless, it must be conceded that beyond this world God could create a corporeal space and any whatever corporeal substances it pleases Him to create. But we ought not to assume that this is so [just] because of this.'[66]

The extracosmic void considered by the schoolmen was very different from the vacuum or non-being proposed by the Greek atomists. It was often conceived to be a spiritual heaven, God's abode, and therefore something which could not be confined to a finite world. Among God's many attributes were that he was omnipotent, transcendent, and infinite (in some non-spatial sense). It was sometimes suggested that an infinite world—or a finite material universe and an infinite void space—would be more consonant with God's power than a finite world. The eminent Oxford mathematician and natural philosopher Thomas Bradwardine identified infinite void space with God's immensity. Although his void space had neither extension nor dimensions, he nonetheless argued that it was real. Bradwardine followed Aristotle's arguments against a void a long way, but did not find them irrefutable. God could make a void anywhere he wished, within this world or outside it. 'Truly, even now, there is in fact an imaginary void place outside of the world, which I say is void of any body and of everything other than God.'[67] Oresme held a similar view.

It should be clear from this brief review that most of the cosmological problems discussed by medieval philosophers had very little to do with the business of the astronomers. The scholastic disputes about cosmology and cosmogony took place in a framework based on Christian theology and Aristotelian philosophy. What mattered was the delicate balance between these two pillars of insight, and in this context astronomical observations and calculations were of little or no relevance.

1.3.4 *New perspectives: Buridan to Cusanus*

The condemnations of 1277 helped create an intellectual climate where Aristotle's writings could be discussed more freely and critically. 'The philosopher' continued to be held in great esteem, but his system of natural philosophy was far from beyond criticism. We have an important example of this in the discussion of the Earth's immobility in the thirteenth century. Although no one drew the conclusion that the Earth actually moved, the arguments for a potentially moving Earth were impressive and demonstrated the willingness of some philosophers to depart from Aristotelian tradition.

Jean Buridan discussed the possibility of a daily rotation of the Earth around 1350. He pointed out that it was a problem of relative motion and that the motion of the stars could equally well be explained on this basis as on the traditional assumption that the stellar sphere revolved around the immobile Earth. In support of the hypothesis of a rotating Earth, he applied arguments based on the simplicity and economy of nature. 'Just as it is better to save the appearances through fewer causes than through many, if this is possible, so it is better to save [them] by an easier way than by one more difficult.'[68] Wasn't it more reasonable to assume that the relatively small Earth rotated with a fairly low speed than that the vast celestial spheres rotated with what must be an incredible speed? In addition to this argument, he added that rest was nobler than motion. As the noblest bodies, the stars therefore ought to be at rest, while the Earth, corruptible and ignoble as it was, ought to be in motion.

However, having presented his arguments in favour of a daily rotation of the Earth, Buridan started, in the spirit of dialectical thinking, to criticize them. He arrived at the conclusion that the Earth does not rotate after all. One of his counterarguments related to the strong wind that we would feel if the Earth rotated at high speed. He realized that supporters of the rotating Earth might 'respond that the Earth, the water, and the air in the lower region are moved simultaneously with diurnal motion,' but did not accept this explanation. At any rate, he adopted the conventional attitude that 'For astronomers, it is enough to assume a way of saving the phenomena, whether it is really so or not.' In the end, he kept to the orthodox Aristotelian view.

Buridan's discussion was further developed by his younger contemporary, Nicole Oresme, in *Le livre du ciel*, one of the classics of fourteenth-century natural philosophy. Here Oresme made the daring suggestion that the laws of terrestrial nature might be valid also for the celestial regions, a first step towards a dissolution of Aristotle's old distinction between the physics of the sublunar sphere and that of the spheres above the Moon. Also in opposition to Aristotle, but less controversially, he denied that the heavens were moved by intelligences (or angels). God had initially placed motive powers into the celestial bodies in such a way that no further application of power, whether animate or inanimate, was needed. Oresme may have been the first to use the metaphor of a clockwork that was later so famous when he wrote that 'the situation is much like that of a man making a clock and letting it run and continue its own motion by itself'.[69]

As far as the Earth's diurnal motion was concerned, Oresme basically discussed the same topics as Buridan, but in more detail and with greater sympathy for the hypothesis. He dismissed the problem of the wind that should constantly blow from the east by noting that the air would rotate along with the surface of the Earth. No experience, he emphasized, was able to dismiss the hypothesis of an axially moved Earth. Like Buridan, he considered the idea of a rotating Earth to be supported by reasons of simplicity as it avoided celestial speeds 'far beyond belief and estimation'. As another bonus, he mentioned that the hypothesis would do away with the generally assumed ninth sphere, which moved only with the diurnal motion:

> If we assume that the Earth moves as stated above, then the eighth heaven moves with a single slow motion and it is consequently unnecessary to imagine a ninth natural sphere invisible and starless; for God and nature would have made this ninth sphere for naught since by another method, i.e., assuming the Earth to move, everything can remain exactly as it is.[70]

Oresme referred to the passage in the Bible (Joshua 10:12–14) where God lengthened the day by commanding the Sun to stand still, and noted that the same dramatic effect could have been achieved much more easily by a temporary cessation of the Earth's rotation. Since God always acted in the most economic way, perhaps this was how he performed the miracle.

Yet Oresme decided that there were convincing theological reasons not to accept the rotating Earth. It was an interesting hypothesis, but not the way nature actually worked. In view of his impressive arguments in favour of a rotating Earth, Oresme's conclusion in *Le livre du ciel* was an anticlimax:

> However, everyone maintains, and I think myself, that the heavens do move and and not the Earth: For God hath established the world which shall not be moved, in spite of contrary reasons because they are clearly not conclusive persuassions. . . . What I have said by way of diversion of intellectual exercise can in this manner serve as a valuable means of refuting and checking those who would like to impugn our faith by argument.

It was one thing to go against Aristotle, quite another to question the authority of the Bible.

Nicholas of Cusa, also known as Cusanus, was a German cardinal and philosopher who wrote widely on a variety of subjects, including theology, mathematics, and natural philosophy.

He was fascinated by the concept of infinity, and in *De docta ignorantia* of 1440 he developed a metaphysical system (the doctrine of 'the coincidence of opposites') which he applied to cosmology, among other areas. The result was a number of bold claims that departed most radically from Aristotelian cosmology. However, it should be pointed out that the Renaissance philosopher Cusanus was essentially a Neoplatonist and Christian mystic, and that none of his arguments referred to empirical observations or were otherwise scientifically based. He stated that the cosmos had no fixed centre and no circumference as it was not bounded by any celestial sphere. His universe was 'relatively infinite' and homogeneous in the sense that any observer anywhere in the universe would observe essentially the same universe. There was no privileged place.

It is impossible for the world machine to have this sensible earth, air, fire, or anything else for a fixed and immovable centre. . . . And although the earth is not infinite, it cannot be conceived of as finite, since it lacks boundaries within which it is enclosed. . . . Therefore, just as the Earth is not the centre of the world, so the sphere of fixed stars is not its circumference. . . . Since it always appears to every observer, whether on the earth, the Sun, or another star, that one is, as if, at an immovable centre of things and that all else is being moved, one will always select different poles in relation to oneself, whether one is on the Sun, the Earth, the Moon, Mars, and so forth. Therefore, the world machine will have, one might say, its centre everywhere and its circumference nowhere, for its circumference and centre is God, who is everywhere and nowhere.[71]

And this was not all, for Cusanus also argued that the Earth was actually in motion. Moreover, he considered gravitation to be a local phenomenon such that each star or planet was a centre of its own gravitational attraction. Going even further than Oresme, he denied that there was any difference at all between celestial and sublunar matter; all celestial bodies, however noble, consisted of the same four elements as found on the Earth. Since there was life on the Earth, and the Earth was but a star, he assumed that there was life all over the universe. He even conjectured that the extraterrestrial beings differed in rank according to their location and that some of them, such as the 'bright and enlightened denizens' of the Sun, were superior to earthlings.

Cusanus' grand and bold cosmological vision anticipated some of the later developments in cosmology, in particular the cosmological principle, which is the claim that the universe is uniform on a large scale. But it should be kept in mind that Cusanus was no scientist and that his aim was not to devise a theory that could account for observable phenomena.

1.4 The Copernican revolution

And new Philosophy calls all in doubt,
The Element of fire is quite put out;
The Sun is lost, and th'earth, and no man's wit
Can well direct him where to looke for it.
And freely men confesse that this world's spent,
When in the Planets, and the Firmament
They seeke so many new; they see that this
Is crumbled out againe to his Atomies.
'Tis all in peeces, all cohaerence gone;
All just supply, and all Relation . . .[72]

This passage from John Donne's *An Anatomie of the World*, published in 1611, expresses a bewilderment and lack of orientation that many men of culture felt was the result of the doubts that natural philosophers raised against the traditional world picture. Foremost among these doubts was the controversial idea that the Earth, hitherto regarded as the immobile centre of the universe, was merely one planet among others, whirling around the Sun at great speed. With the disappearance of the immutable heavens, the comforting sense of order and unity had disappeared too. The revolution in astronomy seemed to confirm 'the frailty and the decay of this whole World'. Donne's better-known contemporary, William Shakespeare, related to the same theme in *Hamlet* II,2:

Doubt Thou the stars are fire
Doubt that the Sun doth move
Doubt truth to be a liar
But never doubt I love.

The controversial part of Copernicus' new world system, as many saw it, was not so much that it removed the Earth from its central position in the universe, for that was not necessarily a dignified position. After all, it was farthest away from the angels and God's eternal heaven. Indeed, it was sometimes argued that the natural place for the Earth, in both a physical and a moral sense, was 'the centre, which is the worst place, and at the greatest distance from those purer incorruptible bodies, the heavens'.[73] It was worse that the Earth had become reduced to a planet, which could be taken to imply that the other planets were inhabited by living and rational creatures as well. If so, the door was open for a host of theological problems.

1.4.1 *A heliocentric cosmology*

Nicolaus Copernicus, born in 1473 at Torun in the north of what is now Poland, received his elementary education at the Jagiellonian University of Cracow and subsequently went to Italy to study at Bologna and Padua. Although his primary field of study was canon law, he also took an interest in medicine and astronomy. In 1503 he returned to Poland, where he settled permanently in Frombork (or Frauenburg), a small town in an isolated corner of Varmia. There he engaged seriously in astronomical studies, the prime result of his studies being the daring hypothesis of a Sun-centred universe. It is unknown when he arrived at this idea, but around 1512 he wrote a brief sketch of the new astronomical system, known as the *Commentariolus*, which circulated in handwritten copies among a small number of scholars.[74] Copernicus had only a single disciple, Georg Rheticus, and it is in a work of his, the *Narratio prima* of 1540 (a second edition appeared in 1541), that we find the first published account of the Copernican system.

After many years of delay, Copernicus' masterpiece *De revolutionibus* was finally published in 1543, the very year of his death. Whatever the reason for the delay, it is most unlikely that it was caused by fear of how the Catholic church would react to the book. In fact, Cardinal Nicolaus von Schönberg had in 1536 urged Copernicus to publish his manuscript, although at the time to no avail. Copernicus knew that his theory might be considered to be theologically controversial, but in his preface to *De revolutionibus*, dedicated to Pope Paul III, he argued that it was not. Only 'by shamelessly distorting the sense of some passage in Holy Writ to suit their purpose' could certain people ignorant of mathematics

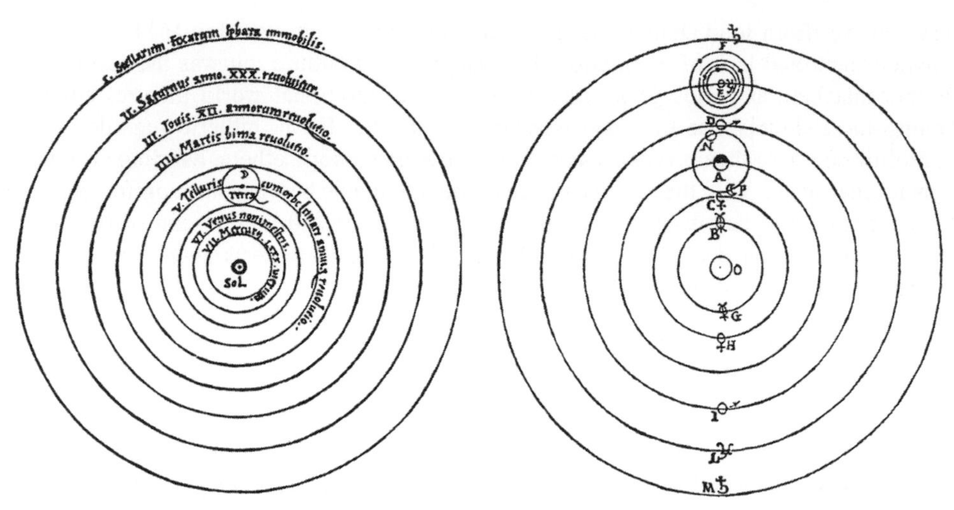

Fig. 1.10 Two historically important reproductions of the heliocentric world system. The picture on the left is from Copernicus' *De revolutionibus* (1543), with the sphere of the fixed stars completing the system. The version on the right is from Galileo's *Dialogo* (1632). The only difference between the two representations is that Galileo included the four moons he had observed moving round Jupiter.

find the work heretical.[75] Much like Ptolemy's *Almagest*, Copernicus' *De revolutionibus orbium coelestium libri sex* was thoroughly mathematical in nature and structure. Written in six books, it was a difficult and technical work, aimed at mathematically informed astronomers and at neither astrologers, philosophers, nor cosmologists. Indeed, Copernicus proudly emphasized that 'Mathematics is written for mathematicians' (*Mathemata mathematicis scribuntur*).

In the *Commentariolus*, Copernicus started by outlining in seven postulates the main features of his alternative to the traditional cosmology. The centre of the Earth was not the centre of the universe, a position which was instead occupied by the Sun. Whatever motion appeared in the firmament did not arise from it, but from the motion of the Earth, and the same was the case for the apparent motion of the Sun. Not only did the Earth rotate around its own axis, it also 'revolve[s] about the Sun like any other planet'. Copernicus further pointed out that his universe was of enormous dimensions: 'The ratio of the Earth's distance from the Sun to the height of the firmament is so much smaller than the ratio of the Earth's radius to its distance from the Sun that the distance from the Earth to the Sun is imperceptible in comparison with the height of the firmament.'[76] The reason for this postulate was the very same problem that Aristarchus had faced in his hypothesis of a heliocentric universe, namely the absence of an observed stellar parallax.

De revolutionibus started with a brief introductory section, in which it was stressed that the sole purpose of astronomy was to devise models that could save the phenomena. The message of this section, apparently written by Copernicus, was that the heliocentric theory was merely a computational model and not one that claimed to be true in a physical sense. It was not written by Copernicus, however, but by Andreas Osiander, a Lutheran theologian who was entrusted with the supervision of the printing of *De revolutionibus*. Copernicus certainly did not share Osiander's opinion, but for a while this was not generally known. Only in 1609 did Kepler reveal that the anonymous introduction was in fact written by Osiander. To believe that Copernicus subscribed to the instrumentalist position outlined in the introduction was 'most absurd', he wrote.

Why did Copernicus find it necessary to turn Ptolemy on his head and develop an astronomical theory that ran counter to tradition and common sense? It is often stated that the Ptolemaic system had grown increasingly complex and that epicycles had to be added to epicycles in order to match observations. This allegedly led to a crisis, which Copernicus responded to with his new world system. However, the contrast between the simplicity of the Copernican system and the complexity of Ptolemy's system of compounded circles is fictitious. There was no state of crisis at the time Copernicus started to develop his alternative.[77]

Copernicus was indeed dissatisfied with the Ptolemaic system, but not because of its number and arrangement of circles or because it failed observationally. His main objection was that the centres of its epicycles did not move with uniform speed on the deferents, but with respect to the fictitious equant. In the opening lines of the *Commentariolus*, Copernicus emphasized that such a system 'seemed neither sufficiently absolute nor sufficiently pleasing to the mind'. He found it to be a betrayal of the fundamental doctrine that uniform circular motion was the only allowed form for motion in the heavens and indicated that it was his desire to remedy this defect that led him to the new theory. Copernicus had also become annoyed that the astronomers had not been able to discover 'the form of the world and the certain commensurability of its parts', a reference to the order and distance of the planets, which had no theoretical justification withing the existing astronomy. As a third reason, he wanted to establish a world system which, methodologically and aesthetically, was simpler than the traditional one, a system which rested only on a few hypotheses. Geocentric astronomers were forced to make use of 'an almost infinite multitude of spheres', whereas Copernicus would rather 'follow the wisdom of nature, which, as it takes very great care not to have produced anything superfluous or useless, often prefers to endow one thing with many effects'.[78]

In addition to these arguments of a methodological nature, there were also arguments that reflected the revival in the Renaissance of Pythagorean and Neoplatonist thought. Thus, in a lyrical passage in *De revolutionibus*, Copernicus conceived the Sun—'this lamp of the very beautiful temple'—to be the most noble of the celestial bodies and for this reason the one which naturally should occupy a central position. And Rheticus stressed how wonderful it was that with Copernicus' innovation, the number of planets was reduced from seven to six. As he pointed out in *Narratio prima*, six was a sacred number: 'For the number six is honored beyond all others in the sacred prophecies of God and by the Pythagoreans and the other philosophers. What is more agreeable to God's handiwork than this first and most perfect work should be summed up in this first and most perfect number?'[79]

Copernicus' system was able to explain in a simple way the retrograde motions of the planets and, equally simply, the limited elongations of Mercury and Venus. These phenomena did not need any special hypotheses, but followed directly from the basic assumption of the Earth's annual revolution around the Sun. In several ways Copernicus' world system resembled that of Ptolemy, only with the Earth and the Sun being interchanged; the celestial spheres were still largely concentric, and Copernicus even had to introduce epicycles in the style of Ptolemy. But when we turn to the structure and dimensions of the Copernican universe we realize how different it was, after all, from the one traditionally accepted.

Table 1.4 Approximate relative distances of the planets according to Copernicus, compared with the modern mean distances. All values in astronomical units.

	Least	Greatest	Mean	Modern mean
Mercury	0.26	0.45	0.38	0.39
Venus	0.70	0.74	0.72	0.72
Earth	0.97	1.03	1.00	1.00
Mars	1.37	1.67	1.52	1.60
Jupiter	4.98	5.46	5.22	5.20
Saturn	8.65	9.70	9.17	9.54

Contrary to the astronomers in the Ptolemaic tradition, Copernicus did not have to guess the order of the planets. He could calculate their distances in terms of the Earth's mean distance from the Sun, the astronomical unit (AU). For this unit distance, he found a value of 1142 Earth radii, which was much too small—the correct value is about 23 600—but Copernicus wisely decided to use the relative planetary distances, as given in Table 1.4.

We first note that Copernicus' planetary universe, as given by the distance to Saturn, is *smaller* by a factor of nearly two than what Ptolemy had found in his *Planetary Hypotheses*. Next, and more interestingly, the planetary spheres are much thinner and do not fill at all the space between the spheres. For example, Mars reaches out to 1.67 AU, far from the minimum distance of Jupiter, which is 4.98 AU. In other words, Copernicus' planetary model did not satisfy the principle of plenitude which was so dear to Ptolemaic astronomy. Even more shockingly, in order to accommodate the unobserved stellar parallax, the distance from Saturn to the sphere of the fixed stars had to be immense. 'The Earth is to the heavens as a point to a body and as a finite to an infinite magnitude', Copernicus wrote.[80] In a cosmological perspective, the Earth was merely an atom. 'It is not at all clear how far this immensity stretches out', he continued, but surely there must be an unoccupied space outside Saturn many times the planet's distance from the Sun. In terms of volume, the Copernican universe was at least 400 000 times as large as that of traditional cosmology! What was the nature and purpose of the space between the celestial spheres? Was it filled with some kind of ethereal substance? Was it a void? Nobody could tell.

When it came to the fixed stars, Copernicus had as little to say as Ptolemy. He seems to have placed all the stars, whatever their magnitude, on the same spherical surface at an immense distance from the Sun. At any rate, he did not indicate that the stellar sphere had any appreciable thickness. In Book I, Chapter 8, he briefly addressed the question of whether there might be something beyond the heavens, or 'If the heavens are infinite, . . . and finite at their inner concavity only'. Copernicus, the mathematical astronomer, did not come up with an answer, and preferred to leave the question to be discussed by the natural philosophers.

1.4.2 *Tycho's alternative*

Copernicus' theory did not immediately attract much attention. It took a couple of decades until its significance and novelty were generally recognized and astronomers began to discuss its merits and defects. A few accepted the heliocentric system, but most of those who studied *De revolutionibus* held a more eclectic attitude: they used what they could use, especially the mathematics of the planetary theory, but without subscribing to the heliocentric theory as physically true. The influential Jesuit mathematician and astronomer Christoph Clavius wrote between 1570 and 1611 a long series of commentaries on Sacrobosco's *De sphaera* in which he critically reviewed alternatives to the traditional Ptolemaic system. He praised many aspects of Copernicus' work, but without accepting its heliocentric cosmology. On the contrary, he objected to Copernicanism with an array of physical, astronomical, and methodological arguments. Clavius' Ptolemaic universe, including the empyreum, consisted of 11 spheres. Although there were neither bodies nor motion in the empyrean heaven, this 'happy seat and home of the angels and the blessed' was no less real than the firmament and the planetary heavens. Clavius stated that beyond the empyreum there might be a kind of infinite space, where God could create other worlds.

Whereas Clavius defended the traditional world picture, the Danish nobleman Tycho Brahe suggested an alternative to both of the existing cosmologies.[81] In 1572, the 26-year-old Tycho observed what appeared to be a new star in the constellation of Cassiopeia, and in his book *De nova stella* of the following year he argued that it was indeed a new, if ephemeral, fixed star. This was of cosmological importance because the interpretation broke radically with the age-old belief that the heavens were perfect and unchanging. Clavius was among those who accepted Tycho's interpretation.

In 1574–75 Tycho gave a series of lectures at the University of Copenhagen in which he introduced the new world system of Copernicus—'the second Ptolemy', as he called him. He had much praise for the theory of the Polish astronomer and stated that he would deal in his lectures with the motion of the planets according to Copernicus and using his parameters; but, significantly, he would transfer them to an Earth at rest. Tycho had lost confidence in the Ptolemaic system, yet he was unable to accept that the Earth really moved around the Sun. After he (and others) had observed the great comet of 1577, he began thinking of an alternative that would accommodate the best of both systems, most likely inspired by 'proto-Tychonic' systems developed by the German mathematician Paul Wittich and others.[82] He realized that if the comet had passed through the spheres of Mercury and Venus, as his data indicated, the spheres could not be solid bodies; hence there was nothing to bar planetary orbits from intersecting, as in the case of Mars crossing the orbit of the Sun.

Tycho published his world system in 1588, as Chapter 8 of his treatise on the great comet, *De mundi aetherei*. According to Tycho, the universe was geocentric, with the Sun and the Moon circling around the immobile Earth; or, perhaps better, it was geo-heliocentric, for all the other planets revolved around the Sun (Fig. 1.11). This was clearly a compromise between the Ptolemaic and the Copernican system, physically closer to the former, whereas mathematically it was closer to the latter. Since the Tychonic system was geometrically equivalent to Copernicus' system, it could match all its predictions except the apparently non-existent stellar parallax. The dimensions of Tycho's world up to Saturn did not differ much from those of Copernicus'. He took the distance of the Sun from the Earth to be about 20 times the Moon's distance, for which he adopted the value 60 Earth radii. For the distance to the farthest planet, Saturn, he ended up with 11,000 Earth radii. In gross contrast to Copernicus, he put the sphere of the fixed stars immediately above Saturn's sphere, at an average distance of about 14 000 Earth radii.

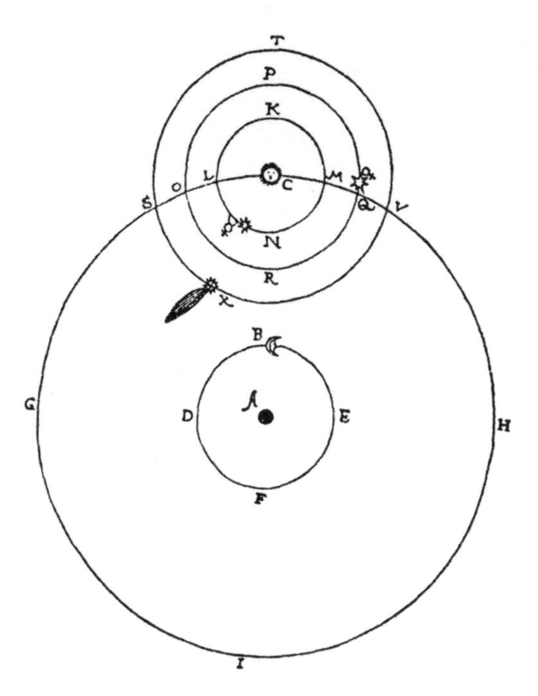

Fig. 1.11 The Tychonic world system, as reproduced in Tycho Brahe's *De mundi aetherei* of 1588. Whereas the Sun (C) revolves around the Earth (A), the other planets encircle the Sun. Only Mercury and Venus are shown in the figure. The object X is a comet, supposed to move in a circular orbit near that of Venus.

Why did Tycho, the self-appointed renovator of astronomy, refrain from going the whole way? Why did he not accept the heliocentric theory as representing the real universe? His reasons were diverse and not particularly original. In some cases they expressed a conservative attitude, as when he used the Bible as evidence against a moving Earth. The German astronomer Christopher Rothmann, with whom Tycho corresponded and who visited Uraniborg in 1590, denied that the Bible held any authority in scientific matters, but Tycho begged to disagree and maintained that the Scriptural evidence against Copernicus' theory must be taken seriously. What was undoubtedly of more importance, in matters of natural philosophy Tycho was at heart an Aristotelian. For this reason he accepted the dichotomy between the world beneath and above the Moon, and he used traditional Aristotelian arguments (already criticized by Buridan and Oresme) to prove the absurdity of a moving Earth.

The missing annual parallax for the fixed stars was another good reason to reject the Copernican theory. Tycho, armed with his excellent instruments, had looked for stellar parallaxes and found none. This he took to mean that the parallax, if there was one, was smaller than 1′ (minute of arc) or that, according to Copernicus' theory, the fixed stars were located at a distance at least 7 million Earth radii away. This enormous void space he simply was unable to accept; it was not only incredible, but also impossible. This uneasiness about the empty space between Saturn and the fixed stars was common at a time when it was generally assumed that the universe had a purpose, that it had been created for the benefit of man. In his *Dialogo* of 1632, Galileo lets Simplicio, the protagonist of the traditional world view, argue against Copernicus as follows:

Now when we see this beautiful order among the planets, they being arranged around the earth at distances commensurate with their producing upon it their effects for our benefit, to what end would there then be interposed between the highest of their orbits (namely, Saturn's) and the stellar sphere, a vast space without anything in it, superfluous, and vain. For the use and convenience of whom?[83]

Later generations of scientists would smile at such teleological rhetoric, but at the time of Tycho and Galileo it was still part of the scientific discourse. Moreover, Tycho shared with other astronomers the belief that the stars had visible diameters, which he found to be between $1'$ and $3'$. This again implied that if the stars were located as far away as required by the Copernican system, they would have impossibly large diameters, several hundred times that of the Sun. Tycho found this to be plainly absurd, and hence a strong argument against Copernicus' mistake, but Copernicans such as Rothmann were not convinced. They typically found refuge in an old theological argument, namely that the vastness of Copernicus' universe reflected the vastness of God's creative power. In response to Tycho, Rothmann wrote:

Or what absurdity follows if a star of the third magnitude equals the entire annual orb? . . . The absurdity of things, which at first glance appear so to the multitude, cannot be so easily demonstrated. Indeed, divine Wisdom and Majesty is much greater, and whatever size you concede to the Vastness and Magnitude of the World, it will still have no measure compared to the infinite Creator.[84]

Although Tycho was mainly interested in devising a planetary system that agreed with observations, he also had an interest in the physics of the heavens. Like most other astronomers, he distinguished between the task of the astronomer (or mathematician) and that of the natural philosopher. As he wrote to Rothmann, cosmology belonged to the realm of philosophy, not astronomy:

The question of celestial matter is not properly a decision of astronomers. The astronomer labours to investigate from accurate observations, not what heaven is and from what cause its splendid bodies exist, but rather especially how all these bodies move. The question of celestial matter is left to the theologians and physicists among whom now there is still not a satisfactory explanation.[85]

On the other hand, the Renaissance holist Tycho was also convinced that astronomy had a non-mathematical, astrophysical, or cosmological side, and that this could not be separated from the studies of terrestrial matter and its changes. A devotee and practitioner of Paracelsian chemistry, he believed that, in a sense, astronomy was the chemistry of the heavens and chemistry a kind of terrestrial astronomy. By studying the heavens, the natural philosopher would get a superior knowledge of processes on the Earth, and he would likewise become a better astronomer if he was well versed in chemistry and alchemy (Fig. 1.12).

Although Tycho followed Aristotle in distinguishing between the sublunar and superlunar regions of the world, he did not admit the distinction to be absolute. He was more inclined to believe that the air gradually became thinner towards the Moon and was then connected to Aristotle's ethereal element (he did not admit fire among the atmospheric

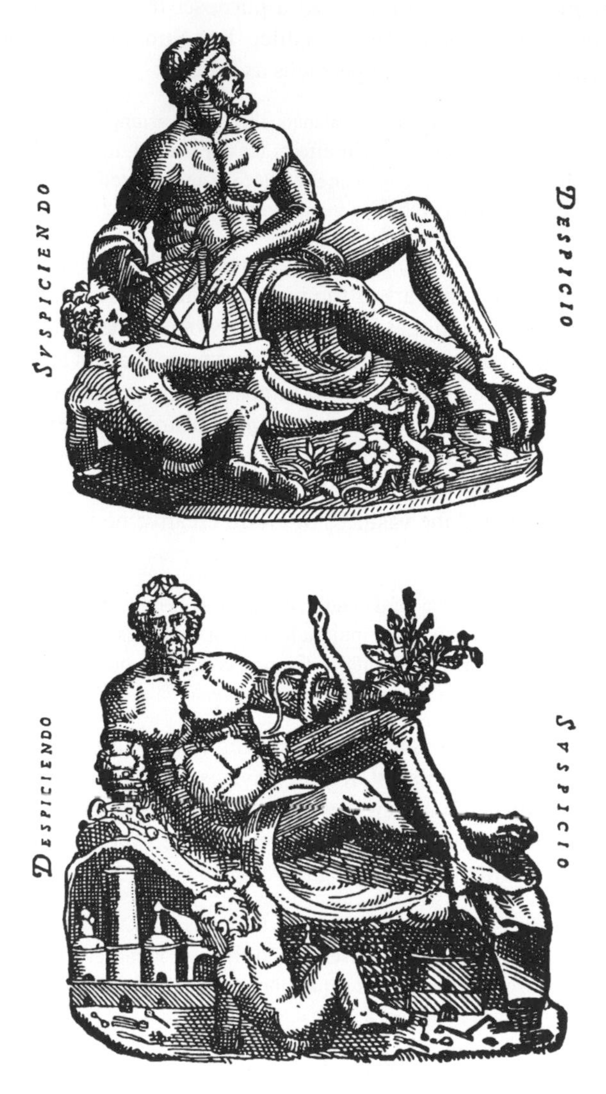

Fig. 1.12 At the two entrances of Uraniborg, Tycho Brahe placed relief sculptures which allegorically represented astronomy and chemistry. The two inscriptions related to the close connections between the two sciences ('By looking up, I see down', and 'By looking down, I see up'). From *Astronomiae instauratae mechanica* (1598).

elements).[86] The heavens were composed of ether, which appeared in a more dense form in the Milky Way and even more densely in the stars. In this way he suggested that the new star of 1572 could be explained as a temporary concentration of ether. Tycho's sketch of a physical cosmology was developed further by his pupil Cort Aslaksen, to whom the celestial ether was material in nature, nothing but air in a highly rarefied state. A representative of what was called Mosaic physics, he pictured the universe as consisting of three heavens, the atmosphere, the space containing the heavenly bodies, and God's eternal heaven. Contrary to Tycho, Aslaksen accepted that the Earth could perform a daily rotation.[87]

In the period from around 1620 to 1660, Tycho's hybrid cosmology received much attention and was to some extent also accepted, especially among Jesuits and other Catholic scholars, who, for theological reasons, could not openly endorse the Copernican theory. For example, the eminent French natural philosopher Pierre Gassendi was a Copernican at heart, but he was also a Catholic priest and, publicly, he defended Tycho's system (and he wrote a biography of Tycho Brahe, the first full biography of a scientist ever). Whereas Gassendi used the Tychonian system to further the cause of Copernicanism, the Jesuit astronomer Giovanni Riccioli believed that Tycho's world model was superior to that of Copernicus (Fig. 1.13). Riccioli's favoured model differed in its details from Tycho's by having Jupiter and Saturn (together with the Sun) circling the Earth, the other three planets revolving around the Sun. During a large part of the seventeenth century, such 'semi-Tychonic' systems were popular and widely discussed.[88] In his important book of 1651, *Almagestum novum*, Riccioli dealt in penetrating detail with the question of the mobility of the Earth. Following the scholastic tradition, he presented arguments for and against, but of course concluded that the Earth was immobile. Characteristically, in this conclusion theological arguments counted as heavily as did arguments based on scientific evidence.

For more than half a century, the Catholic church had no problems with the Copernican system, but in 1616 it was formally banned and after the infamous process against Galileo in 1633 it was impossible for scientists in Catholic Europe to support it. Copernicanism was controversial in the Protestant world as well, if not to the same extent or with the same consequences. Martin Luther allegedly branded Copernicus as a fool who would turn the entire science of astronomy upside down, but the historical basis for this often-quoted judgement is next to worthless. Although Luther was not a pro-Copernican, neither was he an anti-Copernican. For all we know, he may have been indifferent to or perhaps even ignorant about the revolution in astronomy (Luther died in 1546, three years after the publication of *De revolutionibus*).[89]

1.4.3 *Towards infinity*

Copernicus' universe was spherical and no less finite than Ptolemy's and Tycho's. Yet it was immensely larger, and for this reason alone invited a renewal of speculations concerning cosmic infinitude. The possibility of an infinite and infinitely populated universe, as discussed by a few early Copernicans, should be distinguished from the discussion of an infinite void space beyond the cosmos. This latter debate had roots in medieval philosophy and theology, and it continued to be an issue in the sixteenth and seventeenth centuries, largely unaffected by Copernicus' new astronomy. The imaginary, infinite void space was usually conceived to be divine and dimensionless, and hence physically unreal. However, to Otto von Guericke, the famous Magdeburg mayor and pioneer of vacuum technology, the infinite space beyond the material cosmos was real and three-dimensional. In his celebrated treatise of 1672 on the Magdeburg experiments, *Experimenta nova*, he described the infinite nothingness as an active and powerful entity, as what has been called an 'ode to nothing'.[90]

Fig. 1.13 The Italian Jesuit astronomer Giovanni Riccioli published in 1651 a great work with the ambitious title *Almagestum novum*. As is apparent from its frontispiece, he found that a Tychonian system (not quite the same as Tycho Brahe's) should be rated higher than the heliocentric system of Copernicus. The Ptolemaic system is placed on the ground, indicating that it is not considered a worthy competitor. 'I am raised that I may be corrected', Ptolemy utters.

The discussion in the seventeenth century of an infinite universe was indebted to the revival of the ancient atomic theory of matter by natural philosophers such as Francis Bacon in England and Pierre Gassendi in France. The renewed interest in the atomism of Democritus and his followers was primarily of relevance to chemistry, but also included cosmological aspects. In a book of 1675, the Englishman Edward Sherburne summarized what he took to be the essence of atomistic cosmology (see Fig. 1.4): 'The Ancient Philosophers, especially those of *Democritus* and his School, and most of the Mathematicians of those Times, asserted the *Universe* to be *Infinite*, and to be divided into two chief Portions; whereof the One they held to be the World, or rather Worlds, finite as to Bulk and Dimension, but infinite as to Number. The other Part or Portion, they extended beyond the Worlds, which they fancied to be a *Congeries* of infinite Atoms. Out of which not only the Worlds already made received their Sustenance, but new Ones also were produced.'[91]

Apart from the extra-cosmic, more or less theological void, was the world also infinite with respect to celestial bodies? Thomas Digges, an English mathematician and contemporary of Tycho, was among those who observed the new star of 1572. An early adherent of Copernicanism, he tried to prove the new theory by measuring the annual parallax of the fixed stars, but of course he failed. In 1576 Digges added to a book on meteorology written by his father a chapter on cosmology, which included a free translation of the cosmological part (Book I) of Copernicus' *De revolutionibus*. The novelty of 'A Perfit Description of the Coelestiall Orbes' was that Digges did not collect the stars in a sphere, as Copernicus had done, but distributed them throughout an infinite universe (Fig. 1.14) Still, he wrote of the stars as being located in a fixed sphere or orb, albeit one 'reachinge vp in *Sphaericall altitudes* without ende'. Moreover, his infinite starry heaven was 'the gloriouse court of ye great god' and 'the habitable of the elect, and of the coelestiall angelles'. Digges's universe was infinite in this theological sense, but it is more uncertain whether it was also the first infinite Copernican universe in a physical and astronomical sense.[92] Shakespeare may have been acquainted with Digges's works and it has been argued that his world picture enters allegorically in several of the Bard's plays.[93]

The Italian maverick philosopher Giordano Bruno (or Filippo Bruno of Nola, as his name of birth was) was burned at the stake on 17 February 1600 because of his heretical religious views. He was a martyr of intellectual freedom, but not of science—if for no other reason because he was not a scientist. At any rate, his unorthodox and partial support of Copernicanism had little to do with his trial and cruel death.[94] Talented, undisciplined, and influenced in particular by Cusanus and related mystical thought, Bruno dealt with cosmological topics in *The Ash Wednesday Supper* of 1584, in *On the Infinite Universe and Worlds*, also of 1584, and in the Latin poem *De immenso* of 1591.

It is a matter of some dispute whether Bruno was truly a Copernican. Certainly, his understanding of the Copernican system was poor and at least on one occasion he seriously misunderstood it.[95] He had neither an interest in nor suffient knowledge of mathematics to appreciate *De revolutionibus* and frankly stated that 'I care little for Copernicus.' The planetary system that he proposed in *De immenso* had little to do with Copernicus', as he put Venus and Mercury on the same epicycle, which, opposite to it, also carried the epicycle carrying the Earth and the Moon. The proposal lacked any observational evidence, a fact that did not bother Bruno the least. He had only disdain for the astronomers' concern with the number and order of the planets, questions which he

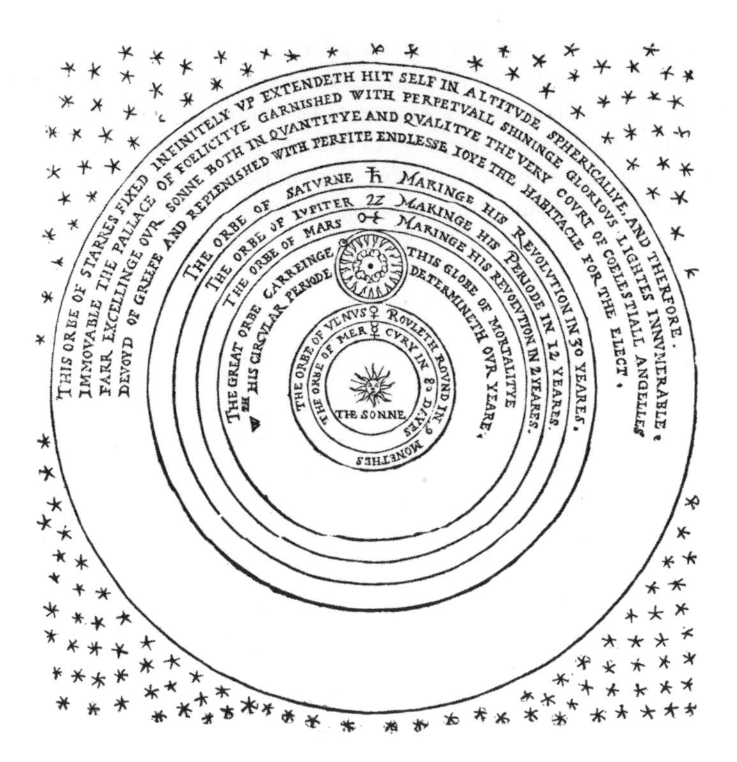

Fig. 1.14 Thomas Digges's Copernican system of 1576.

considered to be unimportant. They were even meaningless, for Bruno was convinced that comets were planets, which implied that the number of planets encircling the Sun could not be known. Given that Bruno's main affinity with Copernicanism was his conviction that the Earth and planets revolved around the Sun, it is doubtful whether he can reasonably be called a Copernican.

At any rate, Bruno saw himself as a reformer of the Copernican system, which in his version was given a different and more grandiose perspective. For one thing, he denied that the orbits of the planets were necessarily circular or reducible to circular motions. For another, he rejected the Aristotelian notion of a fifth element peculiar to the heavens and declared that the celestial bodies were made of the very same elements as those constituting terrestrial matter. As to Copernicus' preservation of the sphere of the fixed stars, he dismissed it as a 'fantasy'. Even more importantly, he emphasized again and again that the universe—the real, physical universe—was infinite in size and in a continual state of change. The Earth was not at the centre of the world; and neither was the Sun, for there was no centre of the universe, only an infinity of local centres. In *On the Infinite Universe and Worlds*, he wrote: 'There are then innumerable suns, and an infinite number of earths revolve around those suns, just as the seven we can observe revolve around the Sun which is close to us.'[96] Each of the infinite number of earths was inhabited. Without going into details, Bruno's

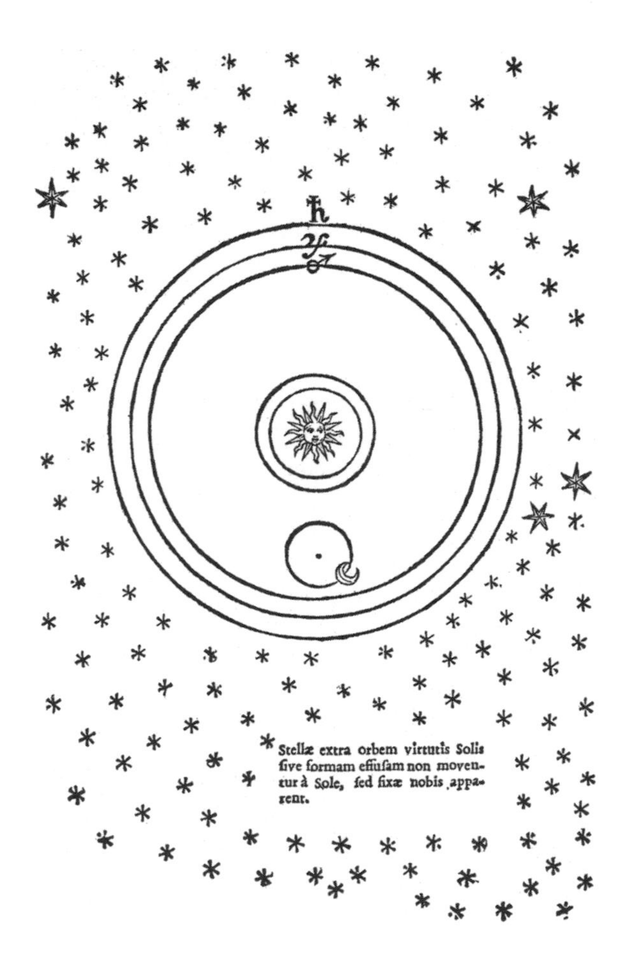

Fig. 1.15 William Gilbert's representation of the universe in his *De mundo sublunari*, published posthumously in 1651.

cosmic vision included many bold proposals of a kind Copernicus would not have admitted. There is no doubt that he went much beyond Copernicus, but then he—contrary to Copernicus and Tycho—could afford the luxury of ignoring observations. The speculative and non-astronomical features in Bruno's poetic vision are further illustrated by his suggestion that there might be other earths revolving around our Sun.

In the years around 1600, elements of Copernicanism appeared in many places outside astronomy. One example was provided by William Gilbert, the English physician who is best known for his pioneering work on lodestones and magnetism, *De magnete* of 1600. Influenced by Bruno and being a Copernican of a sort, Gilbert accepted the diurnal rotation of the Earth, whereas he ignored the more important annual revolution. However, there are reasons to believe that he accepted the system of Copernicus ('a man most deserving of literary honour') at the time he wrote his book on magnetism. He seems to have believed in an infinite world where the fixed stars were distributed at all distances from the Earth. Rhetorically, he asked if the stars had ever been found to reside in a single sphere:

No man hath shown this ever; nor is there any doubt that even as the planets are at various distances from Earth, so, too, are those mighty and multitudinous luminaries ranged at varous heights and at distances most remote from earth: they are not set in any sphæric framework or firmament (as is supposed), nor in any vaulted structure. . . . What then, is the inconceivably great space between us and these remotest fixed stars? and what is the vast immeasurable amplitude and height of the imaginary sphere in which they are supposed to be set? How far away from Earth are those remotest of the stars: they are beyond the reach of eye, or man's devices, or man's thought.[97]

Indeed, the new philosophy called all into doubt. One understands John Donne's worries. Gilbert operated with two cosmic forces, electricity and magnetism, and he suggested that the former was responsible for the aggregation of matter, and hence somehow related to gravitation. His account of gravity was by no means clear, but it did imply that gravity was not a property restricted to the Earth; the other celestial bodies had their gravities, too, a view that contradicted the Aristotelian distinction between the sublunary and superlunary regions of the world.

Renaissance cosmology was a far broader subject than the kind of mathematical astronomy practised by Copernicus and the professional astronomers. Astrology was an integrated and most important part of the period's cosmology, although Copernicus was exceptional in his lack of interest in astral influences. So-called Paracelsianism, named after the Swiss physician Paracelsus (Philippus Aureolus Theophrastus Bombastus von Hohenheim), was an important intellectual force in the second half of the sixteenth century and a source of inspiration for Tycho Brahe, among others. The Paracelsians were primarily interested in chemistry and alchemy, which they used in understanding the cosmos. For example, they explained in detail the creation of the world, as recounted in Genesis, in terms of chemical transformations. Their universe was a living entity where all parts interacted through 'sympathies' and 'antipathies', and it was represented in the microcosmos by means of so-called correspondences. Paracelsus and his allies considered the universe as a vast chemical laboratory, but their interests were largely limited to the Earth and did not include mathematical models of the universe. While it makes sense to speak of Paracelsian cosmology, it was a cosmology of a very different kind from the one cultivated by the astronomers.

Although chemical philosophers of a Paracelsian inclination were strongly anti-Aristotelian, they did not support the Copernican system. One of them, the English physician and mystic Robert Fludd, recognized the primacy of the Sun but nonetheless rejected the views of Copernicus and Gilbert. Fludd's arguments were mainly traditional, including references to the Bible and the lack of an annual parallax. He was convinced that the Earth was the most massive body in the universe and therefore immobile. 'Certainly the reasons of Gilbert are ridiculous,' he wrote in 1617, 'it is impossible to believe that the heavens can be carried around in the space of twenty-four hours because of their boundless magnitude.'[98]

1.4.4 *Galileo and Kepler*

The Copernican revolution was largely completed during the first half of the seventeenth century, not least through the path-breaking works of Galileo Galilei and Johannes Kepler. As a young man, Galileo was in favour of traditional cosmology, but he soon came out in support

Fig. 1.16 The Paracelsian universe as magnificently depicted in Robert Fludd's *Utriusque cosmi* of 1617. The sublunary world of the four elements, governed by the alchemical goddess, is separated from the lower heavenly regions. Beyond the sphere of the fixed stars is the upper celestial world. The ape sitting on the central Earth symbolizes humans' poor reflection of divinity.

of the Copernican world picture, which he tirelessly defended throughout his life. His approach to celestial problems was decidedly physical and in this respect very different from the astronomical approach of Copernicus, Tycho, and Kepler and also different from the philosophical approach of Bruno. This may explain his limited interest in cosmology, a field he seems to have regarded with a mixture of scepticism and indifference. All the same, the sensational discoveries he made with the new optical tube in 1610 and reported in *Sidereus nuncius* did much to change the picture of the universe. When he turned his primitive telescope toward the Milky Way he instantly solved a riddle that had occupied astronomers and natural philosophers for two thousand years. The Milky May, he now realized, was 'nothing but a congeries of innumerable stars grouped together in clusters'. Galileo's discoveries caused great excitement, and the news was rapidly disseminated throughout learned Europe.

Intellectuals and artists excelled in praising the Italian master philosopher, as exemplified by a contemporary poem by Johann Faber, a German–Italian physician and botanist:

Yield, Vespucci, and let Colombus yield. Each of these
Attempts, it is true, a journey through the unknown sea. . . .
But you, Galileo, alone gave to the human race the sequence of stars,
New constellations of heaven.[99]

Galileo also discovered the spots on the Sun, traditionally believed to be a perfect and sacred body, and deduced that the Sun rotated with a period of about 28 days. (The Englishman Thomas Harriot had studied the Sun with a telescope and observed sunspots a little earlier, but without publishing his observations, and Chinese naked-eye observations were made much earlier.) Wherever Galileo directed his telescope, he found crowds of stars invisible to the naked eye, and he discovered that while his instrument could magnify the planets and make them look like discs, it could not do the same with the fixed stars. The stars consequently must be at enormous distances from the Earth, just as Copernicus had claimed. Another strong argument for Copernicanism, and against the Ptolemaic system, came from Galileo's discovery that Venus exhibited phases. The only way to explain the observed phases of Venus was to assume that the planet moved in an orbit round the Sun; the observed change in phases did not fit with the Ptolemaic system.

With regard to the number of stars and their spatial distribution, Galileo was not very clear. He denied that the stars were placed in the same sphere, but without asserting that they were found at all distances with no limit. In the famous *Dialogo*, he denied the infinity of space, and in other of his writings he indicated that it would never be known whether the universe was finite or infinite. During the last years of his life, Galileo, who since the infamous trial of 1633 had lived in forced isolation in his house in Arcetri outside Florence, corresponded with Fortunio Liceti, a professor of philosophy of Aristotelian inclination. From this correspondence, we learn about Galileo's agnostic attitude to cosmology. Concerning the question of the finitude or infinitude of the universe, he wrote:

The reasons on both sides are very clever, but to my mind neither one is necessarily conclusive, so that it always remains ambiguous which assertion is true. Yet one argument alone of mine inclines me more to the infinite [universe] than the finite, this being that I cannot imagine it either as bounded *or* as unbounded and infinite; and since the infinite, by its very nature, cannot be comprehended by our finite intellect, which is not the case for the finite, circumscribed by bounds, I should refer my incomprehension to the incomprehensible infinite [rather] than to the finite, in which there is no necessary reason of incomprehensibility.[100]

In another letter to Liceti, of 1641, he described the question of the centre of the universe as 'among the least worthy of consideration in all astronomy' and went on to state that any search for a centre of space, or for the shape of space, was 'a superfluous and idle task'.[101]

If Galileo expressed reservations with respect to the grand questions of cosmology, his contemporary Kepler did not. On the contrary, the German mathematician was fascinated by such questions and wrote exuberantly about them.[102] His main concern was with the spatial dimensions of the universe but he also had an interest in the temporal dimension. On the basis of Biblical and astronomical evidence, he concluded that God had created the universe in 3992 BC and that Jesus Christ was born in 4 BC.

In his *De stella nova* of 1606, a work discussing a new star that had appeared in the skies two years earlier, Kepler took up the question of the extension of the sphere of the fixed stars. He was aware that Bruno and Gilbert had defended the infinity of the universe, a notion that filled him with a 'secret, hidden horror' and which he was eager to refute. He likewise denied Bruno's version of the cosmological principle, the claim that the world looks the same to every cosmic observer, whatever star is chosen as the vantage point. His arguments in favour of a finite world in which the solar system occupies a privileged position were in part metaphysical, in part based on observations that he thought spoke against a world filled with an infinite number of stars. The Milky Way and the fixed stars limit our space, but is it not possible that beyond the limit there is an infinite space, either a void or a space thinly populated with stars? Kepler discussed the question systematically and his answer was a firm no.

Kepler's early rejection of infinity relied on philosophical reasoning and naked-eye astronomy. The picture of the starry heaven changed with Galileo's telescopic discoveries, yet the change only confirmed Kepler in his conviction of a finite world. This he made clear in *Dissertatio cum nuncio sidereo* of 1610, a hastily composed comment on and summary account of Galileo's *Sidereus nuncius*. In the course of his argument against infinitism, he examined Bruno's idea of an infinite number of worlds, each of them differing from ours. He claimed that in these other worlds the five regular polyhedra—the geometrical basis of his world model described in *Mysterium cosmographicum* of 1596—would not exist in the same form as we know them. To Kepler, this was reason enough to conclude that 'this world of ours is the most excellent of them all, if there should be a plurality of worlds'.[103]

Kepler returned to the question in later works, in particular in the *Epitome astronomiae Copernicanae* published in three instalments between 1618 and 1621. As Galileo had shown, there are numerous stars that cannot be seen with the naked eye. This might be because they were too far away from the Earth, or because they were too small to be seen. Kepler unhesitatingly endorsed the second option, concluding that 'the visible sky is . . . everywhere raised above us by nearly the same distance. There is therefore an immense cavity in the midst of the region of the fixed stars, a visible conglomeration of fixed stars around it, in which enclosure we are.' He believed that an infinite number of stars could be ruled out logically, as the very notion was contradictory—'all number of things is actually finite for the very reason that it is a number'.[104] As to the possibility of a finite world immersed in an infinite space, he rejected it on conceptual grounds, using the Aristotelian argument that there cannot be space without bodies located in it.

In *Epitome*, Kepler not only reconfirmed the finitude of the universe, he also calculated its size.[105] The radius of the sphere of the fixed stars he took to be 60 million Earth radii or 4 million solar radii. From this it follows that the volume of space up to the stars was 64×10^{18} as great as the volume of the Sun. Kepler argued that the volume of the entire stellar sphere was merely 8×10^9 times the volume of the Sun, thus ending up with a cosmos in which the stellar region was of negligible size. The sphere of the fixed stars was curiously thin: its thickness was 6000 times smaller than the radius of the Sun, not more than nine English miles! This implied that the stars were incredibly small bodies, an assertion Kepler thought was supported by telescopic observations (which revealed stars as points, not as round discs). He found this surprising picture of the stellar world to be satisfactory for at least two reasons. For one thing, it refuted Tycho's main objection to the Copernican system; for another, it showed how radically different the Sun was from the fixed stars and how much more impressive the central body was. Kepler was no less a sun-worshipper than Copernicus. 'Of all the bodies in the universe the most excellent is the Sun, whose whole essence is nothing else than the purest light', he wrote. He continued:

> It is a fountain of light, rich in fruitful heat, most fair, limpid, and pure to the sight, the source of vision,. . . called king of the planets for his motion, heart of the world for his power, its eye for his beauty, and which alone we should judge worthy of the Most High God, should he be pleased with a material domicile and choose a place in which to dwell with the blessed angels.[106]

Kepler's universe was indeed heliocentric.

Notes

1. One indication of the strong interest in archaeoastronomy as a research field is the foundation in 1977 of the journal *Archaeoastronomy Bulletin* and in 1979 of *Archaeoastronomy*, published as a supplement to *Journal for the History of Astronomy*. Among the many scholarly works on Stonehenge as an astronomical observatory, North 1996 deserves to be singled out as a most detailed and erudite book on the subject.
2. See Plumley 1975, which I largely follow. See also Frankfort 1959, pp. 51–70.
3. Cited in Plumley 1975, p. 34.
4. Cited here from Jacobsen 1957, pp. 18–19. Other accounts of Mesopotamian cosmology are given in Lambert 1975 and Rochberg-Halton 1993.
5. English translation in Schiaparelli 1905.
6. Gombrich 1975 provides a summary account of ancient Indian cosmologies.
7. Gombrich 1975, p. 115.
8. Needham and Ronan 1993, p. 66.
9. For details and historical evidence, see May 1994. May's assertion that creation out of nothing is not to be found in the Bible has not remained uncontested. Copan and Craig 2004 argue forcefully that the idea is indeed in the Old Testament, if not explicitly stated.
10. Quotations from McKirahan 1994, pp. 11 and 9. Online version: http://sunsite.berkeley.edu/OMACL/Hesiod/theogony.html.
11. Quoted in Whitrow 1998, p. 47. On the idea of the cyclical universe in a cultural and historical perspective, see, e.g., Eliade 1974 and Jaki 1974.
12. Quoted in McKirahan 1994, p. 23. The following quotations are from the same source.
13. A detailed analysis of Empedocles' cyclical cosmos can be found in O'Brien 1969.
14. On Pythagoras and the Pythagorean school, see Riedweg 2002.
15. McKirahan 1994, p. 104.
16. Ibid., p. 326. Much later, this cosmic scenario was taken up by cosmologists of a speculative orientation, including Kant, whose cosmological vision was much indebted to the atomists (see Section 2.2).
17. *De rerum natura* exists in many translations. The quotations are from Lucretius 1997, a reprint of a translation from 1904, pp. 45–46, 93, 96, and 205. An online version can be found on http://classics.mit.edu/Carus/nature_things.html.
18. Quoted in Sambursky 1963, p. 202. See also Jaki 1974, p. 114.
19. Aiton 1981, p. 79.
20. For Plato's possible influence on Eudoxus' world model, see Knorr 1990.
21. Eudoxus' system was reconstructed by Schiaparelli in 1874. However, it is possible that Schiaparelli's clever reconstruction was also to some extent a modernization and that it included features that did not appear in the authentic version.
22. Farrington 1953, p. 279.
23. Quoted in Heath 1959, p. 275.
24. Quoted in Wright 1995, p. 151.
25. For details on Plato's cosmology, see Cornford 1956.
26. *De caelo*, as quoted in Munitz 1957, p. 95.
27. Sambursky 1963, p. 203.
28. Quoted in Freudenthal 1991, p. 50.
29. See Webb 1999 for an overview.

30. Van Helden 1985 provides an excellent survey of distance determinations from ancient Greece to the late seventeenth century.
31. Aristarchus' work is translated in Heath 1959 (first edn. 1913).
32. Heath 1959, p. 353.
33. Van Helden 1985, p. 6.
34. See Gingerich 1985.
35. *The Sandreckoner* is translated in Heath 1953 (first edn 1912). Quotations from pp. 221–221 and p. 232.
36. The similarity between Archimedes' and Eddington's numbers is discussed in Brown 1940, who argues that the number of particles in Archimedes' model universe may be interpreted to be of the same order of magnitude as Eddington's cosmical number.
37. Pliny 1958. Quotations from book II, pp. 171–173 and 177–179. A 'book' denoted a part or large chapter of a work, not a separate text.
38. Ptolemy 1984, pp. 45–46.
39. Ibid., p. 37.
40. Ibid., p. 38.
41. See Goldstein 1967. Accounts of Ptolemy's cosmology can be found in, e.g., Evans 1993 and Aaboe 2001, pp. 114–134.
42. Goldstein 1967, p. 8. Other quotations are from the same source.
43. Quoted in Cohen and Drabkin 1958, p. 118.
44. See Haber 1959 and Dean 1981.
45. The complex story is told in May 1994. But see also Copan and Craig 2004 for a different view.
46. Lindberg 2002, p. 48.
47. Lactantius 1964, III.24
48. Dreyer 1953, pp. 207–219.
49. Cosmas' cosmography was translated into English in 1897. Quotations from Cosmas 1897, pp. 11 and 129.
50. Galileo's *Letter* is reproduced in Drake 1957 (online version at www.fordham.edu/halsall/mod/galileo-tuscany.html). Quotation from p. 184.
51. Hoskin 1999, p. 72. The two historians are Michael Hoskin and Owen Gingerich.
52. Sambursky 1973, p. 135. See also Sambursky 1987, pp. 154–163.
53. Brehaut 1912, III, 32.1, 40.1, and 52.1.
54. Crombie 1953, p. 107. Grosseteste's light (*lux*) should not be understood as ordinary, visible light, but rather as the principle of light, of which ordinary light is only one manifestation.
55. Cosmology during the late Middle Ages is described in detail in Grant 1994.
56. Quoted in Crowe 1990, p. 74.
57. Lindberg 1992, p. 262.
58. Crowe 1990, p. 74.
59. See North 1975, p. 6.
60. On Chaucer's cosmology, see North 1990.
61. Orr 1956, p. 297, which offers a detailed account of Dante's universe.
62. A selection of the condemned articles is presented in Grant 1974, pp. 47–50.
63. From Thomas's *Writings on the Sentences of Peter Lombard*, as quoted in Carroll 1998, p. 88.
64. See the careful account in Dick 1982.
65. Grant 1974, pp. 548–550.
66. Grant 1994, p. 170.
67. Grant 1974, p. 560.
68. Ibid., p. 501.
69. Grant 1994, p. 478.
70. Grant 1974, p. 509.
71. Cusanus 1997, pp. 158–161. Jasper Hopkins' English translation of *De docta ignorantia* can be found online at www.cla.umn.edu/jhopkins/DI-Intro12–2000.pdf.
72. Quoted in Koyré 1968, p. 29. On the Copernican revolution, see also Kuhn 1957.
73. Lovejoy 1964, p. 102, who quotes John Wilkins, *Discourse Concerning a New Planet* (London, 1640). Wilkins, an English natural philosopher who played a leading role in the founding of the Royal Society, was in favour of Copernicanism.
74. *Commentariolus* was only rediscovered and published in the late nineteenth century. I use the English translation in Rosen 1959, which also includes a translation of Rheticus' *Narratio prima*. Rheticus' work can be seen at www.lindahall.org/services/digital/ebooks/rheticus.

75. Copernicus 1995, p. 7. To illustrate the foolishness of his potential enemies, he referred to Lactantius, who, as we have seen, used arguments from the Bible to deny the spherical shape of the Earth.
76. Rosen 1959, p. 58.
77. Gingerich 1975.
78. Copernicus 1995, p. 24. On the basis of the very same metaphysical doctrine of nature's economy, the Copernican system would soon be criticized for its vast gaps in the form of superfluous and useless voids.
79. Rosen 1959, p. 147.
80. Copernicus 1995, p. 14. The sentence is to be understood figuratively, not as a claim that the universe is infinite.
81. On Tycho and his cosmology, see Schofield 1981 and Thoren 1990, pp. 236–264.
82. Gingerich and Westman 1988.
83. Galilei 1967, p. 367.
84. Van Helden 1985, p. 52. Letter of 18 April 1590.
85. Howell 1998, p. 526. Letter of 17 August 1588.
86. Tycho outlined his view of the relationship between the heavens and the Earth in the preface to a book published by one of his assistants in 1591. See Christianson 1968.
87. Blair 2000. A part of Aslaksen's work, *De natura caeli triplicis* from 1597, was translated into English as *The Description of Heaven* (London, 1623).
88. Variants of the Tychonic system are discussed in Schofield 1981.
89. Hooykaas 1972, pp. 121–122.
90. Grant 1969, especially pp. 55–57. Von Guericke's description of the extra-cosmic void has a curious similarity to the quantum vacuum of modern physics.
91. Quoted in Heninger 1977, p. 193.
92. Koyré 1968, pp. 35–39.
93. See Usher 1999, where it is suggested that the true theme of *Hamlet* is the controversy between the three rival world systems, Ptolemy's, Tycho's, and Digges's. Usher argues that Shakespeare was probably a Copernican.
94. Singer 1950; Koyré 1968, pp. 39–54. It is commonly recognized that Bruno's crimes were theological and political, rather than related to his espousal of the Copernican system. See, e.g., Lerner and Gosselin 1973.
95. McMullin 1987.
96. Singer 1950, p. 304, which includes a translation of *On the Infinite Universe and Worlds*. Available online at www.positiveatheism.org/hist/bruno00.htm.
97. Gilbert 1958, pp. 319–320. On Gilbert's magnetic cosmology, see Freudenthal 1983.
98. Quoted in Debus 1977, p. 244. Fludd became involved in a bitter dispute with Kepler concerning the correct application of mathematics to natural phenomena; see pp. 256–260.
99. Quoted in Cohen 1985, p. 74. Amerigo Vespucci was the Italian seafarer who sailed to the Americas about 1500 and after whom America is named.
100. Quoted in Drake 1981, pp. 405–406. See also Koyré 1968, pp. 95–99.
101. Drake 1981, p. 411.
102. Koyré 1968, pp. 58–87.
103. Rosen 1965, p. 44, which includes a complete translation of Kepler's *Dissertatio*.
104. Koyré 1968, pp. 81 and 86.
105. Van Helden 1985, pp. 87–90.
106. Quoted in Burtt 1972, p. 48.

2

THE NEWTONIAN ERA

2.1 Newton's infinite universe

During the seventeenth century, the road to Copernicanism often went through Cartesianism. The famous French philosopher, mathematician, and physicist René Descartes developed an ambitious theory based on matter and motion that purportedly explained all natural phenomena, including those in the heavens. The proud motto of later Cartesian natural philosophers was 'Give me matter and motion, and I will construct the universe.' Cartesian astronomy and cosmology became hugely popular, but at the end of the seventeenth century Descartes's theory was challenged by Isaac Newton's very different system of natural philosophy. Although Newton's physics celebrated its greatest triumphs in celestial mechanics, it was also applied to cosmology and provided, for the first time, the field with a measure of scientific authority based on the universal law of gravity. The Newtonian universe, as it appeared in the early years of the eighteenth century, consisted of a multitude of stars spread out over infinite space. While the law of gravitation governed Newton's cosmos, the true governor was God, who was never absent from the mind of Newton and his contemporaries.

2.1.1 *Celestial vortices*

Starting in 1629, Descartes was preparing a comprehensive work on his mechanical cosmology when he learned about the condemnation of Galileo's *Dialogo*. In a state of shock, he decided to withhold from publication *Le monde*, a cosmological work firmly founded upon Copernican principles. 'I wouldn't want to publish a discourse which had a single word that the Church disapproved of', he piously confided to Marin Mersenne, his learned friend who was not only a chief scientific intelligencer but also had sympathy for Copernicanism.[1] Nonetheless, Descartes did publish the main part of his cosmology (if only anonymously) in his famous *Discours de la méthode* of 1637 and also in *Principia philosophiae*, published in 1644. *Le monde, ou traité de la lumière* appeared posthumously in 1664. He claimed that the relativity of motion made the Copernican theory acceptable from a formal point of view, whereas as a physically true theory it had to be rejected.

Descartes's physics was nothing but geometry and motion, and so was his cosmology. In *Principia philosophiae* he ambitiously sought to understand nature in purely mechanical terms. He argued that space (or extension) and matter were identical, a doctrine that had important consequences. First, if space itself is meaningless without matter, there can be no genuine vacuum. The world is necessarily a plenum. Second, since space cannot vary in density, neither can matter. Yet we do not experience a completely uniform world, but a world with differences between one part and another. The differences, as they appear as structures and objects, are to be explained as different states of motion, which is all that distinguishes a planet or a star from its surroundings. Third, if space is infinite, the same must be the case with the material universe. Descartes believed that the world was infinite in the sense that it was impossible to conceive any limits to the matter of which it consists. Because only God was truly infinite, he preferred to speak of an indefinite rather than an infinite world, but this was merely a tactical manoeuvre. 'Because it is not possible to imagine such a great number of stars that we do not believe that God could have created still more, we shall suppose their number to also be indefinite,' he wrote.[2]

The cosmic machinery of the Cartesian world was driven by mechanical actions of matter particles on other particles, giving rise to vortical motions of any size and kind. Although Descartes's matter was made up of particles, he was no atomist, as the particles were indefinitely divisible. He distinguished between three elements, luminous, transparent, and opaque. The Earth and planets were made of the third element, the Sun and stars of the first; the second, the ethereal or celestial element, filled, together with particles of the first element, the spaces between the cosmic bodies but could also penetrate the pores of terrestrial matter. The large celestial vortices, whirlpools of subtle matter, carried the planets with them and were also the mechanism behind other celestial phenomena, such as comets.

As one of the fathers of the very idea of natural laws, Descartes believed that law-governed mechanical processes had to replace teleology. As he saw it, cosmogony and cosmology were products of matter in motion and nothing but. God had of course installed

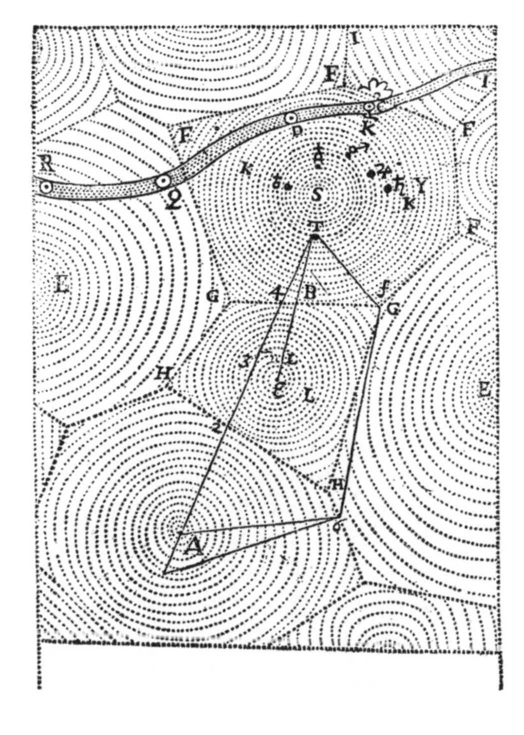

Fig. 2.1 Descartes's universe, as described in *Le monde*, consisted of an indefinite number of contiguous vortices, each with a sun at its centre and with planets revolving around it. The symbol *S* marks our Sun with its six planets. The path above the solar system represents a comet.

the laws, as he had created the material world, but what followed was strictly a consequence of the laws of motion and the initial conditions. In *Discours* and elsewhere, Descartes made it clear that the laws were all-important, not the initial conditions. What would happen, he asked, if God created a new world and 'if He agitated in diverse ways, and without any order, the diverse portions of this matter, so that there resulted a chaos as confused as the poets ever feigned, and concluded His work by merely lending His concurrence to Nature in the usual way, leaving her to act in accordance with the laws which He had established?'[3] Descartes answered that the mechanical laws would eventually lead to the very same world that we inhabit. Cosmic development was strictly determined by the laws of nature, which were of such a nature that even if God had created other worlds, these would be governed by the same laws that we observe. God had imposed his laws on an original chaos, and any kind of chaos would do, as it would lead to the same world.[4]

Cartesian cosmology was a great success, if perhaps more socially and ideologically than from a scientific point of view. It was presented in an elegant and compelling fashion by Bernard le Bouvier de Fontenelle, permanent secretary of the Royal Academy of Sciences in Paris, in his *Entretiens sur la pluralité des mondes*. Fontenelle's work, published in 1686, became an instant classic and did much to disseminate Descartes's world picture. Also, many scientists found his cosmology attractive and sought to develop it further. As presented by Descartes, his theory was qualitative and speculative, for which reason later scientists such as Huygens, Malebranche, and Leibniz strove to turn it into a quantitative, mathematically formulated version. However, the popularity of Cartesian cosmology could not hide the fact that scientifically it was plagued by great problems. By 1740 at the latest, it was no longer considered a scientifically viable theory, not even in France.

The great Dutch scientist Christiaan Huygens, who was much influenced by Descartes's system of natural philosophy, preferred to deal with limited problems rather than questions of a cosmological nature. In his posthumous *Cosmotheoros* of 1698—perhaps best known for its detailed description of extraterrestrial life—he severely criticized Kepler's Sun-centred cosmology and emphasized, as others had done before him, that the Sun was just one star among many. There surely was a gigantic number of stars, but Huygens refrained from entering 'that intricate dispute of infinity'.[5] Although he tended to believe that the universe was infinitely extended, he realized that this was nothing but a belief. Knowledge of the universe at large was beyond man's faculty, he thought.

2.1.2 *Newton's cosmos*

As a young man, Isaac Newton studied Descartes's theory of celestial vortices only to reach the conclusion that it was unsatisfactory as it could not account for Kepler's planetary laws (Descartes, who died in 1650, seems to have been unaware of the laws). It was Kepler's third law and the analysis of circular motion by means of centrifugal force that set Newton off on the long travel that resulted in the universal law of gravitational attraction. Only around 1680 did he encounter Kepler's area law, which to him implied that the Sun must exert a force on the planets that varied with the inverse square of the distance. Some years later he started preparing his masterpiece *Philosophiae naturalis principia mathematica*, which he completed in less than three years of intense labour and which was published in the summer of 1687.

From a cosmological point of view, *Principia* is important mainly because it offered the first mathematically formulated explanation of all known celestial phenomena based on a single set of physical laws.[6] The core of Newton's system of the world was the universally valid law of gravitation, in modern terminology that any two masses (or mass points) m and M separated by a distance r attract each other with a force F given by

$$F = G\frac{mM}{r^2},$$

where G is a constant of nature. It had been suggested earlier that the same laws governed terrestrial and celestial phenomena, but Newton's claim was superior because it included a definite law that could be applied mathematically and had an impressive explanatory range. The action that Newton postulated between material bodies was a *force* acting over distances, something very different from the impact actions in the Cartesian plenum.

The meaning and nature of the gravitational force soon became controversial, and Newton wavered over the years about how to conceive it. To many natural philosophers in about 1700, his concept of gravity seemed to be an 'occult quality', a quasi-Aristotelian notion that had to be rejected. Forces or actions were supposed to be mechanically explainable, yet Newton had dismissed Descartes's vortices and he had nothing mechanical to offer as a substitute. In *Principia* he adopted an instrumentalist attitude and famously wrote: 'I have not as yet been able to deduce from phenomena the reason for these properties of gravity, and I do not feign hypotheses. . . . And it is enough that gravity really exists and acts according to the laws that we have set forth and is sufficient to explain all the motions of the heavenly bodies and of our sea.'[7]

In Book II of *Principia*, Newton subjected Descartes's theory to devastating criticism. First, he demonstrated that there could not exist a cosmic material medium, such as the one postulated by the Cartesians, as it would have catastrophic consequences for the motions of the planets; second, Descartes's vortices were assumed to be self-sustaining, a claim that Newton refuted; third, vortical motion did not yield the relations stated by Kepler's three laws. The visualizable celestial vortices were chimera with no scientific justification, he concluded. The heavens, far from being a plenum, were almost completely devoid of matter: 'The hypothesis of vortices . . . serves less to clarify the celestial motions than to obscure them.'[8]

Newton's rejection of the Cartesian system went back some twenty years before the publication of *Principia*. In an unpublished manuscript from about 1667, called *De gravitatione*, 25-year-old Newton argued that space and matter were unconnected. Space was infinite, he held, 'For we cannot imagine any limit anywhere without at the same time imagining that there is space beyond it.' While we can imagine 'that there is nothing in space, yet we cannot think that space does not exist, just as we cannot think that there is no duration, even though it would be possible to suppose that nothing whatever endures.'[9] At that time Newton thought of the universe as a limited sidereal system surrounded by an infinite space. He explicitly mentioned that empty space—such as known from 'the spaces beyond the world, which we must suppose to exist'—was something different from the void. Although he did away with Descartes's plenum, he did not replace it with a vacuum but with an ethereal substance which was universally diffused through absolute space. However, later in life he denied the existence of a subtle matter that carried the celestial bodies in their orbits. In a letter to Leibniz of 1693 he emphasized that interplanetary space was devoid of matter and that gravity alone was responsible for the motions of the planets.

Contrary to Descartes, Newton insisted that the universe could not be fully understood by the laws of mechanics alone. He found it 'unphilosophical . . . to pretend that it [the world] might arise out of a Chaos by the mere Laws of Nature',[10] a direct repudiation of Descartes. The wonderful uniformity in the planetary and sidereal systems was only possible because they were constructed and maintained by an intelligent agent. In the 'General Scholium' of the *Principia*, Newton noticed that 'the light of the fixed stars is of the same nature with the light of the Sun'. This was an important astrophysical observation, but for Newton it had wider implications, as he considered it a theological argument for monotheism. The uniformity of nature, as illustrated by the Sun and the stars being like bodies emitting the same light, surely demonstrated that all of nature's objects are 'subject to the dominion of One'.

Newton's universe was mechanical but neither deterministic nor free of vital principles and spirits. On the contrary, such non-mechanical principles were all-important to keep the universe going, for 'Motion is more apt to be lost than got, and is always on the decay.' This comment of Newton's reflected an attitude shared by almost all British writers and natural philosophers at the time, the effectiveness of nature's denudational forces.[11] The topography of the Earth was gradually decaying, a process which was as natural as it was inevitable. The concept of denudation was primarily applied to the surface of the Earth, but there was no reason why it should not be applicable to the universe in its entirety.

In a passage which may bring to mind much later discussions concerning possible counter-entropic processes (see Section 2.4), he wrote: 'If it were not for these principles, the Bodies of the Earth, Planets, Comets, Sun, and all things in them, would grow cold and freeze, and become inactive Masses; and all Putrefaction, Generation, Vegetation and Life would cease, and the Planets and Comets would not remain in their Orbs.' Also contrary to Descartes, Newton believed that the laws of nature might well have been different from what they are, and that there might in principle be other worlds where these alternative laws were realized. This was possible because of God's omnipotence, as he explained in *Opticks*:

> And since Space is divisible *in infinitum*, and Matter is not necessarily in all places, it may also be allow'd that God is able to create Particles of Matter of several Sizes and Figures, and in several Proportions to Space, and perhaps of different Densities and Forces, and thereby to vary the Laws of Nature, and make Worlds of several sorts in several Parts of the Universe.

It needs to be emphasized that Newton's cosmos was far from a perfect machine or a clockwork universe. The clockwork metaphor was popular in the seventeenth century and frequently employed by Leibniz, among others, but Newton never mentioned the image of the clock. Leibniz and other critics believed that a perfect God would have created a perfect world, at least in the sense of being the best of all possible worlds, a machine in no need of maintenance. In 1715, Leibniz wrote about the 'very odd opinion' of the Newtonians:

> According to their doctrine, God Almighty wants to wind up his watch from time to time: otherwise it would cease to move. He had not, it seems, sufficient insight to make it a perpetual motion. . . . According to my opinion, the same force and vigour remains always in the world, only passes from one part of matter to another, agreeably to the laws of nature, and the beautiful pre-established order.

But Newton and his protagonists found such a world view dangerously close to deism and crypto-materialism. Samuel Clarke, Newton's spokesman in the controversy with Leibniz, retorted: 'The notion of the world's being a great machine, going on without the interposition of God, as a clock continues to go without the assistance of a clockmaker, is the notion of materialism and fate, and tends (under the pretence of making God a *supra-mundane intelligence*) to exclude providence and God's government in reality out of the world.'[12]

Leibniz disagreed with Newton on a number of issues, including the nature of space and time, but he shared his belief in a materially infinite universe, which he found to be 'more agreeable' to God's wisdom. While there could be no spatial bounds, he admitted the possibility of a temporal bound, a beginning of the universe. Yet Clarke maintained that if one accepted Leibniz's relational view of space and time one would also have to accept temporal infinity. This Leibniz denied. With a reference to the discussion in the Middle Ages, he added that even if the world had existed for an eternity it would still depend upon God, and hence be created. Leibniz further speculated that God's mind necessarily contained an infinity of worlds, each of which was possible in the sense of being logically consistent. However, God had actually created only our universe, which was the best possible world, if not a perfect one. This idea has a certain similarity with later hypotheses of multiple worlds, except that Leibniz did not conceive his alternatives as actually existing. His entire discussion of other possible worlds belonged to philosophy and theology, not to cosmology in a physical or astronomical sense.

During the seventeenth century, cosmology not only was an astronomical science but also was associated with geological attempts to understand the formation of the Earth and the changes of its surface. The century witnessed a number of speculative cosmogonical scenarios, some of which relied on assumptions about the universe at large. In view of the static conception of the universe, it is remarkable that these scenarios were thoroughly evolutionary. Descartes offered a mechanical history of how the solar system had come into being, and Thomas Burnet, a contemporary of Newton, proposed another cosmogony which appealed as much to the Old Testament as to the new science. In his *Telluris theoria sacra* of 1681 Burnet reconstructed the Earth's history, starting from a primeval chaos and going through a sequel of six cosmic phases. The current world was only a transitory stage in the development, as it would give way to a global conflagration after which paradise would be restored. Burnet's evolutionary scenario caused much debate, but it was not the only one of its kind. In 1696 it was followed by William Whiston's *New Theory of the Earth*, a work which was more in line with Newtonian theory and which was received favourably by Newton. Also, Leibniz was engaged in the science of evolutionary cosmogony, although his *Protogaea* only appeared in 1749, many years after his death.

2.1.3 *Correspondence with Bentley*

As indicated, Newton believed that the world, if left to itself, would gradually dissolve. But dissolution was not necessarily the end, for he also believed, at least in his later years, that the decay might be counterbalanced by a replenishing of activity in the cosmos, possibly including the formation of new planets from the light of the Sun. Even if the world were destroyed, turned into a chaos, a new world might emerge out of the ashes of the former.[13] This kind of cyclical world view had roots in Greek antiquity and was entertained by several of Newton's British contemporaries, with regard to either the Earth or the universe at large. It also appeared, if only briefly, in the important correspondence of 1692–93 that Newton had with the young theologian Richard Bentley, later Master of Trinity College, who at the time was preparing for publication the first series of Boyle Lectures.[14] In his third letter, Newton considered the possibility 'that there might be other Systems of Worlds before the present ones, and others before those, and so on to all past Eternity'. Or, alternatively, 'that this System had not its Original from the exhaling Matter of former decaying Systems, but from a Chaos of Matter evenly dispersed throughout all Space'. However, he was careful not to defend the theologically dangerous idea of a phoenix universe, which he dismissed as 'apparently absurd'.

Newton's universe was a dynamic system kept in careful balance between disruptive and constructive forces. Every part of it interacted dynamically with every other part in order to keep the delicate balance that, ultimately, could be traced back to God's benevolent design. This was the case for the universe at large, just as it was the case for the solar system, as he told Bentley in his first letter. Jupiter and Saturn were much greater and more massive bodies than the other planets, and they were also placed very far away. According to Newton, this was no coincidence, for these

> qualifications surely arose, not from their being placed at so great a distance from the Sun, but were rather the cause why the Creator placed them at great distance. For, by their gravitating powers, they disturb one another's motion very sensibly, . . . and had they been placed much nearer to the Sun and to one another, they would, by the same powers, have caused a considerable disturbance in the whole system.

In his correspondence with Bentley, Newton abandoned the notion of a limited sidereal universe contained in an infinite space and instead adopted a space uniformly filled with stars. He argued that if the world was finite, all matter would eventually coalesce into one huge central mass, and therefore went on to consider an infinite sidereal system. Even in this case, his favoured model of the universe, it was hard to imagine how the stars could be 'so accurately poised one among another, as to stand still in a perfect Equilibrium' (he did not consider the possibility of proper motions among the stars). Yet, in spite of his insight that such a universe might be gravitationally unstable, he believed it was possible, 'at least by a divine power'.[15] In his *Confutation of Atheism* of 1693, Bentley confirmed what Newton had taught him: 'The continuance of this Frame and Order for so long a duration as the known ages of the World must necessarily infer the Existence of God. For though the Universe was Infinite, the Fixt Starrs could not be fixed, but would naturally convene together, and confound System with System.'[16]

This is the first appearance of the so-called gravitation paradox, in a more modern formulation the problem that in an infinite universe of uniform mass density the gravitational potential is undefined. This problem was to play an important role in theoretical cosmology until the time of Einstein, and we shall meet it again in later chapters. Newton intended to investigate the problem more closely, but nothing came of it. In one of his last comments on cosmology, in the 1726 edition of *Principia*, he stated that God had placed the fixed stars at enormous distances from each other in order that they should not collapse gravitationally, hardly a convincing argument. He was unable to provide a physical explanation of the stability of an infinite universe uniformly populated with stars.

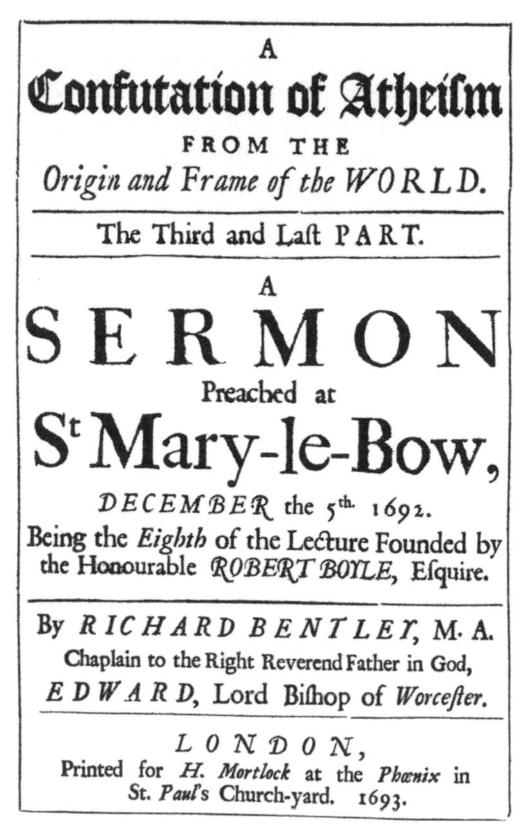

A

Confutation of Atheiſm

FROM THE

Origin and Frame of the WORLD.

The Third and Laſt PART.

A

SERMON

Preached at

St Mary-le-Bow,

DECEMBER the 5th. 1692.

Being the *Eighth* of the Lecture Founded by the Honourable *ROBERT BOYLE*, Eſquire.

By *RICHARD BENTLEY*, M. A.

Chaplain to the Right Reverend Father in God,

EDWARD, Lord Biſhop of *Worceſter*.

LONDON,

Printed for *H. Mortlock* at the *Phœnix* in St. *Paul's* Church-yard. 1693.

Fig. 2.2 Richard Bentley's *Confutation of Atheism* did much to disseminate Newton's view of an infinite universe governed by the law of gravity.

Edmund Halley, for one, did not recognize the insight that Newton had taught Bentley. In a paper read to the Royal Society in 1720, Halley stated that in an infinite universe each star would be attracted by equal forces in any direction and therefore be in a state of equilibrium. But this was precisely the misunderstanding that Newton had corrected in his correspondence with Bentley. Two infinities do not cancel. To Halley and his contemporaries—not to mention later generations—an infinite number of stars was a most strange notion. It implied that there were more stars than any finite number, and how can such a number be construed? Yet Halley did not consider the objection fatal to an infinite universe filled throughout with stars, for 'by the same Argument we may conclude against the possibility of eternal Duration, because no number of Days, or Years, or Ages, can compleat it'.[17]

As Newton's natural philosophy was gradually accepted, first in England and later on the Continent, his view of an infinite stellar universe governed by gravitational forces also won increasing approval. By the mid eighteenth century many scientists also accepted Newton's argument that if gravity remains unchanged (and if divine intervention is excluded), the universe becomes more and more unstable and will eventually come to a catastrophic end.

At least one Newtonian astronomer, James Ferguson, realized that the scenario could be used as an argument against the eternity of the world. In a book of 1757, he explained:

For, had it [the world] existed from eternity, and been left by the Deity to be governed by combined actions of the above [Newtonian] forces or powers, generally called Laws, it had been at an end long ago. . . . But we may be certain that it will last as long as was intended by its Author, who ought no more to be found fault with for framing so perishable a world, than for making men mortal.[18]

More than a century later, Ferguson's argument for a finite-age universe would be reformulated in a thermodynamic context, with entropy replacing mechanical decay, a theme to which we shall return in Section 2.4. Although dressed here in terms of Newtonian mechanics, the argument relied on a more general line of thinking and can be seen as a late version of the concept of denudation, so popular in the second half of the seventeenth century. Indeed, in 1677 Matthew Hale, a British judge, published *The Primitive Origination of Mankind*, in which we find a similar kind of argument: 'That if the World were eternal, by the continual fall and wearing of Waters all the protuberances of the Earth would infinite Ages since have been levelled, and the superficies of the Earth rendred plain, no Mountains, no Vallies, no inequalities would be therein, but the Superficies thereof would have been as level as the Superficies of the water.'[19]

2.2 Enlightenment cosmologies

Cosmology in the age of reason was characterized by two trends of different nature. On the one hand, with the construction of larger and better telescopes, astronomers penetrated farther out in the universe and discovered that it was inhabited by new and strange objects, such as nebulae. It was on the basis of astronomical observations that William Herschel proposed his cosmological theory at the end of the century. The other trend had little to do with observation and much to do with philosophy of a speculative kind. Inspired by progress in stellar astronomy, but in no way restricted by it, scientists and philosophers developed a number of theories of the entire universe. A few of these theories, such as Kant's, included an evolutionary perspective of the universe.

It is often stated that acceptance of an infinite universe followed the victory of Newton's physics in the eighteenth century. But it is not generally the case that the Enlightenment marked a transition from the closed world to the infinite universe. On the contrary, infinity continued to be seen as a strange and unwelcome concept that one could not attach any physical meaning to and therefore had to be avoided in cosmological reasoning. Kant and a few others disagreed, but they were exceptions.[20]

Although the discovery of the finite velocity of light was generally accepted by the early eighteenth century, astronomers were curiously reluctant to recognize that when we look far out in space we also look far back in time. This was known, but it only made an impact on cosmological thought with William Herschel at the end of the century. In order to illustrate the 'prodigious vastness' of the stellar universe, the Englishman Francis Roberts wrote in 1694 that 'Light takes up more time in Travelling from the Stars to us, than we in making a West-India Voyage (which is ordinarily performed in six Weeks). . .[and] a Sound would not arrive to us from thence in 50,000 Years, nor a Canon-Bullet in a much longer time.'[21] For the speed of light, Roberts adopted Newton's value from *Principia*, corresponding to a travelling time from the Sun to the Earth of 10 minutes.

2.2.1 *Advances in astronomy*

With improved telescope technology, which involved not only more powerful telescopes but also sophisticated use of the filar micrometer, in the eighteenth century the fixed stars moved to the forefront of astronomical research.[22] Since the days of Hipparchus, the stars had been regarded as fixed—occupying the same positions relative to one another—a notion that Newton still took for granted. The first to question the orthodoxy was Halley, who in a communication of 1718 compared modern observations of the stars with those reported by the Greeks, such as in the star catalogue of the *Almagest*. He came to the conclusion that the only way to account for the discrepancies was to ascribe to three of the brightest stars (Aldebaran, Sirius, and Arcturus) a southerly motion; although his data for the fainter stars did not allow a similar conclusion, he believed that they, too, performed proper motions.

Halley's claim, although confirmed with respect to Arcturus by Jacques Cassini in 1738, was not generally accepted until the question was taken up for systematic examination by the Göttingen astronomer Johann Tobias Mayer. Rather than using as comparison data Greek observations, he compared modern observations with those made by the Danish astronomer Ole Rømer only half a century earlier. In his *De motu fixarum proprio* ('On the Proper Motion of the Stars') of 1760, Mayer reported incontrovertible evidence that 80 of the brightest stars had changed their positions relative to the equatorial frame of reference. After Mayer's important work, the proper motions of the stars became a reality and 'fixed stars' merely a name of historical convention.

The search for proper motion was mixed up with the search for observational evidence that the Earth revolved around the Sun. By the time of Newton's death no astronomer seriously questioned (at least not privately) that this was the case, but solid proof was lacking. Such proof, or something close to it, came with the discovery of the aberration of light. In an attempt to measure the parallax of the star Gamma Draconis, James Bradley, a young vicar who would eventually become Astronomer Royal, was puzzled by the star's extraordinary movements. After much confusion he arrived at the conclusion that what he had discovered was not a stellar parallax but an effect of the Earth's motion about the Sun. As he explained in his paper 'A new discover'd motion of the fixed stars', published in *Philosophical Transactions* in 1728, the precise position of a star would depend on the ratio between the Earth's orbital velocity and the finite velocity of light.

Bradley's discovery had profound implications, not only for astronomy but also, if more indirectly, for cosmology. For one thing, it amounted to a near proof of the fundamental assumption of the Copernican system: after 1728, the only way to resist the annual revolution of the Earth was to deny Rømer's discovery of 1676, that light propagates with a finite speed. Moreover, since all stars showed the same aberration, it followed that the speed of light was the same, irrespective of the distance it had travelled through space. The speed of light appeared to be a constant of nature, a quantity like Newton's constant of gravitation. Using the value 137.73 million km for the astronomical unit, Bradley found for the speed of light $c = 279\,939$ km/s, only seven per cent smaller than the modern value.

Bradley believed that he was able to determine changes in position as small as 1″, and since he had found no parallax for Gamma Draconis he concluded that the distance of the star must be greater than 400 000 AU. His result agreed nicely with Newton's argument, based on the method of magnitudes, that the nearest stars were at a distance of about one million astronomical units. The two methods were entirely independent and therefore provided convincing evidence that interstellar distances had to be counted in millions of astronomical units. Needless to say, this was a result of the greatest significance.

As mentioned, the discovery of the aberration of light amounted to a proof of the Earth's revolution around the Sun, but it was an indirect one only. What was lacking was still the old problem of stellar parallaxes, a phenomenon that had been looked for since the days of Tycho Brahe (and even earlier) but which seemed to defy resolution. Numerous were the reports of astronomers announcing that now, finally, a stellar parallax had been detected; and equally numerous were the disappointments when it turned out on closer inspection that this was not the case after all. It was only in the late 1830s—nearly three hundred years after Copernicus' *De revolutionibus*—that the thorny question was resolved. The Russian–Baltic astronomer Friedrich Wilhelm Bessel studied the double star system 61 Cygni with an instrument (a heliometer) that allowed resolution of the components, only 1.2″ apart. In 1838 he announced that he had found for the star an annual parallax of 0.3136″, corresponding to a distance of 657 000 AU.[23] This time the claim was not retracted.

Although the discovery of the first stellar parallaxes was of great importance, it was for reasons other than what originally had motivated the search for parallaxes. After all, the Copernican system had long ago been accepted as a fact, scarcely in need of final confirmation. The result of Bessel and his contemporaries was not a cosmological sensation, but it did prove that stellar distances could be determined by observational means. The distances to the nearest stars turned out to be in fair agreement with what Newton had predicted and what Bradley had estimated. Of course, it was assumed that the distances to the fainter stars were much greater, but by how much nobody could tell.

2.2.2 *From Wright to Buffon*

The age of reason was a great era for cosmologies of a speculative and philosophical kind, grand schemes of the universe which paid lip service to physics and astronomy but rarely more than that. Yet some of these schemes included brilliant insights that changed the course of cosmological thinking. Among the early philosophical cosmologies of the century was *Cosmologia generalis*, a book of 1731 written by the influential Leibnizian philosopher Christian Wolff. Whatever its scientific merits, *Cosmologia* is worth mentioning because it was the first book that included in its title 'cosmology' in a sense not limited to theology. The Englishman Thomas Wright, an author and teacher, began at the same time to speculate about the composition of the universe and its relation to God. In 1750 he published *An Original Theory of the Universe*,[24] in which he postulated that all the stars comprising our Milky Way, including the Sun, are in orbital motion around a centre. This centre was divine, the place from where God's infinite power emanated. Wright further stated that there existed in the universe other star systems, similar to our Milky Way, and that these surrounded their own local divine centres.

In order to bring into agreement a spherically arranged stellar universe with the apparent concentration of stars around a plane in the Milky Way, Wright suggested that the shell of stars was thin and, because it was so vast, only slightly curved. By hypothesis, the solar system was part of the spherical shell and so, when we looked along the tangent plane, we would see numerous stars with the appearance of the Milky Way. (When we looked in a direction perpendicular to the plane, we would only see few, bright stars.) Contrary to what has often been claimed, Wright did not propose that the Milky Way was a disc-shaped conglomerate of stars with the Sun near its centre, a picture which was wholly foreign to his theologically based cosmology. Whereas Wright's cosmology as described in *An Original Theory* was largely stable and static, in a later draft essay, *Second or Singular Thoughts upon the Theory of the Universe*, he argued that the universe was in a state of evolution. In his new version the universe was Sun-centred, the Sun being likened to a circulating fire which was responsible for the dynamic features of the world. Whatever the version, it is important to realize that Wright's cosmology was thoroughly moralistic–theological and that its purpose was not primarily to account scientifically for the observed universe.

An Original Theory was not widely read, but it did exert some influence on the course of cosmology, mainly through Immanuel Kant, the famous philosopher. Kant, who at the time was an unknown *Privatdozent* at the University of Königsberg in Prussia, never read the book, but he was acquainted with its main content from a detailed review that appeared in a Hamburg journal in 1751. By his own account, the review inspired him to develop a cosmological system of his own, which appeared in a small book of 1755, *Allgemeine Naturgeschichte und Theorie des Himmels*.[25] Kant was no newcomer to cosmological

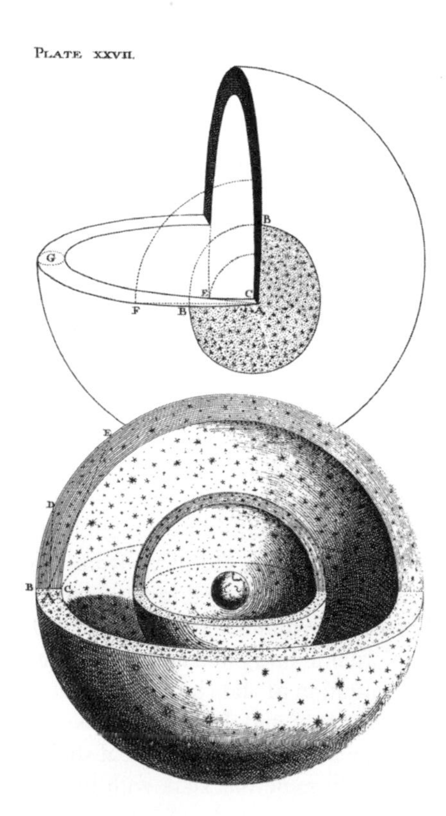

Fig. 2.3 Thomas Wright's universe of 1750. The Sun and the Earth are located in the outer shell together with most of the fixed stars. In the centre is the divine 'sacred throne' in the direction of one of the poles of the Milky Way. From Wright 1971, plate 27.

Allgemeine

Naturgeschichte

und

Theorie des Himmels,

oder

Versuch

von der Verfassung und dem mecha-
nischen Ursprunge

des ganzen Weltgebäudes

nach

Newtonischen Grundsätzen

abgehandelt.

* * * * * * * * * * * * * *

Königsberg und Leipzig,

bey Johann Friederich Petersen, 1755.

Fig. 2.4 The title page of Kant's treatise on the constitution and evolution of the world. The complete title is *Universal Natural History and Theory of the Heavens, or an Essay on the Constitution and Mechanical Origin of the Whole Universe Treated According to Newton's Principles.*

speculations, as witnessed by his very first book of 1749, where the 25-year-old philosopher claimed from theological arguments that there must exist universes with any number of spatial dimensions, not only the three-dimensional world we live in. Kant's 1755 book marked a new phase in the history of cosmology, primarily because it presented a thoroughly naturalistic and evolutionary account of the universe in its totality. It was a cosmogony in the tradition of the ancient atomists, a grandiose attempt at a *Universal History and Theory of the Heavens*, as the title reads in English. Although Kant referred frequently to God and presented his theory as theistic, in reality the references to the Creator were largely rhetorical except for the original creation of matter. Contrary to Newton, but in agreement with Descartes and Leibniz, he found no place for divine miracles in the universe. 'A world-constitution, which without a miracle does not maintain itself does not have the character of steadyness which is the hallmark of God's choice,' he wrote.[26]

Kant started with a primeval, divinely created chaos of particles at rest, distributed throughout an infinite void. This initial chaos was unstable, he said, and the denser particles would begin to attract the more tenuous, and thus form condensations. With Descartes he claimed that, given the laws of nature, the primary chaos must necessarily evolve into regular and orderly structures—a definite cosmos. In order to avoid the possibility that all the particles coalesced into one mass, Kant introduced a repulsive force acting between the particles. However, he did not provide any law for the repulsion and never explained how the mechanisms of attraction and repulsion (to which he added collision) could produce the angular momentum of a rotating system. He merely claimed that 'Through this repulsive force, . . . the elements sinking toward their points of attraction become directed sidewise in all sorts of ways and the perpendicular fall issues in circular motions which surround the center of sinking.'[27] Although Kant's system violated the law of conservation of angular momentum, it would be anachronistic to dismiss it on that count: by 1755, the law had not yet been generally formulated and accepted.

Whatever the details, Kant claimed to be able to explain the formation of the solar system, and went on to generalize his system of formation to still larger structures. However large these structures were, they rotated about a common centre, which contradicted another of Kant's basic assumptions, that the universe was infinite. For how could there be a centre in an infinite universe? Kant acknowledged that, geometrically, there could be no central point, but asserted that, from a physical point of view, the original seed of cosmic evolution 'can have the privilege to be called the centre'.

One of Kant's great insights was that the Milky Way had a disc-like structure, that it was a flattened conglomerate of a multitude of stars encircling a galactic centre. Even more innovatively, he suggested that the nebulous stars were not individual stellar objects, but vast congeries of stars of the same type and structure as the Milky Way. He stated clearly that the nebulae were 'just universes and, so to speak, Milky Ways'. The world was an island universe, the islands being nebulae floating in an immense sea of void space. But the islands were not isolated, for they were themselves members of even larger structures, the hierarchical arrangement continuing indefinitely throughout the infinite depths of space. Kant's insight was to some extent foreshadowed by the French polymath Pierre-Louis Moreau de Maupertuis, who in his *Discours sur les différentes figure des astres* of 1732 discussed the shapes of the nebulous stars. Maupertuis likened the objects to celestial millstones formed by a revolving fluid material, and he also pointed out that they might appear in shapes varying from a sphere to a flattened oval, depending on their position relative to the line of sight. However, he did not recognize that the nebulae might be large aggregations of stars.

Infinitude, evolution, and creation were key notions in Kant's dynamic cosmology. He found it imperative that the world must be infinite in space, as only such a universe accorded with the attributes of God. But God had not created the universe in its present state, it had slowly developed from the primeval chaos governed by the laws of nature: 'One can therefore posit with good basis that the ordering and arrangement of the world-edifice should gradually happen from the supply of the created nature-stuff.' Kant's cosmic creation was anything but creation once and for all. He wrote of the creation process as a sort of wave propagating from a central area of the (infinite!) universe, bringing life, activity and organization with it:

> The sphere of developed nature is incessantly busy in expanding itself. Creation is not the work of a moment. . . . millions and entire mountains [bundles] of millions of centuries will flow by, within which

always new worlds and world-orders form themselves one after another in those reaches [so] distant from the center of nature and reach perfection. . . . Creation is never completed. Though it has once started, but will never cease. It is always busy in bringing forth more scenes of nature, new things and new worlds.[28]

In Kant's vision, destruction was no less important than creation. Entire worlds perished and were 'devoured by the abyss of eternity', but at the same time destruction was counteracted by creative processes from which new cosmic formations resulted. It was in the very nature of finite things, however big, that they would eventually decay. 'We must not, however, bemoan the decay of a world-edifice as a true loss of nature. . . . The infinity of creation is great enough to view in relation to it[self] a world or a Milky Way of worlds [in the same way] in which man looks upon a flower or an insect in comparison with the Earth.'[29] The constructive and destructive forces of the universe were subject to definite laws, for example that the celestial bodies nearest to the centre of the universe would perish first. Kant was not very precise with respect to these laws, which were at any rate untestable, but asserted that the developed world was bounded in the middle between the ruins of nature that had been destroyed and the chaos of nature that was still unformed.

Kant even speculated that the entire world, or parts of the world, might return to a chaotic state and then re-emerge, possibly an infinity of times. 'Can one not believe that nature, which was capable of placing herself from chaos into a regular order and a skilful system, will not be in the position to restore herself from the new chaos, into which the diminution of her motions had lowered her, just as easily and to renew the first combination?' Kant had no problem with believing such a scenario of 'this phoenix of nature, which burns itself out only to revive from its ashes rejuvenated, across all infinity of times and spaces.'[30]

Allgemeine Naturgeschichte remained little known for a long time, and Kant himself seems to have had second thoughts about his cosmogony. At one point he referred to it as a 'weak sketch'. At any rate, when he wrote his famous *Kritik der reinen Vernunft*, published in 1781, he concluded that the notions of age and extent were meaningless when applied to the universe as a whole. The universe was not an object of possible experience, not something that existed objectively, but a regulative principle of merely heuristic value. In his so-called first antinomy, Kant proved by means of a *reductio ad absurdum* argument the thesis that 'The world has a beginning in time, and is limited also with regard to space.' He then went on to prove the anti-thesis, that 'The world has no beginning and no limits in space, but is infinite, in respect both to time and space.' Since the concept of the world is thus contradictory, it cannot cover a physical reality.[31] In his argument, Kant relied crucially on assumptions of Newtonian physics, such as mechanical determinism and Euclidean space. Although these assumptions would turn out to be unwarranted, in Kant's time they were taken for granted and remained so for more than a century. Kant's *Kritik* provided a penetrating analysis of cosmological problems, but it was 'cosmology' in the philosophical and not the scientific sense.

In *Cosmologische Briefe* of 1761 the polymath Johann Heinrich Lambert, an accomplished German mathematician and philosopher, proposed a cosmological theory which in some respects had similarities to Kant's. The book was an honest if by no means clear attempt to deal scientifically with the entire system of the world, but of course it included heavy doses of theology and philosophy as well. On the basis of a mixture of Newton's physics and the teleological philosophy of Leibniz and Wolff, Lambert adopted a picture of the nebulae similar to that proposed by Kant, and he also argued for a hierarchically organized universe in which the same disc-like pattern was displayed on all its levels. However, although his universe was inconceivably vast, it was finite nonetheless. Also in contrast to Kant's dynamic cosmos, it did not evolve in time, for 'the heavens are made to endure, and the things of the Earth to pass away'. In the concluding chapter of *Briefe*, Lambert recapitulated: 'The Sun being of the number of fixed stars, revolves round the centre like the rest. Each system has its centre, and several systems taken together have a common centre. Assemblages of their assemblages have likewise theirs. In fine, there is a universal centre for the whole world round which all things revolve. Those centres are not void, but occupied by opaque bodies.'[32]

Apart from those already mentioned, several other Enlightenment natural philosophers took up cosmological questions. One of them was the Croatian–Italian astronomer and physicist Roger Boscovich, a Jesuit scholar, who in 1758 published his main work *Theoria philosophiae naturalis*. Although best known for its contribution to dynamical atomism and matter theory, the book also included considerations of a cosmological nature. For example, Boscovich imagined that, apart from our space, there might exist other spaces with which we are not causally connected. His conception of the universe was relativistic, such as illustrated by a passage from the end of *Theoria*, which may bring to mind much later cosmological ideas:

> If the whole Universe within our sight were moved by a parallel motion in any direction, & at the same time rotated through any angle, we could never be aware of the motion or the rotation. . . . Moreover, it might be the case that the whole Universe within our sight should daily contract or expand, while the scale of forces contracted or expanded in the same ratio; if such a thing did happen, there would be no change of ideas in our mind, & so we should have no feeling that such a change was taking place.[33]

Boscovich imagined all matter to consist of point-atoms bound together by Newtonian-like attractive and repulsive forces. If no forces were present, a body might pass freely through another without any collision (after all, points have no extension in space). The possibility led him to a daring cosmological speculation: 'There might be a large number of material & sensible universes existing in the same space, separated one from the other in such a way that one was perfectly independent of the other, & the one could never acquire any indication of the existence of the other.'[34] Boscovich did not elaborate. Here we have, in 1758, a new version of the many-universe scenario: not different universes distributed in space and time, but coexisting here and now. It was surely a scenario that harmonized in spirit with ideas that some cosmologists would propose more than two hundred years later.

Perhaps the most important innovation in Kant's cosmology was that he provided the universe with an evolutionary perspective. The emphasis on time and development was novel and can be found in many authors in the second half of the eighteenth century, whether dealing with astronomy, geology, or natural history. If the universe had a history, and if it was created (as of course it was), it also must have an age.[35] Traditionally this age had been inferred from the Bible and estimated to be about 6000 years, but this timescale soon came to be seen as wholly inadequate. Benoît de Maillet, a French diplomat, argued in about 1720 that the Earth had once been covered by water, and by estimating the rates of decline and evaporation he reached the result that the Earth must have existed for more than two billion years. De Maillet prepared a manuscript with his arguments, and *Telliamed*, as the book was titled (the author's name spelled backwards), appeared in 1748, anonymously and posthumously.

The great French naturalist Georges Louis Leclerc, Comte du Buffon, had no patience with Biblical chronology and in his *Époque de la nature* of 1778 he decided to determine the age of the Earth experimentally. Assuming that the Earth was originally formed in a hot molten state, he performed a large number of experiments with red-hot balls of iron and stone and determined the time until they had cooled to room temperature. Extrapolating from his experimental globes to the real Earth, Buffon arrived at the staggering figure of about 75 000 years as the age of the Earth. Privately, he concluded that this figure was much too low and that the age of the Earth was probably closer to a couple of million years. This may not sound much to modern ears, yet at the time it was an age difficult to grasp, scarcely different from the 'Millions and whole myriads of millions of centuries' that Kant had visioned. Buffon was neither an astronomer nor a cosmologist, but his empirically established value for the age of the Earth nonetheless had cosmological implications. The universe must evidently be older than (or as old as) the Earth, and so Buffon's value established for the first time a scientifically based lower limit for the age of the universe.[36]

2.2.3 *Olbers' paradox*

The darkness of the night sky, if combined with the assumption of an infinite space uniformly filled with stars, may present a cosmological problem. Although the light received from a star varies with the inverse square of the distance, and so is negligible for distant stars, the number of stars at a certain distance increases with the square of the distance. Consequently, we should receive so much starlight that the sky at night is as bright as on a sunny day. It is not, of course, and this is the essence of what is known as Olbers' paradox.[37] This puzzle or paradox has a curious history. Although it has been known for nearly four hundred years, for most of the time it was considered neither paradoxical nor particularly interesting. Today the darkness of the night sky counts as an important cosmological fact, but it is only recently that Olbers' paradox has been situated within the domain of cosmology. Before the First World War, the paradox was rarely seen as relating to the universe as a whole.[38]

Kepler seems to have been the first to note the paradox. As we have seen, he was convinced that the universe was finite, and among his arguments in *Dissertatio cum nuncio sidereo* of 1610 was that 'in an infinite Universe the stars would fill the heavens as they are seen by us'. William Stukeley, a physician and antiquarian, became acquainted with Newton in early 1718, and in one of their first conversations they discussed questions of astronomy and cosmology. Stukeley raised the question of an infinite space populated throughout with stars and pointed out that, in this case, 'the whole hemisphere would have had the appearance of that luminous gloom of the Milky Way. We should have lost the present sight of the beauty and the glory of the starry firmament.'[39] It may have been from Stukeley that Halley came to appreciate the paradox, which he examined in a brief paper of 1720, 'Of the Infinity of the Sphere of Fix'd Stars', published in *Philosophical Transactions*. A convinced infinitist, Halley's purpose was to demonstrate that it was not a valid objection against an infinite stellar universe. His demonstration, brief and somewhat obscure, rested on the claim that the intervals between the stars in an infinite system decreased linearly with distance while the intensity of light from them decreased with the square of the distance.

Jean-Philippe Loys de Chéseaux, a 26-year-old Swiss astronomer, published in 1744 an essay on a comet that had recently appeared. In an appendix, he analysed in considerable detail the paradox of the night sky. Following Halley in conceiving the stars to be situated in concentric spherical shells, he stated that the quantity of light emitted from each shell was proportional to the sum of the squares of the apparent diameters of the stars in the shell. He did not consider the possibility of shielding, i.e. that stars may fall on the same line drawn from the Earth (neither did Halley). Chéseaux concluded that the paradox would arise not only if the number of stars was infinite, but also if it was exceedingly large (but finite), namely larger than a value fcorresponding to 76×10^{13} concentric shells. His solution to the paradox was new. Rather than opting for a finite, relatively small stellar universe, he suggested that the intensity of starlight decreased at a greater rate than given by the squared-distance law. Chéseaux was familiar with Pierre Bouguer's work on photometry and light absorption in media, and his answer was that there existed in interstellar space an ethereal fluid which absorbed starlight. He argued that the problem would be solved if the medium was only 33×10^{16} times more transparent than water (that is, if the intensity of light were to diminish by 3% when passing through a layer with a thickness equal to the diameter of our solar system).

The postulate of a light-absorbing medium, first proposed by Chéseaux, would become the standard answer to Olbers' paradox in the nineteenth century. However, in the previous century Chéseaux's work was by and large met with silence. One of the few Enlightenment scientists who did respond to the problem of the dark night sky was Lambert, who was as confident that the stellar universe was finite as Chéseaux was confident that it was infinite. Lambert pointed out that the paradox rested on the basic assumption that the stars were uniformly distributed in space, an assumption that need not be accepted and that Lambert denied.

'Is space not endless? Is it possible to conceive limits of it? And is it conceivable that the creative omnipotence should have left empty that infinite space?'[40] While the first two questions were rhetorical—it was generally accepted that space was unlimited—the third was not. However, at the time when Olbers wrote these lines, the pendulum had swung in favour of an infinite space filled throughout with stars. Heinrich Wilhelm Olbers, an eminent astronomer who earned his living as a physician, realized that the question could not be resolved on observational grounds, but for philosophical reasons he strongly believed in an infinite, Newtonian universe. That the context of his paper of 1826 (which was submitted three years earlier) was cosmological is evidenced by its introduction, a lengthy quotation from Kant's *Allgemeine Naturgeschichte*. Olbers' motive was the same as Chéseaux's, to demonstrate that the riddle of the night sky was not truly paradoxical. His formulation of the riddle was that 'should there really be suns in the whole infinite universe . . . the whole sky should be as bright as the Sun'. To defuse the paradox he relied on the same assumption that Chéseaux had made, that interstellar space was not perfectly transparent, an assumption which he found was most natural. What came to be known as Olbers' paradox was not, in Olbers' view, a paradox at all.

Although the Chéseaux–Olbers hypothesis of interstellar absorption was widely accepted— for example, in 1837 it received support from Friedrich Struve—it was not without problems. In 1848, at a time when the first law of thermodynamics had been fully formulated and accepted, John Herschel pointed out in the *Edinburgh Review* that radiant heat from the stars, when absorbed, would heat up the interstellar medium until it reached a state of thermal equilibrium and itself became radiant.[41] On the other hand, Herschel realized that the infinite universe could be reconciled with the dark night sky even without assuming light absorption, if the stars were arranged in a suitable, non-uniform way. Nothing is easier, he wrote, 'than to imagine modes of systematic arrangement of the stars in space . . . which shall strike away the only foundation on which [the problem] can be made to rest, while yet fully vindicating the absolute infinity of space'.[42] Herschel did not elaborate his proposal, but a cosmological model of the kind he may have envisaged was constructed by the Swedish astronomer Carl Charlier sixty years later.

What if light absorption is abandoned and a hierarchical universe is considered too artificial? Even in that case there is a solution to Olbers' paradox, an alternative first proposed by the German astronomer Johann Mädler, who in 1858 called attention to the following argument:

> The world is created, and hence is not eternal. Thus no motion in the universe can have lasted for infinite time; in particular, this applies to a beam of light. In the finite amount of time it could travel before it reached our eye, a light beam could pass through only a finite space no matter how large the speed of light. If we knew the moment of creation, we would be able to calculate its boundary.[43]

Clearly, if we only receive light from stars within a certain horizon, Olbers' paradox need not arise. Mädler repeated his suggestion three years later in a popular book, but it failed to make an impact on his colleagues in astronomy. However, in 1872 it turned up again in an essay 'On the Finiteness of Matter in Infinite Space' written by the German astrophysicist Karl Friedrich Zöllner, and in 1888 a similar suggestion was made by George Ellard Gore, an Irish author on astronomy. Zöllner, whose essay is of considerable cosmological interest, mentioned the possibility, but without endorsing it. For philosophical reasons he found the solution unsatisfactory, and instead he came up with another and highly original solution to Olbers' paradox, namely to modify the standard assumption that space is Euclidean. Acquainted with the works of Gauss, Riemann, and other pioneers of non-Euclidean geometries, Zöllner suggested that cosmological space might be positively curved. Olbers' paradox would dissolve, he wrote, 'if we ascribe to the constant curvature of space not the value zero but a positive value, however small'.[44] This is, to my knowledge, the first time that non-Euclidean geometry was applied to cosmology.

Zöllner's proposal went as unnoticed as that of Mädler. When William Thomson (then Lord Kelvin) took up the matter in 1901, he was unaware of the contributions of Mädler and Zöllner and may not even have known of Olbers' work.[45] Yet his route to solving what was in fact Olbers' paradox was similar to that proposed by Mädler. Thomson considered a stellar Milky Way universe uniformly filled with stars up to the enormous distance of 3.3×10^{14} light years and took into consideration that each star has a finite lifetime. He believed that a star could shine for at most 100 million years, which meant that the time it took for light to travel from one of the stars farthest away would be about 3.3 million times the lifetime of a star. To 'make the whole sky aglow with the light of all the stars' would require that the periods of time during which the various stars shone were very precisely correlated to their distances from the Earth. This was completely unlikely and for this reason Thomson's reference to a finite, relatively small stellar lifetime provided a solution to Olbers' pradox.

Because of the belief in interstellar absorption, Olbers' paradox was not taken very seriously in the nineteenth century. Only during the early part of the new century did evidence indicate that space was much more transparent than had been assumed, and then Olbers' paradox did become paradoxical, if still not widely noticed. In 1917, the American astronomer Harlow Shapley formulated the paradox in essentially the same way that Halley had done two centuries earlier, albeit with a different conclusion: 'Either the extent of the star-populated space is finite or "the heavens would be a blazing glory of light". . . Then, since the heavens are not a blazing glory, and since space absorption is of little moment throughout the distance concerned in our galactic system, it follows that the defined stellar system is finite.'[46]

2.2.4 *The construction of the heavens*

The discovery of proper motions of the stars inevitably raised the question of whether the solar system itself was moving through space relative to the neighbouring stars. And, if it did, with what speed and in which direction? As Bradley had clearly perceived in 1748, the problem was to disentangle the true proper motions of the stars from the apparent motion caused by a moving solar system carrying the Earth with it. In 1760, Tobias Mayer explained that if the solar system were moving towards some region (the apex), 'all the stars which appear in that region would seem to be gradually separating from each other one by one, and those which are in the opposite part of the sky would seem to be joining up.'[47] Could such a pattern be discerned? Mayer looked for it in his data of proper motions, but found nothing. He concluded pessimistically that it would take centuries until the problem might be resolved. The great French astronomer Joseph-Jérôme Lalande agreed in a memoir of 1779, yet only four years later William Herschel found just such a pattern from a smaller number of proper motions.

William Herschel, one of the greatest astronomers ever, never received formal training in astronomy or any other science. He was born in Hanover as Friedrich Wilhelm Herschel, and when he came to England as a young man he earned his living by teaching what he knew best, which was music. (He had been an oboist in the Hanoverian Guards, was a skilled organ player, and composed symphonies and choral works.) In England, Herschel developed an all-absorbing interest in astronomy, a subject he taught himself, in part by reading and in part by making his own, excellent telescopes. In the construction of his large mirror telescopes, and in his scientific work in general, he received invaluable help from his sister, Caroline Herschel, herself a talented amateur astronomer.[48] William Herschel's fate changed when, in 1781, he discovered a new celestial object, which he first thought was a comet but soon turned out to be a planet—Uranus. Following in the wake of this momentous discovery, he moved to the neighbourhood of Windsor Castle, where he became a kind of personal astronomer to the king, George III. He was also elected a Fellow of the Royal Society and promptly awarded its Copley Medal.

Herschel's astronomical interests were broad and he contributed to most parts of the science, including cosmology. In his 1783 paper 'On the Proper Motion of the Sun and Solar System', he determined the direction of the Sun's course through space, concluding that the solar apex was in the constellation Hercules, close to the star Lambda Herculis.[49] His remarkably precise sky coordinates were largely verified by Pierre Prévost later in 1783 and more fully by Georg Simon Klügel in 1789. In later studies from 1805–06

Herschel returned to the problem, now aiming at determining the speed of the Sun, but without arriving at a convincing result. For a period there continued to be considerable uncertainty with regard to the proper motion of the solar system. Only in 1837 did the German astronomer Friedrich Argelander unequivocally demonstrate that the Sun moves toward a point 6° north of Lambda Herculis, a position not far from that found by Herschel 54 years earlier.

Whereas Herschel's work on proper motions was theoretical, during the following years he concentrated on an ambitious observational programme of scanning or 'sweeping' the night sky for stars and nebulae with his new 20-foot reflector. In order to determine the structure of the Milky Way, which was one of his chief aims, he made two assumptions: first, that the stars were distributed nearly uniformly throughout the space covered by the Milky Way; and, second, that his telescope could reach to the very limits of the star system, and in all directions. These assumptions implied that the more stars he saw in a given direction, the deeper was the extent of the sidereal system in that direction. From this there followed a method of estimating relative distances, given by the cube root of the number of stars seen in the field of the telescope. From such considerations Herschel pieced together the architecture of the Milky Way. He concluded, as Kant and a few others had done earlier, that our Milky Way was only one nebula among many others. Herschel was not the first to propose an island universe theory, but he was the first to support the theory with observational evidence.

In 'The Construction of the Heavens', Herschel's important paper of 1785, he started with some general considerations on the stability of star systems which were very much in the tradition of Newton, Bentley, and Halley. Just as Newton had reassured us, Herschel was confident that 'the great Author' had constructed the system in such a way that gravitational collapse would be avoided or would only occur in an indeterminable future. On the other hand, even if large-scale destruction took place in the universe, this might be God's way to keep it eternally alive: 'We ought perhaps to look upon such clusters, and the destruction of now and then a star, in some thousands of ages, as perhaps the very means by which the whole is preserved and renewed. These clusters may be the *Laboratories* of the universe, if I may so express myself, wherein the most salutary remedies for the decay of the whole are prepared.'[50] This passage is strikingly similar to what Wright and Kant had stated earlier. (Herschel was unaware of Kant's *Allgemeine Naturgeschichte* and, although he owned a copy of Wright's *Original Theory*, it did not influence him.)

As has become clear, Herschel's universe had a history. 'I have looked further into space than ever human being did before me', he told a friend, and mentioned that he had observed stars at a distance of more than two million light years. 'If those distant bodies had ceased to exist millions of years ago, we should still see them, as the light did travel after the body was gone.'[51] Herschel's analysis resulted in a picture of the Milky Way as a 'very extensive, branching, compound Congeries of many millions of stars' with the solar system near the middle of the giant structure. As mentioned, the picture depended on several presuppositions, of which the most problematical was the assumption that the 20-foot reflector was able to spot the farthest objects of the Milky Way. This, he came to realize, was not the case. When he made use of his new 40-foot telescope many more stars became visible, which made him wonder if the number and distances were a matter of telescopic power. The old Herschel consequently modified his earlier view and concluded, somewhat agnostically, that the Milky Way might be 'fathomless' and its extent unknown.

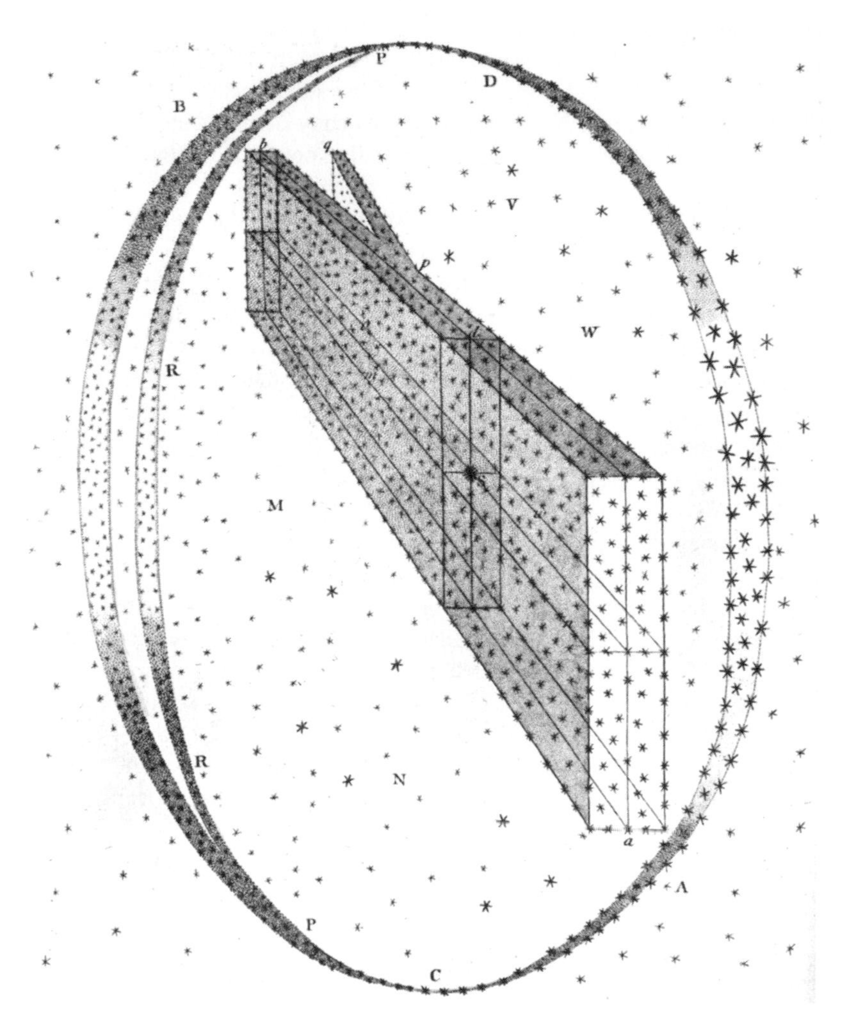

Fig. 2.5 William Herschel's explanation of the Milky Way, as proposed in a paper of 1784. The Sun is located near the centre of the system. From *The Scientific Papers of Sir William Herschel*, vol. 1 (London: Royal Society, 1912), plate 8.

In another paper on the construction of the heavens, published in 1789, Herschel philosophized about how the slow evolution in the heavenly bodies could be recognized empirically. Evidently, the astronomer could not focus on a single nebula and follow its development over time. But he could nonetheless form an evolutionary picture of the universe, namely by collecting data from different parts of it, some far away and others closer to the Earth. Herschel expressed the method in—well—flowery language:

[The heavens] now are seen to resemble a luxuriant garden, which contains the greatest variety of productions, in different flourishing beds; and one advantage we may at least reap from it is, that we can, as it were, extend the range of our experience to an immense duration. For, to continue the simile I have borrowed from the vegetable kingdom, is it not almost the same thing, whether we live succesively to witness the germination, blooming, foliage, fecundity, fading, withering, and corruption of a plant, or whether a vast number of specimens, selected from every stage through which the plant passes in the course of its existence, be brought at once to our view![52]

From the mid nineteenth century onwards, it became common to compare the myriads of stars to the molecules of a gas, an analogy which eventually would be extended to galaxies. The analogy relied on a picture of gases consisting of molecules in motion but can be found even earlier. John Dalton, the founder of chemical atomism and a younger contemporary of William Herschel, did not think of a gas as made up of molecules in swift motion, yet he knew that the molecules were tiny and separated from each other at 'a respectful distance'. In his main work, *A New System of Chemical Philosophy* of 1808, we find, probably for the first time, a picture of stars as the molecules of the universe. 'When we attempt to conceive the number of particles in an atmosphere, it is somewhat like attempting to conceive the number of stars in the universe,' he wrote.[53]

2.3 Astrophysics and the nebulae

Astrophysics, an invention of the nineteenth century, significantly changed the course of astronomy. Moreover, it also had important consequences for cosmology, although these were only fully recognized in the following century. The emergence of the field was closely related to the introduction of spectroscopy, which from its very beginning was applied to the study of the stars. Indeed, for a couple of decades astrophysics and astrospectroscopy were nearly synonymous terms.

Until about 1800 light from the celestial bodies was thought of as just light signals, without structure and therefore with no particular information hidden in them. Thanks to advancements in optical instruments, however, it turned out that the spectrum of light extended to invisible rays at both ends of the spectrum and also, even more importantly, that analysis of light received from stars could yield information about the stars' physical and chemical composition. The very notion of a 'star' changed from a geometrical to a physical concept, a change which implied a profound transformation of the astronomical sciences. Since antiquity astronomy had been thought of as an observational science using mathematical methods, and nothing more. It was a science that observed and hypothesized about the motions of celestial bodies, not one that could possibly deal with the nature of those bodies. As mentioned in Section 1.4, in 1588 Tycho Brahe affirmed the traditional view that it was not part of astronomy to investigate 'what heaven is and from what cause its splendid bodies exist'.

As late as 1832, Bessel emphasized in a lecture that the business of astronomy was restricted to precise measurements of the positions and orbits of celestial bodies. In a letter to the great naturalist Alexander von Humboldt, he wrote 'Everything else that one may learn about the [heavenly] objects, for example their appearance and the constitution of their surfaces, is not unworthy of attention, but is not the proper concern of astronomy.'[54] At about the same time, the French philosopher Auguste Comte composed his massive *Cours de philosophie positive*, a pioneering work of positivist philosophy of science. Comte was well versed in astronomy but, like most of his contemporaries, was unable to imagine that the physics and chemistry of the stars could ever be subjects of scientific study. In an often quoted passage, he wrote about the stars:

> We conceive the possibility of determining their forms, their distances, their magnitudes, and their movements, but we can never by any means investigate their chemical composition or mineralogical structure, still less the nature of the organic beings that live on their surface etc. In short, to put the matter in scientific terms, the positive knowledge we can have of the stars is limited solely to their geometrical and mechanical phenomena, and can never be extended by physical, chemical, physiological, and social research, such as can be expended on entities accessible to all our diverse means of observation.[55]

Comte further advised the astronomers to restrict their attention to the solar system, to what he called 'the world'. This solar or planetary world should be separated from the idea of 'the universe', for certain knowledge could only be attained about the former. 'As for those innumerable stars scattered in the sky, they have scarcely any interest for astronomy other than as markers in our observations, their position being regarded as fixed relative to the movements internal to our system, which alone concern us'.[56] It soon turned out that Comte was wrong, and seriously so.

2.3.1 *Astrospectroscopy*

With the benefit of hindsight, one may trace the origin of astrophysics to optical discoveries made in the early years of the nineteenth century. In measurements of the temperature increase in a thermometer placed at various positions in a solar spectrum, William Herschel observed in 1800 that the increase was greatest (9 °F) beyond the red end of the visible spectrum. Apparently the Sun emitted invisible heat rays—infrared rays—in addition to its visible rays. Although Herschel was able to demonstrate that the new rays followed the familiar optical laws of reflection and refraction, for several decades it remained a matter of dispute whether radiant heat was a phenomenon like ordinary light, but only of longer wavelength. In the wake of Herschel's discovery Johann Wilhelm Ritter, a German *Naturphilosoph* and leading electrochemist, suggested the existence of rays beyond the violet part of the visible spectrum. Herschel had found no heating effect in this area, but in 1801 Ritter proved the existence of 'chemical rays', or ultraviolet light, by means of their blackening effect on paper impregnated with silver chloride (that is, a photochemical effect). The extension of the light spectrum demonstrated by Herschel and Ritter did not immediately lead to advances in astronomy, yet in the long run it would prove most important for astronomy and astrophysics.

William H. Wollaston, a London chemist and retired physician, noticed in 1802 seven dark lines in the spectrum of the Sun. However, he mistakenly believed these to be natural boundaries between colour zones in the spectrum rather than lines originating in the illuminating source. It was only in 1814 that the mistake was corrected, in the course of a careful and systematic study of the solar spectrum made by Joseph Fraunhofer, a Bavarian optician and instrument maker.[57] Fraunhofer found, as Wollaston had done earlier, dark spectral lines, but he interpreted them differently and greatly extended the number of them. Although he did not offer an explanation for the mysterious dark lines, of which he charted nearly 600, he was convinced that they were intrinsic to sunlight, that they somehow originated in processes in the Sun itself. That the dark lines, or 'Fraunhofer lines', were not specially connected with sunlight was recognized in the 1830s, when it was shown, first by the Scotsman David Brewster, that they could be produced artificially in the laboratory by passing white light through a gas.

Fraunhofer's discovery was not much noticed at the time, and it took more than thirty years until the enigmatic dark lines attracted wide attention among physicists and astronomers.[58] It was gradually realized that each chemical element was identifiable by its emission spectrum, and also that the emission lines coincided with the dark lines observed when continuous light from a glowing solid or liquid traversed a cooler gas of that element (such as sodium, which in the vaporized state yields a gas that absorbs certain wavelengths of the continuous spectrum: the dark absorption lines coincide with sodium's double yellow emission line). Spectrum analysis was established on a firm basis in 1859–60 by the German physicist Gustav Robert Kirchhoff, partly in cooperation with the chemist Wilhelm Bunsen, his colleague at the University of Heidelberg. In experiments using Bunsen's new gas burner the two German scientists demonstrated that the emission spectra of the chemical elements coincided with their absorption spectra and could be used to identify the elements. Kirchhoff immediately pointed out that this had important astronomical consequences: 'The dark lines of the solar spectrum . . . result from the presence of that substance in the luminous solar atmosphere which produces in the flame spectrum bright lines in the same place. . . . The dark D lines in the solar spectrum allow one therefore to conclude, that sodium is to be found in the solar atmosphere.'[59]

The work of Kirchhoff and Bunsen effectively founded the chemical study of the Sun and the stars, and it also led Kirchhoff to a theoretical study of the thermodynamics of radiant heat, an area of research that eventually would lead to quantum theory. By means of thermodynamic arguments, Kirchhoff investigated the properties of what he called black-body radiation, proving that the radiant energy emitted from an ideal black body equals the energy absorbed by that body. Later in the century, Kirchhoff's black-body radiation became the subject of intense experimental and theoretical studies, culminating in Max Planck's celebrated radiation law of 1900 which heralded the age of quantum theory and modern atomic physics.

The first prism spectroscopes, based on the original design of Kirchhoff and Bunsen, were primarily used to study the chemical composition of the Sun, comets, and other celestial bodies, but it soon turned out that the modest optical instrument had wider areas of application. In 1842 the Austrian physicist Christian Doppler announced the effect named after him, that if a light source is moving relative to the observer with radial velocity v there will be a change in wavelength given by

$$z = \frac{\Delta\lambda}{\lambda} = \frac{\lambda' - \lambda}{\lambda} = \frac{v}{c},$$

where c is the velocity of light and λ' signifies the measured wavelength, redshifted relative to the emitted wavelength λ. As Einstein proved in his paper of 1905 in which he introduced the theory of relativity, if the recession velocity is large the formula becomes

$$1 + z = \sqrt{\frac{1 + v/c}{1 - v/c}}.$$

Einstein's expression reduces to Doppler's in the limit $v \ll c$.

The effect was soon verified for sound waves, whereas Doppler's claim that it was valid also for light remained controversial for many years.[60] Several attempts to detect stellar

Doppler shifts by means of the spectroscope were made in the 1860s, and in 1868 the British gentleman astronomer and pioneer of astrospectroscopy William Huggins announced a positive result. Comparing the H_β line in the spectrum of Sirius with that produced by a Geissler discharge tube filled with hydrogen, he found a shift in wavelength of about one angstrom.[61] On the assumption that this shift was due to a Doppler effect, it implied a recession velocity of Sirius of 29.4 miles per second, a result which was wrong both in amount and sign but nonetheless was widely accepted as proof of stellar Doppler shifts. Huggins realized that his result was controversial because it contradicted the generally held view of a static universe, and consequently he was careful in formulating his claim.

For some time the situation continued to be unclear and it was only in the 1880s that the validity of the optical Doppler effect was firmly demonstrated, not by using stellar motions but by examining the rotation of the Sun. Accurate measurements of radial velocities began with the work of the German astronomer Hermann Vogel, director of the Potsdam Astrophysical Laboratory, who in work with his colleague Julius Scheiner from 1888 to 1891 obtained results with an error of only 3 km/s. With this accuracy, he could easily demonstrate photographically the orbital motion of the Earth around the Sun. It took another twenty years until the Doppler effect for light was detected in the laboratory, first by the German physicist Johannes Stark in 1905.

2.3.2 *Chemistry of the stars*

In a lecture of 1807 the Danish physicist Hans Christian Ørsted, the discoverer of electromagnetism, prophesied that 'some day chemistry will have just as much influence on cosmology as mechanics so far . . . and all natural science will finally become a cosmogony'.[62] Ørsted's prophecy eventually became reality, namely when the spectroscope was used to gain information about the chemical composition of stars and nebulae. Astrospectroscopy provided chemical science with a greater perspective and promised an extension of terrestrial chemistry to a cosmic or celestial chemistry. As the British chemist Henry Roscoe expressed it in a lecture of 1875, '[We] now possess means for extending our knowledge of the chemistry of the universe beyond the narrow limits of our tiny planet . . . by help of the peculiar light which the Sun and fixed stars emit we are able to ascertain their chemical composition, and to lay the foundation of a celestial chemistry.'[63]

The astrochemistry that emerged in the 1860s opened up new and exciting questions to be answered by means of the spectroscope. What is the chemical constitution of the Sun and the stars? Do there exist chemical elements in the stars that are not found on the Earth? Do stellar spectra provide evidence of matter in a primordial state and of the complexity of the chemical atom? These were some of the questions asked by William Crookes, Norman Lockyer, Angelo Secchi, and other researchers within astrophysics and chemistry. A major aim of the astrospectroscopists was, in the words of Huggins, 'to discover whether the same chemical elements as those of our Earth are present throughout the universe'. He concluded that this was indeed the case and that 'a common chemistry . . . exists throughout the universe'.[64] Yet a few scientists, including Crookes and Lockyer, believed that there were elements in the heavens that did not exist on Earth. Spectroscopy was the most important technique for studying cosmic matter, but it was not the only one. It was assumed that meteorites could be analysed as probes of the chemical composition of the universe, and such analyses demonstrated that the elements of meteorites were the same as those of the Earth, but also that the abundance distribution of elements in meteorites differed from that found in terrestrial minerals.

A small group of researchers, in particular in Great Britain, believed that the stellar spectra indicated the complexity of the chemical atom and that there might exist elemental matter in the stars different from that found on the Earth. In an address to the 1886 meeting of the British Association for the Advancement of Science, Crookes brilliantly speculated on the nature and origin of the elements. Anticipating later ideas in cosmology, he invited his audience to 'picture the very beginnings of time, before geological ages, before the earth was thrown off from the central nucleus of molten fluid' and to 'imagine that at this primal stage all was in an ultragaseous state, at a temperature inconceivably hotter than anything now existing in the visible universe; so high, indeed, that the chemical atoms could not yet have been formed, being still far above their dissociation point'. According to Crookes, the elements were formed cosmologically through processes of 'inorganic Darwinism'.[65] Even more daringly, he ventured to look into the distant past before any matter existed:

> Let us start at the moment when the first element came into existence. Before this time matter, as we know it, was not. It is equally impossible to conceive of matter without energy, as of energy without matter; from one point of view the two are convertible terms. . . . Coincident with the creation of atoms all those attributes and properties which form the means of discriminating one chemical element from another start into existence fully endowed with energy.

Crookes was not the only Victorian chemist who associated the supposed evolution of the elements with cosmological speculations. Benjamin Brodie, an Oxford professor of chemistry, developed an unorthodox system of what he called 'ideal chemistry', on which subject he delivered a lecture in 1867. The lecture included a remarkable anticipation of much later ideas of cosmological element formation:

> We may conceive that, in remote time or in remote space, there did exist formerly, or possibly do exist now, certain simpler forms of matter than we find on the surface of our globe. . . . We may consider that in remote ages the temperature of matter was much higher than it is now, and that these other things existed then in the state of perfect gases . . . We may then conceive that the temperature began to fall, and these things to combine with one another and to enter into new forms of existence . . . We may further consider that, as the temperature went on falling, certain forms of matter became more permanent and more stable, to the exclusion of other forms. . . . We may conceive of this process of the lowering of the temperature going on, so that these substances, when once formed, could never be decomposed—in fact, that the resolution of these bodies into their component elements could never occur again. You would then have something of our present system of things.[66]

The power of the spectroscope as a detector of chemical elements was convincingly demonstrated in 1860 when Kirchhoff and Bunsen discovered a new metallic element, caesium, from its blue spectral lines; the success was duplicated the following year with the discovery of rubidium and Crookes's discovery of thallium. Although most spectral lines from the stars could be identified with lines known from laboratory experiments, there were some unidentified lines that might indicate the presence of elements, or states of elements, particular to the stars. On the basis of stellar spectra, scientists claimed the

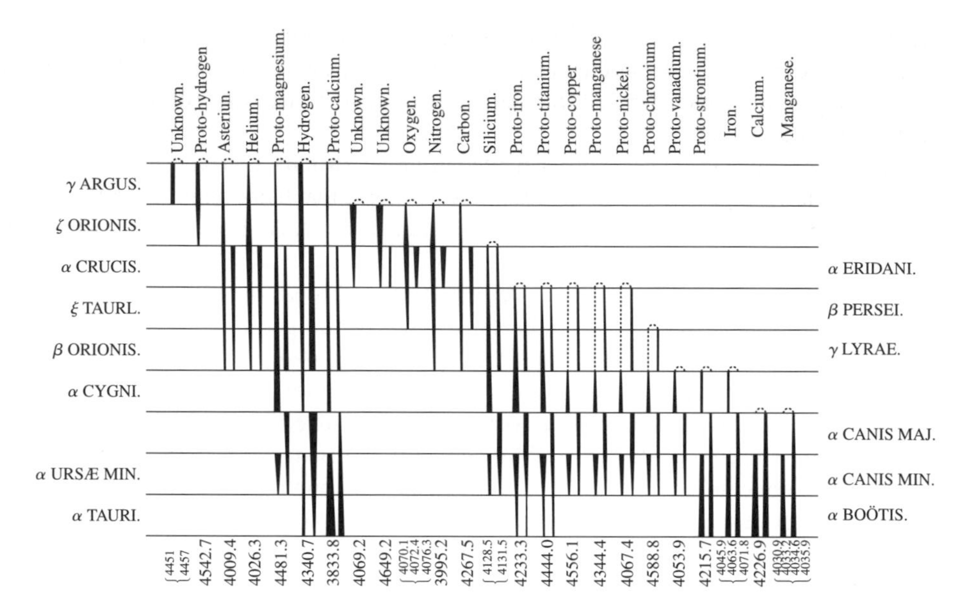

Fig. 2.6 Norman Lockyer's comparison of spectral lines and bands from different stars. He believed that some of the lines were due to elements in forms unknown on Earth (asterium, helium, proto-hydrogen, and several proto-metals). From N. Lockyer, *Inorganic Evolution as Studied by Spectrum Analysis* (London: Macmillan, 1900), p. 62.

existence of several spurious elements, including 'coronium' (1871), 'nebulium' (1898), and 'asterium' (1900). However, not all the claims of celestial elements were wrong.

In 1868, Lockyer studied spectral lines from a solar prominence and noticed a yellow line with a wavelength of 5876 angstroms, smaller than that of the D doublet lines from sodium. The new line, which he named D_3, did not correspond to any line from a known element, nor to any of the Fraunhofer lines in the solar spectrum. Lockyer consequently supposed that he had detected a new element that might exist only in the Sun, and for which he suggested the name 'helium'. For most of two decades helium remained a ghost element that the majority of chemists refused to accept, but in 1895 its status changed abruptly when William Ramsay identified the gas in the terrestrial mineral cleveite. Meanwhile, Lockyer had found the D_3 line in the absorption spectra of certain stars. Helium was originally believed to be exceedingly rare, and since it was chemically inert it was considered little more than a curiosity. At that time nobody could foresee helium's central role in the universe and how importantly knowledge of that element and its cosmic abundance would influence cosmological science.

Lockyer, a self-taught Victorian sage and founder of the journal *Nature*, was a leading proponent of the so-called dissociation hypothesis, the idea that the chemical elements in the stars were broken down to smaller and simpler forms of matter.[67] He first proposed the hypothesis in 1873 and expounded it fully in his *Chemistry of the Sun* of 1887; over the next two decades he continued to defend and develop it. Lockyer reasoned that if chemical compounds dissociated in a gas burner, at the much higher temperatures in the stars the elements themselves might plausibly dissociate. His enthusiasm for element dissociation and the complex atom made him experiment with electrical-discharge decomposition of elements, and in 1879 he announced that he had succeeded in actually decomposing some elements into hydrogen (impurities were the cause of his blunder).

The dissociation hypothesis, in one of its several versions, was popular among British scientists in particular. For example, it was of direct importance to J. J. Thomson in his celebrated experiments of 1897 which resulted in the discovery of the electron, the first known elementary particle. Indeed, Thomson considered the electron a kind of chemical proto-element and referred in his 1897 address to Lockyer's 'weighty arguments' in favour of the composite atom. In 1903 he discussed why the hydrogen atom was the lightest of all known atoms and why there was a limited number of chemical elements. According to Thomson, the elements had gradually evolved during the long cosmic history; atoms lighter than hydrogen had once existed, but later had formed aggregations, the smallest of which happened to be hydrogen. The following year he discussed the opposite hypothesis, that the final stage of the universe would consist of the simplest atoms, which he identified with electrons (or 'corpuscles' as he insisted on calling them).

Speculations about dissociation and evolution of the chemical elements were entertained also by George Ferdinand Becker, an American geologist. Impressed by the discoveries of the electron and radioactivity, Becker suggested in 1908 a periodic system based on 'the cosmic distribution of the elements', probably the first of its kind. Becker distinguished between elements found on the Earth and those found in meteorites, the Sun, the stars, and the nebulae. He concluded that by charting the distribution of the elements in this way 'the plausibility of the evolution hypothesis is increased' and that his table pointed to 'the truth of the hypothesis that elements are evolved, those of highest molecular weight being youngest and confined to cooling stars or planetary bodies'.[68] At about the same time, the British physicist John Nicholson contributed to the astrochemical tradition by suggesting that proto-elements such as coronium and nebulium existed in stellar atmospheres and that astrophysical data provided evidence for the complex structure of atoms. Foreshadowing much later developments in the physics of elementary particles, he suggested that, in the future, fundamental physics would have to rely on astrophysics:

> Astronomy, in the wider interpretation of its scope which is now general, owes much to Physics. . . . A point appears to have been reached in its [astronomy's] development at which it becomes capable of repaying some of this debt, and of placing Physical Science in its turn under an obligation. . . . The reason for the possibility of this position of astrophysics—as an arbiter of the destinies of ultimate physical theories—is . . . [that] the terrestrial atoms are apparently in every case too complex to be dealt with by first principles . . . [but] when an astrophysicist discovers hydrogen in a spectrum, he is dealing with hydrogen in a simpler or more primordial form than any known to a terrestrial observer.

Moreover, referring to the primitive elements that he and others assumed to exist in the stars but no longer on Earth, he wrote, 'The astronomer . . . may have a field of chemical study which is closed to the chemist.'[69] Nicholson was not the first to use the metaphor that the stars were celestial laboratories—as mentioned, it goes back to William Herschel—nor was he the last.

2.3.3 *The riddle of the nebulae*

William Herschel started observations of the nebulae in 1781 and five years later he published his *Catalogue of One Thousand Nebulae and Clusters of Stars*. A major aim of his observation programme was to elucidate the nature of the objects, but no easy answer was forthcoming. On the one hand, he was convinced that he had seen changes in the nebula in Orion, which could therefore not be a star system; on the other hand, he was able to resolve many nebulae into stars and more or less suspected that it was only because of the great distances that not all nebulae could be resolved. For a few years he thought that all nebulae were clusters of stars, although he was cautious not to state this conclusion categorically. And wisely so, for in 1790 he discovered what he called a 'nebulous star' which caused him to conclude that truly nebulous objects—'shining fluids' perhaps comparable to the aurora borealis—did after all exist in the heavens. He continued to hold this view, which became an integral part of his cosmological speculations concerning the evolution of star clusters from nebulae, which he described in his 1811 paper on the construction of the heavens. According to Herschel's version of the nebular hypothesis, some of the nebulae were composed of hot gaseous clouds that would undergo different stages of condensation and end up as stellar objects.[70]

By 1830 the nebular hypothesis had won wide acceptance, not least because it seemed to support and widen the eminent French physicist Pierre-Simon Laplace's nebular theory of the origin of the solar system, first proposed in 1796 in his *Exposition du système du monde*. Although Laplace was not a cosmologist in the sense of either Kant or Herschel—his applications of celestial mechanics were largely restricted to the solar system—his work was highly significant to what was considered cosmology in his own time.[71] For example, Laplace proved that 'the world' was a mechanically stable system and that there was no reason to fear that perturbations or frictional forces would one day cause the system to disrupt, such as Newton had believed.

During parts of the nineteenth century the nebular hypothesis was associated with the fashionable view of nature being in a state of continual evolution, and for this reason it was rated highly by evolutionists many years before Darwin gave a new meaning to evolution. John Pringle Nichol, a professor of astronomy in Glasgow, was among the champions of evolutionary astronomy. Contrary to most of his colleagues, he believed that astronomy should not be concerned merely with what could be observed from telescopes on Earth. His cosmo-evolutionary vision covered not only stars and nebulae, but also the universe at large, and in his case the vision was tightly connected with his religious belief in a progressive and teleological development.[72] Cosmic progressivist as Nichol was, though, he shared the view of many of his contemporaries that the solar system, and possibly the entire universe, was decaying. However, he saw no reason for pessimism in the ultimate dissolution of the solar system, for he convinced himself that evolution was the grand design of the universe and the overall tendency was towards progress. This was the happy message that ran through his popular book *The Architecture of the Heavens* of 1851.

In spite of his interest in the nebular hypothesis and the evolutionary aspects of the universe, Nichol was not a 'cosmologist'. In accordance with the standards of the time, he refrained from treating the universe *in toto* as a subject of astronomical science. This may be illustrated by his successful *Cyclopaedia of the Physical Sciences*, a work of 900 pages which first appeared in 1857. This one-volume encyclopedia included extensive articles on

Fig. 2.7 'The Leviathan of Parsonstown', Lord Rosse's giant reflector, was ready in 1845. For a period of seventy years it was the world's largest telescope.

'Nebulae' and 'Nebular Hypothesis', but the reader would look in vain for entries on 'Cosmology', 'Cosmogony', and 'Universe'.

After William Parsons, the Earl of Rosse, had completed his giant mirror telescope (54 feet in focal length) at Birr Castle, Ireland, the nebular hypothesis began to crumble. Rosse had his telescope ready in 1845 and among the first to use it was Thomas Robinson, an astronomer at Armagh Observatory, near Dublin. Contrary to Nichol, Robinson was a staunch opponent of the nebular hypothesis, which to his mind was associated with dangerous ideas such as evolutionism and materialism. He was therefore happy to conclude from his observations of nebulae that they were all resolvable into stars, a conclusion which agreed with his preconceived dislike of the nebular hypothesis.

Rosse discovered in the same year the spiral structure of the M51 nebula, the first known example of a spiral nebula. In the next year, he found that the Orion Nebula was resolvable into stars. The more nebulae Rosse studied with his huge telescope, known as 'the Leviathan of Parsonstown', the more of them he was able to resolve into clusters of discrete stars. From this it was tempting to infer that all nebulae were star clusters, as Robinson did, but Rosse realized that his observations did not justify the wider claim 'that all nebulosity is but the glare of stars too remote to be separated by the utmost power of our instruments'.[73] Yet this was a claim accepted by most astronomers. For example, in 1849 John Herschel admitted in *Outlines of Astronomy* that there were inductive reasons to abandon the reality of nebulous matter. As he wrote, 'it may very reasonably be doubted whether

Fig. 2.8 With his great reflector, Lord Rosse and his collaborators discovered the spiral structure of many nebulae. These two drawings are based on observations from 1850, reported in the *Philosophical Transactions of the Royal Society of London*. Reproduced from *The Scientific Papers of William Parsons* (London, 1926), p. 111.

there be really any essential physical distinction between nebulae and clusters of stars'. Nichol was led to the same conclusion.

John Herschel was, like most nineteenth-century astronomers, cautious about indulging in cosmological speculations and grand theories about the structure and distribution of the nebulae. Yet he did from time to time expound cosmological views, if only with many qualifications, and usually he did not include them in his publications and public addresses. At the end of his life he corresponded with Richard Proctor, a popularizer of astronomy, and in one of his letters he revealed a glimpse of his vision of an infinite, hierarchical universe not unlike the one that Kant had suggested in 1755. Speaking of the forms of nebulae and clusters of nebulae, John Herschel speculated that,

> if the forms belong to and form part and parcel of the galactic system, then *that system includes within itself miniatures of itself* on an almost infinitely reduced scale; and what evidence then have we that there exists a universe beyond?—unless a sort of argument from analogy that the galaxy, with all its contents, may be *but one* of these miniatures of that vast universe, and so *ad infinitum*: and that in *that* universe there may exist multitudes of other systems on a scale as vast as *our* galaxy, the analog of those other nebulous and clustering forms which are *not* miniatures of our galaxy.[74]

At the time of Herschel's letter the nebular hypothesis had dramatically reappeared as another phoenix from the ashes, a revival that was mainly due to the invention of the spectroscope. But even before this invention there were a few observations in favour of the nebular hypothesis. Thus, in 1852 the British astronomer John Hind found a small nebula in Taurus which over the next couple of years rose in luminosity and then started to fade, to disappear in 1861. Such an appearance strongly suggested that the nebula could not be a huge system of stars.

Huggins knew from Kirchhoff's work that line spectra were produced only by gaseous bodies, whereas hot solids yielded a continuous spectrum. In 1897 he recalled how he had observed the planetary nebula in Draco one August night in 1864. 'I looked into the spectroscope. No spectrum such as I expected! A single bright line only!' To Huggins, there could be no doubt about the significance of the observation: 'The riddle of the nebulae was resolved. The answer, which had come to us in light itself, read: Not an aggregation of stars, but a luminous gas. . . . There remained no room for doubt that the nebulae, which our telescopes reveal to us, are the early stages of long processions of cosmical events, which correspond broadly to those required by the nebular hypothesis in one or other of its forms.'[75]

The origin of the bright green line that Huggins had found (there were a few more lines) was a subject of confusion and controversy, as neither Huggins nor other scientists were able to reproduce the line in the laboratory. For a time he thought it was due to nitrogen, but was forced to abandon the idea. In 1898 he proposed that the unknown nebular gas responsible for the lines was 'nebulium', a hypothetical element that was taken fairly seriously for more than a decade. The nebulium lines were only explained long after Huggins's death. In 1927 the American physicist Ira Bowen used quantum theory to identify the lines as transitions from metastable states of doubly ionized oxygen and nitrogen.[76] Exit nebulium.

The riddle of the nebulae was of crucial importance to another cosmic riddle, of whether or not the nebulae were structures similar to the Milky Way or much smaller objects within it. If all nebulae could be resolved into individual stars, this would amount to a proof of the former view, the island universe theory as proposed by Kant and others. When Huggins in 1864 found bright line spectra in six planetary nebulae, he considered it an argument against the nebular hypothesis and its associated island universe theory, and his further work only confirmed him in this view. The spectroscopic observations of Huggins and others of the gaseous nature of nebulae were a major reason why the island universe view fell in to disfavour by the end of the century. In addition, in 1885 a blazing nova (in fact a supernova) was observed near the centre of the Andromeda Nebula, and the 'new star' was so bright that it was comparable in luminosity to the entire nebula. If the Andromeda Nebula consisted of a myriad of stars, how could a single one of its stars be so bright?

On the other hand, some of the nebulae, including the Andromeda Nebula, showed continuous spectra, which seemed to imply that island universes were at least a possibility. Huggins did not find this a compelling argument, though. In 1889 he preferred to interpret the spectrum of the Andromeda Nebula as coming from single stars physically connected with the nebula rather than a collection of stars outside the Milky Way. He believed that 'the nebula, as a whole, may not be at a distance from us greater than that which we should attribute to such stars, if they occurred alone in the heavens'.[77]

Julius Scheiner, a German astronomer at the Potsdam Astrophysical Observatory near Berlin, succeeded in 1899 in obtaining a spectogram of the Andromeda Nebula. From an examination of the spectrum, he concluded that it was surprisingly similar to the solar spectrum, which made him suggest that the spiral nebulae were huge star clusters comparable to the Milky Way system. 'Since the previous suspicion that the spiral nebulae are star clusters is now raised to a certainty,' he wrote, 'the thought suggests itself of comparing these systems with our stellar system, with especial reference to its great similarity to the Andromeda nebula.'[78] Scheiner's observation obviously supported the island universe theory, but it did not have the effect of significantly changing the balance between this theory and that of the rival Milky Way universe. By about 1910, the theory of island universes had regained some of its former strength, if more because of its grandeur and aesthetic attractiveness than because of observational support. The whole question of island universes remained unsolved until the mid 1920s, and we shall return to it in a later section.

2.4 Thermodynamics and gravitation

Cosmology in the second half of the nineteenth century was not primarily, and certainly not exclusively, an astronomical science. Astrophysical and astrochemical considerations made their impact, and much of the interest in cosmology continued to have its origin in philosophical and theological contexts rather than in contexts of a strictly scientific nature. In contrast to the timidity with which most astronomers approached cosmological problems, physicists, philosophers, and amateur cosmologists addressed fearlessly such problems as the finiteness or infiniteness of the universe, whether in the spatial or the temporal sense. They came up with many interesting speculations, but, perhaps unsurprisingly, with little that could be justified by means of empirical tests. Two fundamental laws of science were at the centre of these discussions, one being Newton's old law of gravitation and the other being the second law of thermodynamics, a product of the mid nineteenth century. These laws concerned different aspects and phenomena, but they had in common that they were believed to be valid for any part of the universe, and perhaps even the universe itself.

2.4.1 *The heat death*

The thermodynamical theory that emerged in the mid nineteenth century was claimed to be universally valid, and for this reason it had important cosmological implications. The two laws on which the theory was founded were supposed to work not only for steam engines and test tube reactions but also for the solar system and perhaps even for the universe at large. The first law stated that the net energy of a closed system remained the same; or, in its cosmological version, that the energy of the universe might change in form but overall it would remain constant. The second law implied a unidirectionality of all natural processes in harmony with the evolutionary world view, and it had consequences for both extremes of the cosmic timescale. If extrapolated to the far future, it indicated that the world would come to an end; and if extrapolated to the far past, it might lead to the conclusion that the world had not always existed but had a beginning in time. Neither of these predictions was beyond criticism or could be tested, but this only made them more popular as subjects of discussion.

The cosmic significance of heat phenomena was occasionally discussed before 1840, notably by Jean-Baptiste-Joseph Fourier in his analytical theory of heat, which he applied to the physics of the Earth and the Sun.[79] The problem of the Sun's heat received a new foundation after the principle of energy conservation (the first law of thermodynamics) was enunciated in the 1840s by scientists such as Julius Robert Mayer, James Joule, and Hermann von Helmholtz. Indeed, the law led Mayer to advance a 'meteoric hypothesis' as early as 1848 and Helmholtz to suggest an alternative theory in 1854 based on the hypothesis of the Sun's gradual contraction. (Both ideas were independently proposed by the Scottish engineer John James Waterston).[80] These were early applications of the law of energy conservation to solar physics, but not to cosmology in the wider sense. It was the second, not the first law of thermodynamics that principally was discussed in relation to the universe at large.

The second law of thermodynamics was formulated only a few years after Helmholtz had written in 1847 the definitive essay on the principle of energy conservation. In his seminal paper of 1850, Rudolf Clausius stressed the natural tendency of heat to equalize temperature, and four years later he reformulated his theory by basing it on a function that in 1865 reappeared under the new name 'entropy'. Armed with his entropy concept, Clausius famously stated the second law of thermodynamics as 'the entropy of the world tends towards a maximum' and similarly expressed the first law globally, as 'the energy of the world is constant'. Although Clausius's formulations referred to the world or universe (*die Welt*), in his later works he only rarely phrased the thermodynamical principles in such global terms.

The cosmological connection was cultivated more fully in William Thomson's alternative route to the second law, the first result of which was 'On the Dynamical Theory of Heat' of 1851. Thomson never used the concept of entropy, but preferred to speak of dissipation of heat or energy, a concept that corresponds roughly to what Clausius conceived as entropy change. In another paper, of 1852, Thomson summarized, 'There is at present in the material world a universal tendency to the dissipation of mechanical energy.'

As a consequence, he stated that 'Within a finite period of time past, the Earth must have been, and within a finite period to come the Earth must again be, unfit for habitation of man as at present constituted, unless operations have been, or are to be performed, which are impossible under the laws to which the known operations going on at present in the material world are subject.'[81]

At the Liverpool meeting of the British Association in 1854, Thomson went a step further. Inviting his audience to trace backwards in time the actions of the laws of physics, he speculated that the source of mechanical energy in the universe might be sought in 'some finite epoch [with] a state of matter derivable from no antecedent by natural laws'. However, such an origin of matter and motion, mechanically unexplainable and different from any known process, contradicted his sense of both causality and uniformitarianism. 'Although we can conceive of such a state of matter,' he wrote, 'yet we have no indications whatever of natural instances of it, and in the present state of science we may look for mechanical antecedents to every natural state of matter which we either know or can conceive at any past epoch however remote.'[82] Here we have the second law used not to predict the far future but to speculate about a singular state in the distant past.

Thomson did not propose a universal 'heat death' in 1852 (although he did suggest a terrestrial heat death), but two years later Helmholtz extended his ideas to the prediction that in the course of time the universe would approach a state of equilibrium and, when this state had been reached, it would be condemned to eternal rest. In the 1860s the heat death scenario was expounded by several leading physicists and entered, either explicitly or implicitly, the physical literature. Clausius, who coined the term 'heat death' (*Wärmetod*), formulated it in terms of entropy in the following way: 'The more the universe approaches this limiting condition in which the entropy is a maximum, the more do the occasions of further change diminish; and supposing this condition to be at last completely attained, no further change could evermore take place, and the universe would be in a state of unchanging death.'[83] Clausius further emphasized that the second law contradicted any idea of a cyclic universe. Not only for Clausius, but also even more so for Thomson and his circle of Christian scientists (which included Maxwell and Peter Guthrie Tait), was it an appealing feature of the second law that it countered what they considered the materialistic and un-Christian notion of a cyclical world.

The claim of the heat death did not go uncontested. Far from it. Many scientists, as well as non-scientists, felt it unbearable that life in the universe (if not the universe itself) should one day cease to exist, and they came up with various suggestions to avoid the scenario, either by devising counter-entropic processes or by questioning the premises upon which the heat death prediction rested. Remarkably, astronomers took very little part in this discussion, possibly a reflection of their reluctance to deal with such a metaphysical concept as 'the universe'. It is striking that works on astronomy and astrophysics rarely included references to the second law of thermodynamics and its cosmological consequences.[84]

As early as 1852, before the heat death hypothesis had been fully formulated, the Scottish engineer and physicist William Rankine suggested that radiant heat might under certain circumstances allow a reconcentration of energy (and hence physical activity) to go on endlessly. He conjectured that radiant heat was conducted by a bounded interstellar medium, and that outside this medium there was nothing but empty space. In that case,

Fig. 2.9 The fate of the Earth and the universe was much discussed in the last part of the nineteenth century, when it was realized that the cosmos and its constituent bodies, including life, might not last forever. The theme is illustrated here in a plate from Camille Flammarion's *Astronomie populaire* of 1880.

when the radiant heat reached the boundary it would be reflected and eventually reconcentrate in one or more focal points. If one further imagines one of the extinct celestial bodies to pass such a focal point, 'it will be vaporised and resolved into its elements', and part of the heat radiation would be converted into chemical energy and wake the body alive. It was thus conceivable that 'the world, as now created, may possibly be provided within itself with the means of reconcentrating its physical energies, and renewing its activity and life'. Dissipative and constructive processes might eternally go on together, 'and some of the luminous objects which we see in distant regions of space may be, not stars, but foci in the interstellar ether'.[85] Clausius would have nothing to do with Rankine's brilliant but contrived speculation, and answered it in 1864 in a long paper in which he made use of Kirchhoff's recent work on black-body radiation. Clausius unambiguously concluded that radiant heat was no exception to the second law and consequently it could provide no escape from the heat death.

Less discussed than the heat death, but of no less cosmological importance, the second law of thermodynamics was also taken to indicate that the universe had a finite age. The 'entropic argument' is simple, if not necessarily convincing: according to Clausius's law, the entropy of the world increases continually towards an equilibrium state, but our present world is obviously far from this state and so cannot be of infinite age. To some commentators in the Victorian era this implied that the world was created supernaturally, which explains why the argument attracted much attention among theologians and Christian scientists (and, for different reasons, among atheist scientists). The entropic argument was enunciated in the late 1860s, in public first by the Würzburg physiologist and physicist Adolph Fick, who in a lecture series of 1869 presented it as the following dilemma:

> Either we have overlooked some important points in our highest, most general and most fundamental abstractions of science, or—if these abstractions are strictly and generally valid—the world cannot have existed for an eternity but must have come into existence at a time not infinitely far from today in an event which cannot be understood as part of the chain of natural courses; that is, a creative act must have taken place.[86]

In Great Britain, this line of reasoning was taken up independently by Maxwell and Tait, among others. In an address to the British Association in 1871, Tait concluded that 'the present order of things has *not* been evolved through infinite past time by the agency of laws now at work, but must have had a distinctive beginning, a state beyond which we are totally unable to penetrate, a state, in fact, which must have been produced by other than the now acting causes'.[87] Tait's other agency was of course God. For the next fifty years or so the entropic argument for a universe of finite age was much discussed, if more by theologians, philosophers, and social critics than by scientists trained in physics and astronomy. I am not aware of any astronomer of repute who referred to the argument.

If entropy was a problematic concept to use in arguing for a finite-age universe, after about 1900 there was the possibility of replacing it with another cosmic clock, the newly discovered radioactivity. In a popular lecture of 1911 the Austrian physicist Arthur Haas suggested that the laws of physics indicated a finite, unbounded universe, such as allowed by Riemannian geometry. In that case he could avoid Olbers' paradox and also the gravitational paradox without modifying Newton's law of gravitation. As to the timescale, he pointed out that radioactive elements such as uranium and thorium were still present in the

Earth's crust, in spite of their lifetimes being finite (if very long). How could there still be radioactive elements if the world had existed for an eternity? Haas's suggestion of linking radioactivity and cosmology in this way may not have been the first of its kind, but it was one of the first examples of an argument that would attract attention in the context of later and more scientifically based ideas of a finite-age universe.

2.4.2 *The universe—finite or infinite?*

Apart from denying the validity of the second law altogether (which no physical scientist dared to propose), two arguments were mounted against the heat death: there might be processes in the universe that reduced the entropy, and the second law might not be applicable to an infinitely large universe. Although no proof was offered, it was widely believed that the second law of thermodynamics could be meaningfully applied to the universe, if it could be applied at all, only if the universe contained a finite amount of matter. For this reason the cosmo-energetic discussion involved some of the classical questions of cosmology, namely whether the universe was finite or infinite in space, matter, and time. In this discussion, however scientifically it was framed, matters of ideology and faith played an important role.

In spite of being one of the chief architects of the second law of thermodynamics, William Thomson did not accept the universal heat death as physically real. His argument was that matter was distributed throughout endless space, and in this case the law of energy dissipation did not hold. How did Thomson know that space was infinite? He did not, of course, his only argument being that it was impossible to conceive a limit to space. In a popular lecture of 1884 he said, '*finitude* is incomprehensible, the infinite in the universe *is* comprehensible'. His illustration of the claim was less than convincing: 'What would you think of a universe in which you could travel one, ten or a thousand miles, or even to California, and then find it come to an end? Can you suppose an end of matter or an end of space?'[88] Thomson could not. The Swedish astronomer Carl Charlier, a professor at the University of Lund, favoured the opposite combination of views. In a paper of 1896 he argued that the universe was spatially finite, but temporally infinite: 'A finite time is a contradiction, . . . An infinite time may be difficult to conceive, but it is not contradictory.'[89] This kind of subjective argument—based on what individual scientists could comprehend or not comprehend—coloured much of the debate.

The Viennese physicist and philosopher of science Ernst Mach attacked the heat death and its corollary, the entropic creation argument, from the perspective of positivist methodology, arguing that the concepts were scientifically meaningless. In a lecture published in 1872, he stated that it was illusory to apply thermodynamics to the universe because no meaningful statements could be attached to the universe in its entirety. Scientific statements about the universe 'appear to me worse than the worst philosophical theorems' was the verdict of the physicist–philosopher.[90] Expressions such as Clausius's 'the energy of the world' and 'the entropy of the world' made no scientific sense because they did not cover measurable quantities. Mach's critique was shared by other scientists and philosophers of a positivist orientation, such as John Stallo in the United States and Georg Helm in Germany. From their point of view, there could be no physics of the universe, only a metaphysics, a view which made physical cosmology impossible. The French physicist, historian, and philosopher Pierre Duhem, who was influenced by Machian positivism, similarly denied the validity of the heat death, if by somewhat different arguments. He understood the entropy law as a statement that the entropy of the world increased endlessly, not that it had any lower or upper limit. If so, there was no need to accept the heat death or any other cosmological inference from thermodynamics.

Orest Chwolson, a respected Russian physicist who served as a professor at the University of St Petersburg, took up the question in papers of 1908 and 1910. Distinguishing between the observable 'world' and the much larger 'universe', he concluded (like Thomson) that a finite, bounded universe was an impossibility; but (contrary to Thomson) he was not therefore led to infinitism, for an infinite universe was nothing but 'a meaningless combination of empty words'. He also denied that the laws of physics, including the entropy law, were valid throughout the entire universe or could have any meaning in an infinite universe. Chwolson vehemently insisted that the domain of science was strictly the world, or what were parts of the world: 'Physics has nothing to do with the universe; it is not an object of scientific research as it is not accessible to any observation. . . . When the physicist speaks of the "world", he means his limited world. . . . To identify this world with the universe is a proof of either *thoughtlessness or madness*, and in any case *lack of scientific understanding*.'[91]

Whereas Mach and his allies attacked the heat death on methodological grounds, the Swedish Nobel Prize-winning chemist and physicist Svante Arrhenius recognized the force of the second law even on a cosmological scale. Yet he was convinced that the universe was infinite and self-perpetuating, in a steady state of eternal evolution. As he wrote in his best-selling *Worlds in the Making*, 'My guiding principle . . . has been the conviction that the Universe in its essence has always been what it is now. Matter, energy, and life have only varied as to shape and position in space.'[92] In a paper of 1909 Arrhenius argued that the universe at large was not only spatially infinite but also uniformly populated with stars and nebulae, a claim he justified by analogy to the observed part of the universe (or 'world' in Chwolson's terminology).[93] He strongly believed that both the heat death and the creation scenario must be wrong—'absolutely inconceivable', he stated—and therefore was led to look for entropy-increasing processes, such as Rankine had done more than half a century earlier. In works between 1903 and 1913 he developed a theory based on radiation pressure that would compensate entropy increase and allow continual cosmic development. However, Arrhenius's suggestion did not win approval, and after Henri Poincaré had shown that it was unable to counter the second law of thermodynamics, little more was heard of it.[94]

The American mathematician, physicist, and philosopher Charles Sanders Peirce seems not to have believed in a global heat death, and neither did he believe in immutable laws of nature. He thought that the universe had evolved from a chaotic primordial state, completely characterized by chance and spontaneity, to later states increasingly governed by law-bound regularities. Peirce did speak of an end of the universe, but it was a state of maximum complexity and thus very different from Clausius's heat death.[95]

As a final example of a late-nineteenth-century physicist who was led to cosmological questions from thermodynamics, consider Ludwig Boltzmann, the eminent Austrian theorist who founded thermodynamics on the basis of statistical mechanics. Whereas the laws of mechanics are symmetric in time, the principle of entropy increase expresses an irreversible feature in nature, leading to an apparent contradiction, which was much discussed in the 1890s. Boltzmann realized that there was a theoretical possibility that the second law of thermodynamics could be reversed (at some time and in some region in the universe), and in 1895 he developed a remarkable scenario of anti-entropic pockets in a universe which as a whole was in thermal equilibrium. Because of the probabilistic nature of entropy, there was a non-zero probability that our world would be in its present, low-entropy state even though the universe at large was in thermal equilibrium:

> But can we imagine, on the other side, how small a part of the whole universe this world is? Assuming the universe great enough, the probability that such a small part of it as our present world be in its present state, is no longer small. If this assumption were correct, our world would return more and more to thermal equilibrium; but because the whole universe is so great, it might be probable that at some future time some other world might deviate as far from thermal equilibrium as our world does at present.[96]

Boltzmann continued for some time to think over his scenario of entropy fluctuations in an otherwise high-entropy universe. In the second volume of his classical textbook on gas theory, *Vorlesungen über Gastheorie* of 1898, he included a chapter in which he repeated and amplified his ideas of many worlds, entropy fluctuations, and time reversal. He fully realized that these cosmological considerations were highly speculative, but found them consistent, and valuable enough to include them in his book. Astronomers ignored them, but many years later this kind of many-worlds or 'multiverse' thinking would occupy a central position in theoretical cosmology.

Not all late-nineteenth-century cosmological speculations were related to issues of thermodynamics. The British physicist Arthur Schuster suggested in 1898 that there might exist a hitherto unknown form of matter—he called it 'anti-matter'—with the property that it would be repelled gravitationally by ordinary matter. 'Worlds may have formed of this stuff, with elements and compounds possessing identical properties with our own, undistinguishable in fact from them until they are brought into each other's vicinity.' The fact that no such anti-matter had been detected did not count as a compelling argument against the hypothesis, 'for had it ever existed on our Earth, it would long have been repelled by it and expelled from it'. Schuster speculated that atoms and anti-atoms might enter into chemical combinations, with the short-range attractive forces dominating over the gravitational repulsion. 'Large tracts of space might thus be filled unknown to us with a substance in which gravity is practically non-existent, until by some accidental cause . . . unstable equilibrium is established, the matter collecting on one side, the anti-matter on the other until two worlds are formed separating from each other, never to unite again.' As Schuster was well aware, this was nothing but a speculation, 'a holiday dream', and he did not pursue it.[97]

The discovery of X-rays and radioactivity in the 1890s stimulated speculations that all matter might be unstable, in the process of transforming into the imponderable ether from where it had once originated. According to the French psychologist and amateur physicist Gustave LeBon, matter and energy represented two different stages in a cosmic evolutionary process, the end result of which would be a pure ethereal state. His was a vision of a cosmic death, but not the heat death justified by thermodynamics. In *The Evolution of Matter*, a hit that sold 44 000 copies, LeBon summarized his cosmic scenario, which started with 'a shapeless cloud of ether'. Through 'forces unknown to us', this primordial ether was organized in to the form of energy-rich atoms. However, these would be unstable and radioactive, and slowly release their energy: 'Once they have radiated away all their store of energy in the form of luminous, calorific, or other vibrations, they return . . . to the primitive ether whence they came. This last, therefore, represents the final nirvana to which all things return after a more or less ephemeral existence.'[98] Here we have another early case of physical eschatology.

Speculations along such lines attracted considerable attention from physicists, many of whom found LeBon's scenario fascinating, even reasonable. For example, the respected British physicist Oliver Lodge held views not widely different from LeBon's. Another Englishman, the chemist Frederick Soddy (a Nobel laureate of 1921), likewise thought that radioactivity was of cosmological importance, but he favoured a cyclical scenario, namely that 'matter is breaking down and its energy being evolved and degraded in one part of a cycle of evolution, and in another part still unknown to us, the matter is being built up with the utilisation of waste energy'. Thus, 'in spite of the incessant changes, an equilibrium condition would result, and continue indefinitely'.[99]

2.4.3 *Gravitational paradoxes*

In his correspondence with Bentley, Newton argued in qualitative terms that an infinite, uniform stellar universe was possible. Although the two infinite gravitational forces that acted on a particular mass did not cancel (as Bentley believed), he asserted that the mass would nonetheless be in a state of equilibrium. Curiously, it took almost precisely two hundred years until Newton's argument was subjected to rigorous examination.

In 1895 the German astronomer Hugo von Seeliger, a professor in Munich and secretary of the German Astronomical Society (and from 1896 its president), proved that an infinite Euclidean universe with a roughly uniform mass distribution could not be brought into agreement with Newton's law of gravitation. He showed that calculation of the gravitational force exerted on a body by integration over all the masses in the infinite universe does not lead to a unique result, as the integral diverges. Hence the conclusion, 'Newton's law, applied to the immeasurably extended universe, leads to insuperable difficulties and irresolvable contradictions if one regards the matter distributed through the universe as infinitely great.'[100] Seeliger's concern was not really to save Newton's infinite stellar system, for he rejected the notion of an actual infinity and tended to believe that the universe was finite.

The following year Seeliger framed the gravitation paradox differently, by showing that the Newtonian universe allowed motions that start with finite speed and accelerate to infinitely great speeds in a finite time. As he pointed out, such motions are no less inadmissible than a collapsing universe. In a popular presentation, he summarized that whatever the mass distribution in the universe, there must occur infinitely great accelerations in it. From this followed 'motions which, starting from finite velocities, would lead within a finite time, to infinite velocities'. The conclusion 'contains within itself either an absurdity, or a direct contravention of the theory of mechanics'.[101] Seeliger therefore suggested that Newton's law should be modified at very large distances. A body moving in a gravitational field of a central mass M will, according to Newton, experience a gravitational potential $\phi(r) = -GM/r$. Seeliger therefore suggested that Newton's law should be modified at very large distances. A body of mass m moving in the gravitational field of a central mass M will, according to Newton, experience a gravitational pull given by GmM/r^2. Seeliger suggested that for very large distances, the body would move as if there was a repulsive force in addition to the attractive gravitational force. The suggestion amounted to introducing an attenuation factor of the form $\exp(-\Lambda r)$, which leads to the new law of force

$$F(r) = \frac{GmM}{r^2} e^{-\Lambda r}.$$

Seeliger contemplated that the constant Λ might be important in planetary astronomy but did not seriously pursue the idea. The modified force law was essentially ad hoc and also arbitrary, since many other modifications might resolve the gravitation paradox in a similar way. The idea of modifying Newton's inverse-square law was not, by itself, very original, as many such modifications were proposed in the nineteenth century. The exponential correction factor can be found in 1825 in Laplace's *Mécanique céleste*, which can hardly have avoided Seeliger's attention. However, what was original in Seeliger's approach was that he used it in a cosmological context and not, as in most other proposals, to solve problems of planetary astronomy (such as Mercury's anomalous revolution around the Sun).

William Thomson, apparently unaware of Seeliger's work, arrived at essentially the same results in two papers published 1901–02. In the first paper he proved that in an infinite universe with a non-zero density of matter, 'a majority of the bodies in the universe would each experience infinitely great gravitational force'. Considering a homogeneous model universe of radius 3×10^{16} km, he concluded that the number of stars, assumed to be of solar mass on average, must be neither too great nor too small. He found it 'highly probable' that there were fewer than two billion stars within the sphere, and more than one hundred million. Thomson also calculated the time it took for the stellar system to collapse to zero radius. This collapse time turned out to be independent of the initial size of the universe and to depend only on its density ρ_0:

$$t_{\text{collapse}} = \frac{1}{4}\sqrt{\frac{3\pi}{2} G\rho_0}.$$

For a universe containing one billion stars, the collapse time would be 17 million years, a figure of the same order of magnitude as that which Thomson had found for the age of the Earth from thermodynamical reasoning (his favourite age at the time was 20 million years). Contrary to Seeliger, Thomson did not present his investigation clearly as a problem for the infinite Newtonian universe, and he did not propose a way out of the problem.

There were other ways to escape gravitational collapse than modifying Newton's law of gravitation. One could leave the law intact and change some of the cosmological assumptions of the Newtonian universe, such as the homogeneous distribution of matter. This is what Richard Proctor did in his widely read *Other Worlds than Ours* of 1870, although he developed his model in the context of Olbers' optical paradox and not the gravitational paradox. Proctor conceived of a hierarchic model universe in which the higher star systems were separated by increasingly larger distances from the lower ones; in that case, the contributions of light from stars lying in successive shells would not be equal, but successively less, and the total amount of light received from the infinite number of stars could be quite small.[102]

In 1908, Carl Charlier developed in mathematical detail a hierarchic model in a paper titled 'Wie eine unendliche Welt aufgebaut kann' (i.e., 'how an infinite world can be constructed').[103] This was not the first time he expressed an interest in the grander aspects of cosmology. Unusually for professional astronomers at the time, Charlier transcended the divide between scientific astronomy and philosophical reflections on the more speculative aspects of cosmology. In 1896 he argued that Olbers' and Seeliger's paradoxes indicated that the universe must be finite, a solution he preferred to Seeliger's modification of Newton's law of gravitation. However, he also pointed out that the paradoxes rested on the assumption of uniformly distributed stars, an assumption that might be questioned. If the distribution of stars followed a law such that 'the density of the stars decreases faster, as we move out in space', the paradoxes did not need to arise.[104] As mentioned, this idea, as far as it is related to Olbers' paradox, goes back to John Herschel in 1848 and was revived by Proctor in 1870.

Charlier's early advocacy of a finite universe was contradicted by his later work, where he concluded that the infinity of the world could be defended after all, at least in the sense that the gravitation paradox could be avoided. His idea was a fractal, hierarchic universe built up in a particular way from spherically arranged nebulae and clusters of nebulae. Let the Milky Way S_1 be composed of N_1 stars and let N_2 Milky Way galaxies form a second-order galaxy S_2; N_3 galaxies of type S_2 form a third-order system S_3, and so forth. The system S_i has a radius R_i. Charlier showed that if the mean density of matter decreased in such a way that the inequality $R_{i+1}/R_i \geq N_{i+1}$ was satisfied, Seeliger's gravitation paradox would disappear and there would be no infinite velocities.His ideas received support from Franz Selety, a Viennese physicist, who in 1922 developed them as a Newtonian alternative to the relativistic cosmology that Einstein had introduced five years earlier. Selety criticized Einstein's theory and argued that if matter became diluted in a suitable way with distance (faster than $1/r^2$), Newtonian theory allowed an infinite universe completely filled with matter. A brief debate followed between Einstein and Selety in the pages of *Annalen der Physik*, but after his first reply in 1922 Einstein chose not to respond to Selety's papers.

One year after Charlier's paper had appeared, another Swedish scientist, Arrhenius, examined the gravitation paradox in a work which has been referred to above. As mentioned, Arrhenius was an ardent advocate of a homogeneous and infinite universe, and consequently he felt it necessary to criticize Charlier's hierarchic model. One might have thought that he would be sympathetic to Seeliger's solution, but this was not the case; he could see no problem with the Newtonian universe and therefore found Seeliger's work to be irrelevant. (As Seeliger was quick to point out, Arrhenius had partly misunderstood his work.) 'There really is no weighty reason why the world would not be sown uniformly with stars', Arrhenius concluded.[105] Arrhenius's denial that there was a problem at all hardly convinced anyone except himself.

2.5 The Via Lacta

At the end of the nineteenth century it was generally recognized that if an observationally based cosmology was ever to be established, a first step would be to understand the size and structure of the Milky Way. Space was usually thought to be infinite, but there was no consensus at all about the distribution of stars and nebulae in the universe—many astronomers hesitated to admit it as a problem that could be solved scientifically. In a book of 1878, Proctor gave voice to the dilemma by writing that the 'only question for us is between an infinity of occupied space and an infinity of vacant space surrounding a finite [material] universe'.[106] Some years later, the astronomer and astronomy writer Agnes Clerke raised what by then had become a crucial question, whether the nebulae were located inside or outside the Milky Way; or, to put it differently, whether everything visible belonged to our galaxy. She confidently asserted:

No competent thinker, with the whole of the available evidence before him, can now, it is safe to say, maintain any single nebula to be a star system of coordinate rank with the Milky Way. A practical certainty has been attained that the entire contents, stellar and nebular, of the sphere belong to one mighty aggregation, and stand in ordered mutual relations within the limits of one all-embracing scheme—all-embracing, that is to say, so far as our capacities of knowledge extend. With the infinite possibilities beyond, science has no concern.[107]

Not all astronomers would agree, but there is little doubt that the majority shared the view that the Milky Way was approximately identical to the material universe. In an essay of 1906 on unsolved problems in astronomy, the leading American astronomer Simon Newcomb addressed the same question, of whether the universe was populated with stars all over, or whether they were largely contained in the system of the Milky Way. This question, he wrote, 'must always remain unanswered by us mortals . . . Far outside of what we call the universe might still exist other universes which we can never see.' For all practical purposes, the Milky Way 'seems to form the base of which the universe is built and to bind all the stars into a system'.[108] In the spirit of positivism, Newcomb asserted that theories and hypotheses were put forward to explain facts, and hence no theory would be required where there were no facts to be explained: 'As there are no observed facts as to what exists beyond the farthest stars, the mind of the astronomer is a complete blank on the subject. Popular imagination can fill up the blank as it pleases.'[109]

2.5.1 *The Milky Way universe*

The first spiral nebula had been discovered by Rosse in 1845, and with the many observations of spiral-shaped nebulae that followed during the subsequent decades it became natural to ask whether the Milky Way itself might not be a spiral. The first such suggestion was made as early as 1852 by an American astronomer, Stephen Alexander, but it took nearly half a century until the idea became widely known.

Cornelis Easton was a Dutch journalist, science writer, and respected amateur astronomer who around the turn of the century published several works on stellar astronomy. Having initially thought of the Milky Way as an annular structure, in a review article in the *Astrophysical Journal* of 1900 he proposed a new theory, in which he compared the Milky Way with some of the spiral nebulae. He placed the solar system containing the Earth in the centre of the system, and argued that the convolutions of the 'galactic spiral' were situated in two planes forming an angle of about 20°. Easton stated that his drawing of the Milky Way 'indicates in a general way how the stellar accumulations of the Milky Way might be distributed so as to produce the galactic phenomenon, in its general structure and its principal details, as we observe it'.[110] In a later article, published in 1913, he revised his picture by placing the Sun halfway between the centre and the edge of the spiral system. Easton's modern-looking picture and conclusion that the Milky Way was shaped like a spiral nebula did not mean that he advocated the island universe theory. On the contrary, he described the other spirals as nothing but 'small eddies in the convolutions of the great one' and thought it safe to assume that 'the great majority of the small spiral nebulae, if not all, form part of our galactic system'.[111]

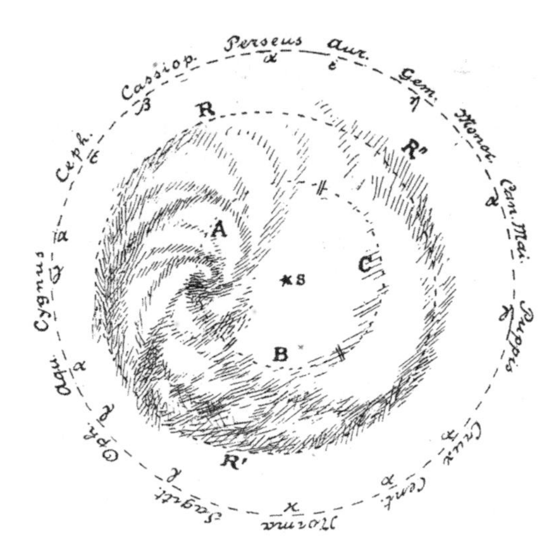

Fig. 2.10 Cornelis Easton's representation of 1900 of the Milky Way as a spiral nebula. In 1913 he suggested a similar, but more elaborate model. From Easton 1900.

Cosmology around 1900 was a field that appealed as much to amateurs as to professional astronomers, or even more so. Alfred Russell Wallace is best known as Darwin's rival, a brilliant naturalist and pioneer evolutionist, but he also had an interest in astronomy, in part because he found the science relevant to his spiritual convictions. In 1903, at the age of 80, he published *Man's Place in the Universe*, in which he presented his view of a stellar universe centred on the solar system. Drawing on authorities such as Newcomb, J. C. Kapteyn, and John Herschel, he concluded that the Sun was centrally located in a globular cluster of stars situated in the centre of a finite, bounded universe, the Milky Way. He took Olbers' paradox to be an 'altogether conclusive' proof that the stellar universe was of limited extent, and estimated its diameter to be a mere 3600 light years. Wallace's universe was Sun-centred and unusually small, but apart from this it did not differ greatly from the view held by many astronomers.

Professional astronomers trained in observational techniques and mathematical methods were not impressed by Easton's visualizable model of the universe and preferred to ignore Wallace's. Seeliger and his contemporary, the leading Dutch astronomer Jacobus C. Kapteyn of the University of Groningen, attempted to obtain a scientifically founded picture of the Milky Way by a laborious analysis of the proper motions and apparent magnitudes of the stars. Although they worked independently and used different methods—Seeliger's approach was mathematical and analytical, Kapteyn's empirical and numerical—their conclusions had much in common, and for our purpose we need not make a sharp separation between their theories.[112] Seeliger started his research programme in 'statistical cosmology' in 1898, based on advanced mathematical analysis of star counts and stellar magnitudes. For a long time the lack of good data prevented conclusive results, but in 1920 he finally published 'Untersuchungen über das Sternsystem' ('investigations of the star system'), which contained the mature form of his cosmology. The stellar system he arrived at was ellipsoidal, extending about 33 000 light years in the Milky Way plane and 3900 light years toward the galactic poles. Seeliger's work benefitted from mathematical improvements by Karl Schwarzschild, a former student of his. Schwarzschild's work on stellar statistics around 1910 resulted in a picture of the Milky Way as a flattened, Sun-centred disc with dimensions in rough agreement with Seeliger's later conclusion.

Kapteyn's favoured method was based on his discovery of 'star streaming', which he announced in 1904. By examining the proper motions of a large number of stars, he found that they did not move randomly but tended to drift in two streams that pass through each other as they move in opposite directions in the Milky Way plane.[113] The extensive and data-demanding work of the Dutch astronomer bore fruit in 1920 when Kapteyn published a paper together with his former student Pieter van Rhijn. The stellar data led to a density distribution and a corresponding model for the stellar system, the so-called 'Kapteyn universe', a term coined by James Jeans in 1922. Kapteyn's universe was, like Seeliger's, an ellipsoidal, Sun-centred stellar system in which the star density diminished with increasing distance from the centre. In the galactic plane it covered a distance of 59 000 light years and towards the galactic poles about 7800 light years; at a distance of 26 000 light years from the centre the density was one-hundredth of the value in the solar region, or about four stars in a volume of ten cubic light years.

In 1922, shortly before his death, Kapteyn revised his model in another paper in the *Astrophysical Journal*.[114] Whereas he had previously assumed that the Sun was at the centre of the system, he now stressed that this was 'infinitely improbable' and found a way to estimate the true position of the Sun. Kapteyn now concluded that the Sun lay in the galactic plane, but was probably located some 2100 light years from its centre. This was still suspiciously near the centre, which many astronomers considered to be a problematic, un-Copernican feature of the model. In his 1922 paper, Kapteyn estimated the mass density of the Milky Way, on the basis of the number of luminous stars, to be about 10^{-23} g/cm^3. He was aware that dark matter might increase the real density, but concluded that 'this mass [from dark matter] cannot be excessive'. Seeliger and Kapteyn both assumed that starlight was not absorbed by interstellar matter. Although Kapteyn realized that there might be a slight absorption, in the absence of conclusive results he chose to ignore the possible effect. As it would turn out later, light is in fact attenuated on its journey from the stars, which was

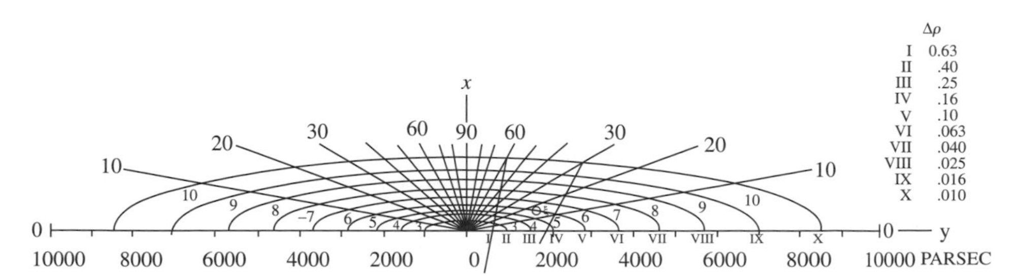

Fig. 2.11 The final version of Kapteyn's universe, from the *Astrophysical Journal* of 1922. The Sun is placed about 1800 parsecs to the right of the galactic centre. The lines refer to surfaces in the ellipsoid with constant matter density.

a major reason why the stellar universe of the statistical cosmologists had a size of only a few tens of thousands of light years.

2.5.2 *Discoveries in stellar and galactic astronomy*

There are essentially two kinds of interstellar absorption caused by diffuse matter. General absorption relates to the overall reduction in intensity of the received light, whereas selective absorption is a reddening effect. The question of whether or not there existed light-absorbing matter in interstellar space, and how much it affected the apparent magnitude of a star, was important to nineteenth-century astronomers discussing Olbers' paradox, as mentioned on p. 84. It was also highly relevant to the models of the Milky Way developed in the first decades of the twentieth century, which was the main reason why Kapteyn took such a strong interest in the absorption problem. However, observations were confusing, and Kapteyn and most other astronomers wavered in their attitude; around 1910 many were ready to accept the evidence for absorption, yet few wanted to commit themselves.[115]

It was only in 1917, when Harlow Shapley published his results of studies of globular clusters, that the situation changed. After studies at the University of Missouri, Shapley had gained his PhD in Princeton in 1913 and the following year he went to the Mount Wilson Observatory, where he established his reputation as one of the period's most brilliant astronomers. In his work from 1917 Shapley found that globular clusters were distributed asymmetrically and he concluded that there was no selective absorption. His arguments were accepted by most astronomers, including Kapteyn. However, there still was room for dissent, and not all agreed with Shapley's interpretation. For example, the young Jan Oort at Leiden Observatory concluded from his studies of the Milky Way's rotation that the Sun must be located much farther from the galactic centre than was implied by Kapteyn's model. In an important paper of 1927 he suggested that the discrepancy between his and Kapteyn's results could best be explained by interstellar absorption. By that time many astronomers had begun to reconsider the reality of absorption, and a few years later the increasing evidence for the phenomenon was turned into a definite proof of interstellar obscuring matter.

The proof was due to the Swiss-born Robert Trumpler at Lick Observatory, whose comprehensive study of open clusters was published in 1930. Trumpler argued convincingly that his data could be explained only if it was assumed that there was general as well as selective absorption in stellar space. For the interstellar absorption effect, he found that it amounted to an average change in apparent magnitude of 0.67 per kiloparsec (1 parsec = 3.26 light years). With the confirmation of a significant absorption in space, Kapteyn's theory of the stellar universe was further undermined. However, Trumpler failed to recognize this implication and erroneously believed that his work showed the Milky Way to be spiral-shaped and with the Sun at its centre. He wrote about his work that it was 'in good agreement with the results of the statistical investigations of Seeliger, Kapteyn and others who describe the stellar system as a flattened lens shaped system 10,000–15,000 parsecs in diameter and 3,000–4,000 parsecs in thickness, with the stars concentrated toward the center and thinning out toward the edge'.[116] Although he mentioned Oort's results, he did not try to reconcile his own data with those obtained by Oort.

In order to obtain a reliable picture of the Milky Way, and of the relationship between it, the globular clusters, and the spiral nebulae, it was all-important to have a method of measuring distances over very long stretches of space (where the parallax method was of no help). Such a method was developed in the second decade of the twentieth century, and was based on a kind of variable stars known as cepheids. The prototype of the cepheid variables is the star Delta Cephei, whose regular light variation was discovered in 1784 by the English amateur astronomer John Goodricke.

That cepheids can serve as cosmic yardsticks was shown in 1908, when Henrietta Swan Leavitt of Harvard College Observatory examined 16 of the stars she had identified in the Small Magellanic Clouds. She noticed that the brighter the cepheid, the longer its period, but it was only four years later that she followed up on the matter and showed that it implied a new distance indicator.[117] In her paper of 1912, now based on 25 cepheids, she demonstrated that the cepheids' periods P were related logarithmically to their maximum (or minimum) brightness by $\log P + 0.48m = \text{constant}$, where m denotes the maximum (or minimum) apparent magnitude. The Small Magellanic Clouds are so far away that all the stars in them have nearly the same distance from the Earth, and Leavitt therefore concluded that the relation was also valid for absolute magnitudes. In terms of average magnitudes,

$$\langle M \rangle = a + b \log P,$$

where a and b are constants to be determined from observations. From the standard relationship $M = m - 5 \log r$, it follows that if the absolute magnitude can be determined, so can the distance r. On the assumption that the period–magnitude relationship held for all cepheids, Leavitt's finding thus implied that astronomers would be able to determine the relative distances of objects containing cepheid variables. In order to find absolute distances, which would be of much greater interest, the constants a and b would have to be fixed, but unfortunately Leavitt did not know the distance to the Small Magellanic Clouds. The first to attempt a calibration was the Danish astronomer Ejnar Hertzsprung, but his 1913 determination of the distance to the Small Magellanic Clouds was imprecise (he got 10 000 parsecs, as compared with the modern value of 60 000 parsecs).

Where Leavitt and Hertzsprung had shown the way, Shapley succeeded in turning the period–luminosity relation into an operational method for determining galactic distances. His interest in cepheids as distance indicators was related to his successful attempt to understand the cause of their light variation. Cepheids had traditionally been believed to be eclipsing binaries, but according to Shapley's theory of 1914 they were individual stars pulsating in size and brightness. Inspired by Hertzsprung's paper, Shapley proceeded to improve the method, and in an important paper of 1918 he argued that the absolute magnitude for a certain cepheid with $P = 5.96$ days was -2.35. Shapley reported that the formula

$$\langle M \rangle = -0.60 - 2.10 \log P$$

reproduced the logarithmic relationship discovered by Leavitt.[118] His conclusion of 1918 was criticized for being too confident and resting on questionable assumptions (such as ignoring interstellar absorption), and the calibration constants were subsequently changed. However, the important thing was that Shapley had provided the astronomers with a practical way of using cepheids as standard candles. After 1918, if a cepheid variable could be unambiguously identified in some faraway object, the distance to that object could be determined. This was a major step ahead in observational cosmology.

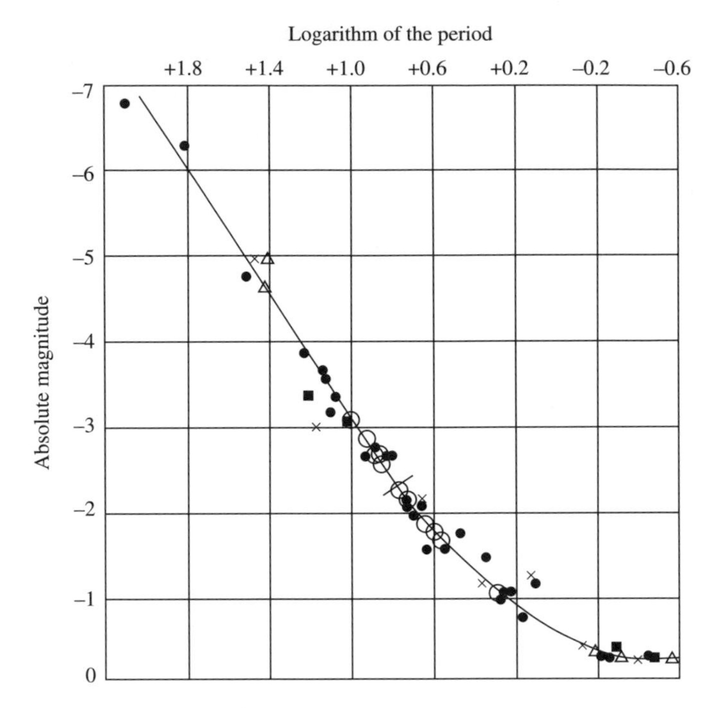

Fig. 2.12 Shapley's 1918 graph of luminosity versus period of variation for cepheids. The various symbols designate variables from seven different systems. From Shapley 1918, p. 104.

The discovery of nebular redshifts, of unprecedented importance to modern cosmology, belongs to Vesto Melvin Slipher, a graduate from the University of Indiana who spent his entire research career (1901–52) at the Lowell Observatory in Flagstaff, Arizona. A self-taught specialist in stellar spectroscopy, Slipher was looking for evidence of rotation of spiral nebulae when he found in 1912 that the spectral lines of the Andromeda Nebula were shifted towards the blue end of the spectrum. If interpreted as a Doppler effect, this implied that the spiral nebula approached the Sun with the exceptionally high velocity of 300 km/s. Two years later Slipher could announce to a meeting of the American Astronomical Society that he had obtained spectral shifts for 13 more spirals and that most of them were redshifts rather than blueshifts. In a review article of 1917, 'A Spectrographic Investigation of Spiral Nebulae', he reported results for 25 spirals with radial velocities between 300 km/s and 1100 km/s. Of these, only four were approaching the Sun, indicating that recession was the rule to be found among the spiral nebulae. In 1921 he reported a recessional velocity of 1800 km/s, indicating the fastest-moving celestial body known at the time.

Slipher was not only the discoverer of spiral redshifts, he was also practically alone in cultivating the new area of research. He continued his research programme until 1925, by which time he had collected data for 45 nebulae, of which 41 were redshifts and all but five were measured by himself. What did the redshifts imply? Initially, Slipher tended to believe that the data indicated that the Milky Way moved relative to the spirals with a velocity of about 700 km/s, and in 1917 he was still not convinced that recession was the normal pattern. He cautiously suggested that when more nebular spectra were obtained, more spirals would be found to have approaching motions. Other astronomers were more impressed by the large velocities and the preponderance of redshifts, and they suggested that the spirals could not be part of the Milky Way. As Hertzsprung wrote to Slipher in March 1914, 'It seems to me, that with this discovery the great question, if the spirals belong to the system of the Milky Way or not, is answered with great certainty to the end, that they do not.'[119] Hertzsprung was right, yet his suggestion was premature. Slipher came to agree that his findings probably indicated that the spirals were extragalactic, but his main concern was with measurements, not interpretation. Referring to the island universe theory, he wrote in his 1917 paper, 'This theory, it seems to me, gains favor in the present observations.'[120]

The discovery of large radial velocities among spiral nebulae was generally seen as support of the island universe theory. However, in the absence of knowledge of the distances to the spirals, the phenomenon was of evidential nature only and the matter was far from settled. More was soon to come.

2.5.3 *Shapley's universe and the 'Great Debate'*

'This is a peculiar universe.' So wrote Shapley to his mentor, the eminent Princeton astrophysicist Henry Norriss Russell, in a letter of 31 October 1917, referring to a new model of the Milky Way he had recently formulated but not yet published.[121] Whereas a couple of years earlier he had accepted a Milky Way of the size of Kapteyn's universe and been sympathetic to the island universe theory, he now believed that the system of stars and nebulae was about 300 000 light years in diameter and 30 000 light years in thickness, that is, it had increased in linear dimensions roughly tenfold. In early 1918, Shapley summarized his galactocentric theory to Hale, describing the Milky Way as an 'enormous, all-comprehending galactic system . . . The diameter of the system is some 300,000 light years in the plane.' He stated confidently that the picture did not conflict with any known facts and added, in a speculative mood, 'There is no plurality of universes of which we have evidence at present.'[122]

This amazing model of a universe made up of one enormous galaxy, published in 1918 as 'Globular Clusters and the Structure of the Galactic System', was based on studies of the distribution of globular star clusters. Shapley suggested that these clusters formed a vast, nearly spherical system enveloping the plane of the Milky Way on both sides and that the clusters belonged to the extended system; they were physically associated with the Milky Way, not independent star systems. He placed the solar system about 65 000 light years away from the centre of the Milky Way. Determinations of the distances to the clusters were crucial to Shapley's model, and he used the newly developed period–luminosity relation to find the distances to the nearer ones. For the distant clusters, where cepheids could not be discerned, he used more indirect methods, which led him to conclude that some remote clusters were located more than 200 000 light years away from the Sun.

As one might expect, Shapley found his immense galactic system to be incompatible with the island universe theory (and of course also with the Kapteyn universe). For one thing, if the spiral nebulae were external galaxies comparable in size to the Milky Way—and this is what most advocates of the island universe theory assumed, at least implicitly—they would be at inconceivably great distances. This created several difficulties, one of which was related to the apparently reliable results for rotations of spiral nebulae that the Dutch astronomer Adriaan van Maanen had announced, beginning 1916. His measurements indicated velocities that were incredibly great if the spirals were at distances of one million light years or more. For example, in 1920 he concluded that if the galaxy M33 was of the same size as Shapley's Milky Way system it must be so far away that it rotated at the speed of light. This was out of the question, and van Maanen considered it a strong argument against the island universe theory. So did Shapley, who accepted van Maanen's data and interpretation. In 1921 Shapley wrote to his Dutch colleague: 'Congratulations on the nebulous results! Between us we have put a crimp in the island universes, it seems,—you by bringing the spirals in and I by pushing the galaxy out. It is certainly nice of those nebulae to have measurable motions.'[123]

Shapley's daring model of the Milky Way attracted wide attention and, expectedly because of its novelty and radical features, also some hostility. Among the early critics were his compatriots Heber Curtis[124] and Walter Adams, but their opposition was more than balanced by the support Shapley received from leading astronomers such as Eddington, Russell, and Hale. Whether for or against Shapley's alternative, it was realized that the question of the size and structure of the Milky Way remained open. The opposing views of the universe became the subject of what has passed into the literature as the 'Great Debate', constituted by a public meeting of the National Academy of Science and the subsequent publication of two articles with the same title ('The Scale of the Universe') but widely different conclusions. The discussants at the Washington meeting of 26 April 1920 were Curtis, defending the island universe theory, and Shapley, who argued the cause of a much larger Milky Way that makes up almost the entire material universe.[125]

The debate between Curtis and Shapley was asymmetrical in the sense that whereas Curtis focused on the spiral nebulae, Shapley had little to say about these and instead focused on the clusters and the structure of the Milky Way. The lack of symmetry was highlighted by differences in style, with Curtis's technical article forming an odd contrast to Shapley's much more elementary and popular presentation. Because of the asymmetry and also because none of the discussants could supply their arguments with new observations, nothing new emerged from the debate, which merely provided a pedagogical overview of the two alternative conceptions of the world. The essence of the debate can be summarized in Curtis's words from his published article:

Present Theory	*Shapley's Theory*
Our galaxy is probably not more than 30,000 light-years in diameter, and perhaps 5,000 light-years in thichness. The clusters, and all other types of celestial objects except the spirals, are component parts of our own galactic system. The spirals are a class apart, and not intra-galactic objects. As island universes, of the same order of size as our galaxy, they are distant from us 500,000 to 10,000,000, or more, light years.	The galaxy is approximately 300,000 light-years in diameter, and 30,000 or more, light-years in thickness. The globular clusters are remote objects, but a part of our own galaxy. The most distant cluster is placed about 220,000 light-years away. The spirals are probably of nebulous constitution, and possibly not members of our own galaxy, driven away in some manner from the regions of greatest star density.

The Washington debate did not end the controversy, which continued for a few more years. Both parties had good arguments and could cite observational support for their views, and there were no observations that unambiguously spoke for one theory and against the other. As is turned out, the debate was in a sense misguided as it presented an either–or situation. A decade later it became clear that there was no winner and no loser in the debate; or perhaps both debaters were winners and both losers. (We do live in an island universe, as Curtis claimed, but we also live in a Milky Way that is closer in size to what Shapley claimed.) Two years after the Washington meeting, the young Estonian astronomer Ernst Öpik estimated from the rotation velocities of the Andromeda Nebula that its distance might be as great as 1.5 million light years and that its mass was about 4.5 billion solar masses. Öpik, who was clearly in favour of the island universe, concluded that his result 'increases the probability that this nebula [M 31] is a stellar universe, comparable with our Galaxy'.[126]

Edwin Powell Hubble, born in Missouri in 1889, was in favour of the island universe even before he found the cepheid in the Andromeda Nebula that practically settled the matter. After studies at the universities of Chicago and Oxford (where his fields were law and Spanish), he returned to the University of Chicago, from which he received in 1917 a doctorate in astronomy. His astronomical career was interrupted when the United States entered the First World War. Shortly after having passed his final examination at Yerkes Observatory, he joined the army and, with the rank of major, he was sent to France in September 1918. However, the war ended before he went into combat. After a stay in England, Hubble joined the staff of the Mount Wilson Observatory and began the series of observations that would make some astronomers compare him to Galileo and William Herschel.[127]

In the autumn of 1923, Hubble began a study of novae in the spiral nebulae, and in the course of this work he found in the Andromeda Nebula two objects that exhibited the same variation in brightness as cepheid variables. The first and clearest of these he initially believed was a nova, but he soon corrected his mistake and in February 1924 he reported his discovery to Shapley. Rather than rush into print, the cautious Hubble waited for nearly a year until he made the discovery public. This happened on 1 January 1925, at the Washington meeting of the American Astronomical Society, where Russell read his paper (Hubble did not attend the meeting). Later in 1925, Hubble announced his observations of a large number of galactic cepheids in the journals *Observatory* and *Popular Astronomy*, but without emphasizing the cosmological implications of his discovery.

But, of course, Hubble was fully aware of these implications. The main reason for Hubble being so cautious in publishing was the conflict between his findings and the results of van Maanen. As he wrote to Russell, 'The real reason for my reluctance in hurrying to press was, as you may have guessed, the flat contradiction to van Maanen's rotations.'[128] Using the period–luminosity relation and Shapley's calibration, he found the distance to the Andomeda Nebula to be about 930 000 light years, which meant that it could not possibly be part of the Milky Way system. (Because of an error in Shapley's calibration, Hubble underestimated the distance, which is actually about 2.2 million light years.) Hubble's discovery drastically changed the attitude in the astronomical community in favour of the island universe theory, which now became accepted by Russell, Jeans, and most others. However, not all astronomers were convinced, primarily because of the glaring disagreement with van Maanen's measurements of motions in the nebulae. The most

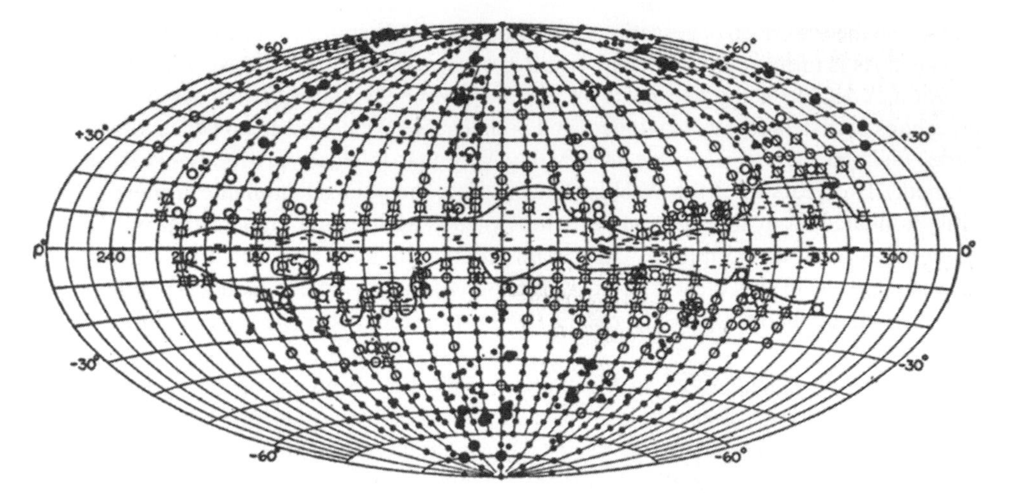

Fig. 2.13 Hubble's 1934 survey of the distribution of galaxies. The horizontal line represents the central plane of the Milky Way, with its 'zone of avoidance' marked. The blank parts on the right and left of the diagram represent the southern skies, which could not be observed by the Mount Wilson telescopes. From Hubble 1936, p. 62.

persistent of the sceptics was van Maanen himself, and he was followed by the Harvard (and later Oxford) astronomer Harry H. Plaskett, who as late as 1931 argued that Hubble's work on cepheids did not vindicate the island universe theory.[129] The opponents of the island universe were, however, few and did not succeed in reversing the consensus view that the spiral nebulae were extragalactic worlds with a scale and structure comparable to the Milky Way.

The developments that led to the island universe in the first quarter of the twentieth century were observation-driven and firmly located within the classical, Herschelian tradition of astronomy. This was distinct not only from the philosophical tradition mentioned in Section 2.4 but also unconnected with the developments that transformed physics in the period. These developments, in quantum theory and relativity theory, did, however, influence astrophysics and cosmology most profoundly and will form the main content of the following chapter.

Notes

1. Quoted in Gaukroger 1995, p. 291. For a summary account of Descartes's cosmology, see Baigrie 1993. Some of the comments that follow are taken from Kragh 2004.
2. Descartes 1983, p. 14.
3. Descartes 1996, p. 26.
4. This so-called indifference principle can be found also in modern cosmological theory, for example in chaotic cosmology and models of the very early, inflationary universe. On the indifference principle, see McMullin 1993. According to the physicist Julian Barbour, Descartes's ideas of the creation of order from chaos 'have an uncanny similarity to many an introduction to many papers on inflationary cosmology'. Barbour 2001, p. 432.
5. *Cosmotheoros* was translated into English as Huygens 1722 (second edition), where the comments on infinity appear on pp. 155–157.

6. On Newton and his cosmological views, see, e.g., Koyré 1965, Westfall 1980, and Harrison 1986.
7. Newton 1999, p. 943.
8. Ibid. p. 790.
9. An English translation of *De gravitatione* can be found in Hall and Hall 1962, pp. 121–156. Quotation from pp. 137–138.
10. Newton 1952, p. 402. The following quotations are from the same source, pp. 399–404.
11. See Davies 1966.
12. See Alexander 1956, pp. 11–12, 14. For an analysis of the Leibniz–Clarke correspondence, see Vailati 1997.
13. Kubrin 1967, reprinted in Russell 1973, pp. 147–169.
14. The letters to Bentley were first published in 1756. They are reproduced in Cohen 1978, pp. 279–312.
15. Cohen 1978, p. 287. In his first letter to Bentley, Newton described the divine power as 'very well skilled in Mechanicks and Geometry'.
16. Cohen 1978 includes a reprint of Parts II and III of *A Confutation of Atheism*. Quotation from p. 351.
17. Halley 1720–21, p. 23.
18. Ferguson 1778 (sixth edition), p. 84.
19. Quoted in Davies 1966, p. 278. As mentioned in Section 1.2, the argument goes back to Stoic philosophy.
20. The famous *Encyclopédie*, edited by Denis Diderot and Jean d'Alembert, included several articles of cosmological relevance, yet in none of these was the universe described as infinite (Jaki 1990, p. 52).
21. Roberts 1694, p. 103.
22. For details, see Hoskin 1982 and Hoskin 1999.
23. The Scottish astronomer Thomas Henderson, studying the stars on the southern hemisphere, detected at the same time a parallax of just over 1″ for Alpha Centauri. Henderson had completed his analysis before Bessel, but his publication came a little later. The modern values for the parallaxes of the two stars are 0.29″ (61 Cygni) and 0.74″ (Alpha Centauri).
24. In accordance with the standards of the time, the full title was more elaborate: *An Original Theory or New Hypothesis of the Universe, Founded upon the Laws of Nature, and Solving by Mahematical Principles the General Phenomena of the Visible Creation; and Particularly the Via Lactea*. A facsimile edition of the book, provided with a detailed introduction by Michael Hoskin, appears as Wright 1971. Wright's cosmology has attracted a good deal of interest; see Schaffer 1978; Hoskin 1982, pp. 101–116; and Jaki 1990, pp. 81–106. See also the more popular account in Belkora 2003, pp. 35–73.
25. English translation in Kant 1981, which includes a valuable and critical introduction by Stanley Jaki, according to whom Kant's cosmology is overrated and basically unscientific.
26. Kant 1981, p. 152.
27. Ibid., p. 115.
28. Ibid, p. 151 and pp. 154–155.
29. Ibid., pp. 157–158. In *An Original Theory*, Wright had expressed himself similarly. See Wright 1971, p. 76.
30. Kant 1981, pp. 159–160. See also Schaffer 1978.
31. See any edition of *Critique of Pure Reason*, Chapter II, Section II.
32. Quoted from Munitz 1957, p. 263. On Lambert's cosmology, see the introduction to Lambert 1976; Jaki 1990, pp. 39–80; and Hoskin 1982, pp. 117–123.
33. Boscovich 1966, p. 203, a reprint of the first English edition of 1763.
34. Ibid., p. 184.
35. Haber 1959; Gorst 2002.
36. Toulmin and Goodfield 1982, pp. 142–150.
37. Detailed historical accounts of Olbers' paradox can be found in Jaki 1969 and Harrison 1987.
38. The expression 'Olbers' paradox' may have been used first in Bondi 1952, which includes an extensive (but historically flawed) treatment of it.
39. See Stukeley 1936, p. 75 (originally published 1752). Stukeley was among the first to suggest that Stonehenge was an astronomical temple (North 1996).
40. Olbers 1826, p. 111.
41. Strangely, Herschel believed that ordinary light could be absorbed without any temperature increase in the absorbing medium.
42. Quoted in Jaki 1969, p. 147.
43. Mädler, *Der Fixsternhimmel* (Leipzig, 1858) as quoted in Tipler 1988, p. 320. The poet and novelist Edgar Allan Poe said much the same in 1848; compare Cappi 1994.
44. Zöllner 1872, p. 308. See Jaki 1969, pp. 158–164 and Kragh 2004, pp. 24–25.
45. Thomson 1901. Thomson did not refer to Olbers' name in his discussion.

46. Quoted in Berendzen, Hart and Seeley 1976, p. 183.
47. Quoted in Hoskin 1982, p. 56.
48. The remarkable partnership between William and Caroline Herschel is described in Hoskin 2003.
49. Together with other papers by Herschel, this paper is reproduced in Hoskin 1963.
50. Hoskin 1963, p. 85.
51. Lubbock 1933, p. 336.
52. Hoskin 1963, p. 115.
53. Dalton 1808, p. 212. In his pioneering paper of 1917, Einstein compared directly 'the stellar system with a gas in thermal equilibrium'. Einstein et al. 1952, p. 178.
54. Felber 1994, p. 14. The term 'physical astronomy' was used in the late eighteenth century by Laplace and others, but in the sense of celestial mechanics, not astrophysics.
55. Quoted from Crowe 1994, p. 147. The reference to 'physiological and social research', curious to a modern reader, indicates Comte's belief in advanced life elsewhere in the universe.
56. Ibid., p. 148.
57. On the early history of line spectra, see James 1985. Fraunhofer constructed the heliometer with which Bessel detected the parallax of 61 Cygni in 1838.
58. Among the scientists involved in this process were John Draper, William Fox Talbot, John Herschel, David Brewster, and Jean Foucault. The idea that spectral lines might be used for chemical analysis was suggested by Fox Talbot as early as 1826 but only turned into an experimental technique much later. For details, see McGucken 1969 and Hearnshaw 1986.
59. Quoted in Hearnshaw 1986, p. 42.
60. Doppler formulated his law in terms of colour change and did not refer to distinct wavelengths. The first to interpret the effect as a wavelength shift of spectral lines was the Frenchman Hippolyte Fizeau in 1848. Doppler incorrectly argued that the colour of a star would undergo a perceptible change because of its motion relative to the Earth. He suggested that for large stellar velocities the light might be shifted to such an extent that the star would become invisible, a line of reasoning that he used to explain the colour of double stars.
61. Huggins and Huggins 1909, pp. 197–215.
62. Quoted in Kragh 2001a, p. 161. See also Kragh 2000. This section is based on material from these two sources.
63. Roscoe 1875, p. 22.
64. Huggins and Huggins 1909, p. 49.
65. Crookes 1886, pp. 558 and 568. References to Darwinian-like evolution were common in the inorganic sciences.
66. As quoted in Crookes 1886, pp. 559–560.
67. Meadows 1972.
68. Becker 1908, pp. 125 and 145. Becker accepted the existence of coronium, a 'protylic gas' found in the Sun's corona and, he suggested, probably with an atomic weight smaller than that of hydrogen.
69. Nicholson 1913, pp. 103–105. George Hale, the American astronomer, described in 1904 the Sun as an 'enormous crucible' in which 'experiments [occur] on a scale far transcending any that can ever be performed in the laboratory'. Quoted in Kargon 1982, p. 95.
70. See Hoskin 1982, pp. 125–153. For other aspects of the nebular hypothesis, see Brush 1987.
71. Laplace's theory is often linked with Kant's cosmology, but the often-used term 'Kant–Laplace nebular hypothesis' is unfortunate because the two theories have little in common. First of all, whereas Kant suggested a theory of the entire universe, Laplace was concerned only with the origin of the solar system. For an appreciation of Laplace as a cosmologist, see Merleau-Ponty 1977.
72. Scheuer 1997; Kragh 2004, pp. 30–32.
73. Quoted in Hoskin 1982, p. 143.
74. Letter of 1 August 1869, quoted in Hoskin 1987, p. 28.
75. Huggins and Huggins 1909, pp. 106–107. Huggins's retrospective essay on 'The new astronomy', first published in 1897, should be read critically and not as an authoritative account of the history of early astrophysics. See Becker 2001.
76. See Hirsh 1979.
77. Huggins and Huggins 1909, p. 154.
78. Scheiner 1899, p. 150.
79. Fourier's 'cosmological' theory of heat is analysed in Merleau-Ponty 1983, pp. 212–225.
80. On these and other early ideas of solar heat generation, see James 1982.

81. Thomson 1882–1911, vol. 1, p. 514.
82. Ibid., vol. 2, pp. 37–38.
83. Clausius 1868, p. 405.
84. In Agnes Clerke's masterful 1903 account of astrophysics, *Problems in Astrophysics*, there is no mention of the second law, and the law of energy conservation is only referred to in connection with Helmholtz's theory of solar energy. The same is the case with her *History of Astronomy During the Nineteenth Century* published in 1886.
85. Rankine 1881, pp. 200–202.
86. Fick 1869, p. 70. The entropic argument and its relation to cosmology and theology are given a fuller treatment in Kragh 2004, pp. 50–66. See also Landsberg 1991.
87. Tait 1871, p. 6.
88. Thomson 1891, p. 322.
89. Charlier 1896, p. 481. Charlier asserted that the principle of matter conservation ruled out a finite-age universe.
90. Mach 1909, pp. 36–37.
91. Quoted in Kragh 2004, p. 57.
92. Arrhenius 1908, p. xiv.
93. Arrhenius 1909.
94. Poincaré 1911, pp. 240–265.
95. Peirce thought that the universe would end in 'an absolutely perfect, rational, and symmetrical system', a notion clearly at odds with the general interpretation of the entropy law. On Peirce's cosmology, dating from the 1890s, see Reynolds 1996.
96. Boltzmann 1895, p. 415. Boltzmann did not clarify what he meant by 'our world', whether it referred to the solar system, some larger stellar region, or perhaps the entire Milky Way system. For the continual fascination of Boltzmann's argument among physicists and philosophers, see Ćircović 2003a.
97. Schuster 1898. Not only did Schuster introduce names such as 'antimatter' and 'antiatoms', but his speculation also has a vague similarity to the ideas of dark matter and dark energy. He suggested that atoms and antiatoms might combine into 'potential matter' and asked: 'Can we imagine a vast expanse . . . filled with this primordial mixture, which we cannot call a substance because it possesses none of the attributes which characterise matter ready to be called into life by the creative spark? Was this the beginning of the world?'
98. LeBon 1907, p. 315.
99. Soddy 1909, pp. 241–242.
100. Seeliger 1895, p. 132. See the lucid analysis in Norton 1999. On Seeliger and his cosmology, see also Paul 1993. For attempts to modify Newton's law of gravitation, see North 1990, pp. 30–49.
101. Seeliger 1897–98, p. 546.
102. Proctor 1896, p. 285.
103. Charlier 1908. An extended version in English appeared in 1922 in *Arkiv för Matematik, Astronomi och Fysik* (vol. 16, pp. 1–34) (the same journal as where the original paper was published). On Charlier's cosmos, see Jaki 1969, pp. 198–204 and Holmberg 1999, pp. 73–78.
104. Charlier 1896, p. 486.
105. Arrhenius 1909, p. 226.
106. Quoted in Jaki 1969, p. 183.
107. Clerke 1890, p. 368. Charlier referred approvingly to Clerke's view (Charlier 1896, p. 489).
108. Newcomb 1906, pp. 5–6.
109. *The Observatory* **30** (1907), p. 362, in an anonymous review of J. E. Gore, *Astronomical Essays, Historical and Descriptive* (London: Chatto & Windus, 1907).
110. Easton 1900, p. 158.
111. Easton 1913, p. 116.
112. Paul 1993 provides a careful exposition of their work in statistical cosmology. For Kapteyn and his work, see also Van der Kruit and Van Berkel 2000.
113. That the stars move non-randomly in the plane of the Milky Way was first proposed by the German astronomer Hermann Kobold in 1895. The star streaming discovered by Kapteyn is a consequence of the rotation of the Milky Way, as shown by Bertil Lindblad at Uppsala University, Sweden, in 1925.
114. The paper is included (with omissions) in Lang and Gingerich 1979, pp. 542–549.
115. On the history of interstellar absorption, see Seeley and Berendzen 1972 and Berendzen, Hart, and Seeley 1984, pp. 70–99.
116. Trumpler 1930. Reproduced in Lang and Gingerich 1979, pp. 593–604 (on p. 600).

117. Leavitt's paper appeared under the name of Edward C. Pickering, the Director of Harvard College Observatory. See Lang and Gingerich 1979, pp. 398–400. On the development of the period–luminosity relation, see Fernie 1969.
118. Shapley 1918.
119. Smith 1982, p. 22.
120. In Lang and Gingerich 1979, p. 707.
121. Smith 1982, p. 62.
122. Ibid.
123. Berendzen, Hart, and Seeley 1984, p. 116. Russell, too, supported wholeheartedly van Maanen's observational claims. However, in 1925 the Swedish astronomer Knut Lundmark reanalysed van Maanen's photographic plates and concluded that his results were probably spurious and certainly greatly exaggerated. Today it is believed that van Maanen read his expectations into his data and thereby produced an erroneous result. For a full discussion of the case, see Hetherington 1972 and Hetherington 1988, pp. 83–110.
124. Heber Doust Curtis had an unusual academic career. He started in classical languages and was for a period Professor of Latin and Greek at Napa College, California. Having an interest in astronomy, he decided to shift subjects to astronomy and mathematics and in 1902 he joined the staff of the Lick Observatory, where he remained until 1920.
125. The two articles, published in the *Bulletin of the National Research Council*, are reproduced in Crowe 1994, pp. 273–327, and online at http://antwrp.gsfc.nasa.gov/diamond_jubilee/1920/cs_nrc.html. The published versions differed significantly from the oral presentations (Hoskin 1976; Berendzen, Hart, and Seeley 1984, pp. 35–47).
126. Öpik 1922.
127. Christianson 1995 is the most complete biography of Hubble.
128. Letter of 19 February 1925, quoted in Berendzen, Hart, and Seeley 1984, p. 138.
129. Smith 1982, p. 134.

3

FOUNDATIONS OF MODERN COSMOLOGY

3.1 Early relativistic models

Cosmology experienced a paradigm shift with Einstein's general theory of relativity of 1915 and its application to the universe as a whole two years later. Among the conceptual advantages that Einstein's new theory of gravitation had over Newton's was that it incorporated the notion of curved space and thus made it possible to describe scientifically a closed universe without boundaries. Observational astronomy had almost no share in Einstein's breakthrough, where mathematics was far more important than experiments and observations. Although Einstein's knowledge of astronomy was limited, he was sufficiently informed to share the belief that the universe was essentially stable and timeless, a belief that shaped his cosmological field equations of 1917. It took many years to fully understand the amazing richness of the Einstein equations, and for more than a decade the small community of theoretical cosmologists thought that general relativity provided only two cosmological models, both of them describing an idealized static universe. One was Einstein's original solution, and the other was due to his friend and cosmological rival, the Dutchman Willem de Sitter. Whereas Einstein's solution has long been of historical interest only, de Sitter's continued throughout the century to play an important role in cosmology.

From 1917 to the late 1920s, relativistic cosmology was predominantly a mathematical science with very little contact to the work done by observational astronomers. The separation was not complete, though, and with Hubble's discovery in 1929 that the redshifts of spiral galaxies varied linearly with their distance, theory and observation entered a closer and more fruitful relationship.

3.1.1 *Curved space*

The model of the universe that Einstein proposed in 1917, and which heralded a revolution in theoretical cosmology, was based on conceptions of space (and, in Einstein's case, of space–time) with roots in the mid nineteenth century. Space, both in a purely geometrical sense and in the sense of experience, had for centuries been thought to comply with the rules of Euclidean geometry. Kant, for one, believed this was true a priori, which entered into his dismissal of cosmology as truly scientific. In a Euclidean space, the sum of the angles in any triangle equals two right angles (180°), and two parallel lines never meet.

It was mathematical curiosity with regard to the role and status of the parallel axiom within Euclid's system that led to the idea that there were consistent models of space different from the traditional one. With the discovery of new types of space, none of which had a privileged position from the point of view of logic or mathematics, it became natural to ask about the geometry of physical space, that is, the space of the experienced world and the laws of physics. Many decades before Einstein, it was realized that the question could not be answered in purely mathematical terms but needed recourse to experiment and observation. Geometry and physics could not be entirely separated.

The discovery of non-Euclidean geometries—where parallel lines do not remain parallel and the sum of the angles in a triangle differs from 180°—was made independently by Carl Friedrich Gauss in Germany, Nikolai Lobachevskii in Russia (in far-away Kazan), and János Bolyai in Hungary.[1] Lobachevskii and Bolyai published their findings in the early 1830s (unfortunately in obscure publications), whereas Gauss's earlier ideas became generally known only after his death. As early as 1817, Gauss suspected that Euclidean geometry was not necessarily true, but at best a system that happened to be empirically true. Space was not a purely mathematical construct, but had a kind of physical reality, he suggested in a letter to Olbers. 'I become more and more convinced that the necessity of our geometry cannot be demonstrated... Geometry, therefore, has to be ranked... not with arithmetic, which is of a purely aprioristic nature, but with mechanics.'[2] Gauss may even have made an attempt to discover by means of geodetic methods whether physical space was Euclidean or not.[3] As the sides of the triangle he measured—if he did—were of the order of magnitude of 100 km, it is understandable that his experiment led to nothing. Lobachevskii, too, attempted to determine the structure of real space, in his case by astronomical means, namely by measurements of stellar parallaxes. His attempt was premature, as he could only establish that the sum of the angles in his heavenly triangle deviated from 180° by an amount much less than the errors of observation. He hesitatingly concluded that his new geometry was 'without application to nature'.

It took many years until these early ideas of non-Euclidean geometry became generally known among mathematicians, and even longer until they caught the interest of a few physicists and astronomers. Bernhard Riemann, who was a student of Gauss, presented some new and exciting ideas in 1854, but they too failed to attract wide attention. Riemann put the concept of curvature as an intrinsic property of space on a firmer basis and pointed out that there were three geometries with constant curvature, corresponding to flat or Euclidean space (zero curvature), a spherical space (positive curvature), and a hyperbolic space (negative curvature). He further suggested a deep connection between geometry and physics, although it was mostly microphysics and not astrophysics that he had in mind. Yet Riemann briefly noted that an infinite extent does not follow from space being unbounded, for 'if we... ascribe to space constant curvature, it must necessarily be finite provided this curvature has ever so small a positive value'.[4]

The ideas of Bolyai, Lobachevskii, and Riemann circulated slowly. It was only when they were taken up and presented in a better argued and clearer way by the Italian mathematician Eugenio Beltrami in 1868 that non-Euclidean geometries truly entered the world of mathematics. Five years later, the British mathematician William Kingdon Clifford translated Riemann's address into English, which had the effect that the ideas of curved space and of a geometrization of physics became further known, indeed almost popular. As mentioned on p. 85, Zöllner referred to Riemannian geometry as early as 1872 in his discussion of Olbers' paradox. In agreement with Riemann, he wrote: 'The assumption of a positive value of the spatial curvature measure involves us in no way in contradictions with the phenomena of the experienced world if only its value is taken to be sufficiently small.'[5] During the late nineteenth and early twentieth century, non-Euclidean geometries were often mentioned in astronomical texts, if typically in brief and uncommitted ways. In *Popular Astronomy*, a book from 1880, Newcomb considered the subject, but seems to have preferred an infinite Euclidean universe:

Although this idea of the finitude of space transcends our fundamental conceptions, it does not contradict them and the most that experience can tell us in the matter is that, though space be finite, the whole extent of the visible universe can be but a very small fraction of the sum total of space.[6]

The first astronomer who brought up the subject in a serious manner was the versatile Karl Schwarzschild in 1900. His aim was the same as that of Gauss and Lobachevskii before him, to determine the structure of space from observations. In his discussion of hyperbolic space, he concluded that to match observations the radius of curvature of space must be at least four million astronomical units. Should space have a constant positive curvature, the minimal radius must be about 100 million astronomical units (in light years, these two radii are about 64 and 1600). For philosophical reasons, Schwarzschild found it 'satisfying to reason' about whether

. . . we could conceive of space as being closed and finite, and filled, more or less completely, by this stellar system. If this were the case, then a time will come when space will have been investigated like the surface of the earth, where macroscopic investigations are complete and only the microscopic ones need continue. A major part of the interest for me inherent in the hypothesis of an elliptic space derives from this far reaching view.[7]

Not all physicists and mathematicians agreed that the geometry of space could be determined empirically. According to Seeliger, space had no properties at all, from which it followed that measurements could not decide which kind of space we live in. On the basis of his conventionalist conception of science, Henri Poincaré in France argued similarly that experiments and observations were of no value when it came to a determination of the structure of space; they could only tell us about the relations that hold among material objects such as rigid rods. As far as Poincaré was concerned, the geometry of space was not something that could be determined objectively, it was merely a convention. One should choose the geometry that allowed the simplest description of nature, and this geometry, he believed, was Euclidean.

Poincaré's conventionalism turned up in a thought experiment about how we would perceive a hypothetical expansion of the universe. This was not, however, an anticipation of later knowledge, but merely an exercise in conventionalist philosophy of science. His conclusion was that an expansion of the universe would be unobservable because the dimensions of everything, measuring instruments included, would increase in the same ratio: 'The most precise measurements will be incapable of revealing to me anything of this immense convulsion, since the measures I use will have varied precisely in the same proportion as the objects I seek to measure.' Because space was not absolute, but relative, 'nothing at all has happened, which is why we have perceived nothing'.[8] Poincaré was not the first to state that a uniform expansion (or contraction) of all magnitudes in the universe would be unobservable. As mentioned on p. 82, Boscovich pointed out the same thing as early as 1758.

Auguste Calinon, a French scientist and philosopher, was a friend of Poincaré and shared his interest in the foundations of mechanics and geometry. However, contrary to Poincaré, he did not believe that Euclidean geometry was always to be preferred because of its simplicity. In a paper of 1889 on the geometry of space, he suggested that the curvature of space need not be constant, but might vary in time, for example it might oscillate between Euclidean and non-Euclidean forms. The original idea remained, however, on a philosophical level and Calinon made no attempt to place it within an astronomical context. In another paper, he suggested that space might differ from the Euclidean form at very large distances and that Newton's law of gravitation might consequently need to be modified. 'We may therefore very well conceive that at such large distances the law of attraction... could find its simplest expression in another geometric representation of the universe, different from the Euclidean representation.'[9]

3.1.2 *Einstein's theory of general relativity*

In Albert Einstein's special theory of relativity, published in 1905, space and time were connected in a flat space–time continuum. The metric of the space–time, giving the distance between two neighbouring events in space and time, was a four-dimensional generalization of Euclid's distance formula,

$$ds^2 = c^2\,dt^2 - (dx^2 + dy^2 + dz^2).$$

This kind of flat space–time is called Minkowskian, after the German mathematician Hermann Minkowski, who in 1907–08 formulated the special theory of relativity in a four-dimensional version where the space and time coordinates appear in a symmetrical way. (By introducing $c\,dt = i\,dw$, where i is the square root of -1, the metric appears in a more symmetric form.)

The first step towards an extended theory of relativity, aimed at covering gravitation also, was taken in 1907 when Einstein formulated in a generalized way the principle of equivalence. This principle states that no experiment can distinguish between a homogeneous gravitational field and a uniformly accelerated frame in which there is no gravitational field. In the process of developing this idea, Einstein, while a professor in Prague, published in 1911 a paper that included the first generalization of his earlier-stated, restricted principle of relativity. Already in 1907, Einstein understood that the rate of a clock would be slowed down near a large gravitating mass. The clock might be a light-emitting atom, and Einstein showed in 1911 from the principle of equivalence that the light emitted from the atom would experience a gravitational redshift. If a beam of light is subject to the gravitational force from the surface of a spherical mass M of radius R, it will be received redshifted by an amount

$$\frac{\Delta\lambda}{\lambda} = \frac{\Delta\phi}{c^2} = -\frac{GM}{c^2 R},$$

where $\Delta\phi$ is the difference between the gravitational potentials at the points of emission and reception of the light. This kind of redshift is entirely different from the Doppler shift of light emitted by a receding source.

In his paper of 1911, Einstein also demonstrated how the principle of equivalence led to the propagation of light being affected by gravity. This was not to be explained by light being attracted by a gravitating mass, but by the geometry of space being distorted by the presence and distribution of matter. Einstein found in 1911 that for a light ray grazing the

Sun, the deflection would be about 0.83″, too small by a factor of two, as it would later turn out. 'It would be a most desirable thing if astronomers would take up the question here raised,' he ended the paper.[10] Two years later he addressed George Hale at the Mount Wilson Observatory, asking him about the possibility. Hale replied that it might be possible to measure the effect during a solar eclipse, but no immediate action was taken. Unknown to Einstein and Hale, the deflection of light by a massive body had been studied more than a century earlier, by the German astronomer Johann Georg Soldner in a paper of 1801 based on the corpuscular theory of light. Soldner obtained for the Sun almost exactly the same deflection as derived by Einstein, namely 0.84″. However, he concluded that the effect would be unnoticeable and that the perturbation caused by the Sun could therefore be neglected.[11]

Einstein's further development of the theory led him deeply into the study of differential geometry and tensor analysis, branches of mathematics which at the time were unknown to most physicists and were believed to be of mathematical interest only. According to an address he gave in Kyoto in 1922, it was in 1912 that he realized that non-Euclidean geometry might help him in his search for a new theory of relativity: 'Until then I did not know that Bernhard Riemann had discussed the foundation of geometry deeply. I happened to remember the lecture on geometry in my student years by Carl Friedrich Geiser who discussed the Gauss theory. I found that the foundations of geometry had deep physical meaning in this problem.'[12] It was from his friend the mathematician Marcel Grossmann that he learned about Riemann's work and related mathematical topics. In collaboration with Grossmann Einstein developed in 1913 the first tensor theory of gravitation, known as the *Entwurf* theory (from the title of their paper, *Entwurf* meaning 'outline'). It contained the first field equations for gravitation, but these were not generally covariant, meaning that they did not have the same form in every frame of reference. Einstein realized that physical laws should preferably satisfy this requirement, but after much thinking he nonetheless ended up convincing himself that the non-covariant features of the 1913 theory were unavoidable. This was an error, and it took him two years of intense work to come back on to the right track, that of generally covariant field equations.

In two papers of 18 and 25 November 1915, Einstein presented his theory in its final form to the Prussian Academy of Sciences. A full account, published in the *Annalen der Physik*, had to wait until March 1916.[13] The formal core of Einstein's general theory of relativity was the gravitational field equations, also known as the Einstein equations, which first appeared in his paper of 25 November. In the usual tensor notation, they read

$$R_{\mu\nu} - \tfrac{1}{2} g_{\mu\nu} R = -\kappa T_{\mu\nu}.$$

The equations express in a condensed way how the geometry of space–time relates to its content of matter and energy (given by the energy–momentum tensor $T_{\mu\nu}$). The quantity $R_{\mu\nu}$ is a tensor of metric curvature known as the Ricci tensor, and R is a curvature invariant derived from it; the components of the fundamental tensor $g_{\mu\nu}$ are functions of the chosen coordinates. The quantity κ is known as the Einstein gravitational constant and is proportional to Newton's constant ($\kappa = 8\pi G/c^2$). In later literature the terms on the left side were grouped together as $G_{\mu\nu}$, known as the Einstein tensor. As early as the beginning of 1916, Schwarzschild analyzed the field equations for a point mass and derived an exact solution, thereby (unwittingly) laying the foundation for later studies of black holes.[14]

Schwarzschild, who was then serving in the army on the Eastern front, died shortly afterwards from a disease he had contracted at the front.

In the first of his papers of November 1915, Einstein calculated from his theory the motion of Mercury and thereby solved a mystery that had long puzzled astronomers. It had been known since 1859 that the innermost planet did not move around the Sun exactly as it should according to Newtonian mechanics, since its perihelion precessed slowly around the Sun with a speed of rotation in excess of the predicted value. The anomalous precession amounted to a mere 43″ per century, yet this was a large enough discrepancy to constitute a serious problem for the celestial mechanics founded on Newton's law of gravitation. Many attempts to solve the problem were made over the years—including ad hoc modifications of the law of gravitation and the assumption of an intra-Mercurial planet called 'Vulcan'—but it was only with Einstein that it proved possible to account for the anomaly in a way that was based on fundamental theory. His calculated value of the precession agreed almost perfectly with observations. On 10 November, he wrote to his friend Michele Besso, 'My wildest dreams have been fulfilled. *General* covariance. Perihelion motion of Mercury wonderfully exact.'[15]

In addition to this success, Einstein derived from his new theory a value for the predicted gravitational deflection of light that was twice as large as the value he had found in 1911. A light ray passing the surface of the Sun, he now concluded, would be bent by an angle of 1.7″. Whereas the explanation of the Mercury anomaly concerned a phenomenon already known, the calculation of the bending of light in a gravitational field was a proper prediction, relating to a phenomenon that had not yet been observed. Einstein was keenly aware of the importance of light bending as a confirmation of general relativity, as already indicated by his paper of 1911. In his 1916 paper in *Annalen der Physik*, he briefly discussed the three predictions and wrote, 'a ray of light going past the Sun undergoes a deflection of 1.7″; and a ray going past the planet Jupiter a deflection of about 0.02″.'[16]

In the years immediately following 1916, astronomers showed little interest in the general theory of relativity. However, there were exceptions, and these were important. Apart from Schwarzschild, de Sitter, and Eddington, the Berlin astronomer Erwin Freundlich merits attention. Freundlich, a former student of Schwarzschild and of the eminent mathematician Felix Klein, was immediately drawn to Einstein's theories and sought as early as 1911 to confirm the prediction of light deflection. During the war years he served as Einstein's mouthpiece in the world of astronomy, and he announced the astronomical relevance of the theory of relativity in *Astronomische Nachrichten* and other journals read by German astronomers.[17]

Attempts to detect the bending of light as it passed the rim of the Sun started with Einstein's 1911 paper, but it was only with the famous British solar eclipse expedition of 1919, headed by Frank Dyson and Arthur Eddington, that good data were obtained. Photographs were taken at both Principe Island (off the West African coast) and Sobral in Brazil, and when they were analysed they showed a light deflection in fair agreement with Einstein's theory. There is little doubt, the British astronomers reported, that 'a deflection of light takes place . . .[as] demanded by Einstein's generalised theory of relativity'. Measured in seconds of arc, the Principe Island measurements gave 1.98 ± 0.12, and those obtained at Sobral 1.61 ± 0.30. The highly publicized eclipse measurements created a sensation and made the general theory of relativity known to (if not necessarily understood by) scientists and laypersons alike.[18] More than anything else, it was the results from the British expedition that catapulted Einstein into the public limelight and made him the best-known scientist since Darwin.

The third prediction of general relativity—the gravitational redshift—was much harder to verify, and it took many years until it was established beyond doubt that in this case, too, Einstein was right. Measurements of spectral lines from the Sun's atmosphere did not give an unambiguous answer, but in 1924 Eddington believed he had found an alternative way to prove Einstein's gravitational redshift. He argued that the faint companion star of Sirius A was a white dwarf of exceptionally high density and that light from it would therefore have to escape a very strong gravitational field; consequently, Sirius B (as the companion star is called) would be ideally suited to test Einstein's prediction. Eddington predicted from the general theory of relativity that light from Sirius B would be redshifted by what corresponded to a radial motion of 20 km/s, and since the American astronomer Walter Adams found a value of 21 km/s, the Einsteinian redshift seemed to be confirmed.[19]

3.1.3 *A closed, static universe*

Einstein did not rest on his laurels. One of the few scientists outside Germany who in 1916 knew of the general theory of relativity was the Dutch astronomer Willem de Sitter, a 44-year-old scientist equally at home with astronomical observations and advanced mathematical analysis. As a foreign member of the Royal Astronomical Society, he was invited by Eddington (who then served as the society's secretary) to produce an account of the new theory, which he did in three articles in the *Monthly Notices*. These articles introduced Einstein's theory to the English-speaking world and it was on the basis of them that Eddington wrote his *Report on the Relativity Theory of Gravitation* in 1918. In the autumn of 1916 de Sitter discussed the theory with Einstein, and as a result of these discussions Einstein attempted to apply his theory to the universe at large. In doing so, he was faced with conceptual problems of the same kind as Newton had struggled with in his correspondence with Bentley and which, in a much modernized version, had been reformulated by Seeliger in the 1890s. However, at the time Einstein turned to cosmology he was unaware of this historical tradition and was not even acquainted with Seeliger's work. He learned about this only later in 1917.[20]

The fruits of Einstein's thinking about the universe appeared in February 1917 in a paper to the Prussian Academy of Sciences with the title 'Kosmologische Betrachtungen zur allgemeinen Relativitätstheorie' ('Cosmological Considerations on the General Theory of Relativity'). Having sent the paper to Besso, he argued in an accompanying letter that a homogeneous, symmetrical distribution of matter throughout all of infinite space would not be sufficient to produce the stable universe that he presupposed. 'Only the closure of the universe frees us from this dilemma; this also suggests itself *in that the curvature has the same sign throughout because, according to experience, the energy density does not become negative*,' he wrote.[21] He also dismissed the other Newtonian alternative, the picture of the stellar universe as a finite island in an otherwise empty and infinite universe. Likening the stars to the molecules of a gas, he argued that individual stars would eventually escape the gravitational field of the other stars—evaporate, as it were.

Einstein's solution was to circumvent the classical boundary problem, which he did by conceiving the universe as spatially closed in accordance with his general theory of relativity. As he wrote in his paper of 1917, 'If it were possible to regard the universe as a continuum which is finite (closed) with respect to its spatial dimension, we should have no need at all of any such boundary conditions. We shall proceed to show that both the general postulate of relativity and the fact of the small stellar velocities are compatible with the hypothesis of a spatially finite universe; though certainly, in order to carry through this idea, we need a generalizing modification of the field equations of gravitation.'[22]

Einstein thus assumed the universe to be a spatially closed continuum, 'spherical' in four dimensions, although the real, non-uniform universe would be what he called 'quasi-spherical'. The idealized model is also referred to as Einstein's 'cylinder' world: with two of the spatial dimensions suppressed, the model universe can be pictured as a cylinder, where the radius represents the space coordinate and the axis the time coordinate. Of course, Einstein was also guided by the available empirical evidence, or what little he knew of it. This suggested, he believed, that the universe was indeed spatially finite, that it was static, and that it contained a finite amount of matter. In order to keep to what he considered convincing empirical evidence, he was led to the following conclusion: 'The curvature of space is variable in time and space, according to the distribution of matter, but we may roughly approximate to it by means of a spherical space.' Einstein found this model to be logically consistent and natural from the point of view of general relativity, whereas he was less concerned with its agreement with the observed universe: 'Whether, from the standpoint of present astronomical knowledge, it is tenable, will not here be discussed.'[23]

Einstein's interest in Mach's principle, named after Ernst Mach, significantly shaped his cosmological model. According to one version of this famous principle, which dates back to the 1860s, the laws of mechanics should be seen as purely relational, namely, relative to the universe as a whole. Einstein's version was somewhat different, as he understood it in the sense that the space–time metric should be fully determined by the masses in the universe, and thus that the local dynamics was conditioned by the universe at large. Put differently, matter was necessary for space. The name 'Mach's principle' was introduced by Einstein in 1918, when he pointed out that although the principle was not generally fulfilled in general relativity, this was the case when the field equations were supplemented with a cosmological term. With his cosmological model of 1917, he had ensured that his new gravitation theory satisfied Machian ideas, which was an important reason for his adventure into cosmology.[24] Later in his career, Einstein changed his mind with regard to Mach's principle, but in 1917 he was a Machian at heart.

In order to secure a universe static in time, Einstein was led to an important change to his field equations of 1915, the 'generalizing modification' referred to in the quotation above. His change consisted in adding a term proportional to the metric tensor. The factor of proportionality soon became known as the cosmological constant, and is always denoted by the Greek letter lambda (Einstein used the symbol λ in 1917, whereas at present the capital letter Λ is mostly used). With this change, the fundamental equations read

$$R_{\mu\nu} - \tfrac{1}{2} g_{\mu\nu} R - \Lambda g_{\mu\nu} = -\kappa T_{\mu\nu} \,..$$

It is to be noted that the cosmological term appears on the left side of the equations, in conformity with Einstein's belief that it was a property of space–time, not of matter–energy. Einstein admitted in his 1917 paper that the introduction of the cosmological constant

kommen analog ist. Wir können nämlich auf der linken Seite der Feldgleichung (13) den mit einer vorläufig unbekannten universellen Konstante $-\lambda$ multiplizierten Fundamentaltensor $g_{\mu\nu}$ hinzufügen, ohne daß dadurch die allgemeine Kovarianz zerstört wird; wir setzen an die Stelle der Feldgleichung (13)

$$G_{\mu\nu} - \lambda g_{\mu\nu} = -\varkappa \left(T_{\mu\nu} - \frac{1}{2} g_{\mu\nu} T \right). \qquad (13a)$$

Auch diese Feldgleichung ist bei genügend kleinem λ mit den am Sonnensystem erlangten Erfahrungstatsachen jedenfalls vereinbar. Sie befriedigt auch Erhaltungssätze des Impulses und der Energie, denn

Fig. 3.1 Einstein's cosmological field equations in his article of 1917. The cosmological constant (λ) appears on the left side, introduced as a 'for the time being unknown universal constant'.

'is not justified by our actual knowledge of gravitation,' i.e. that it was largely of an ad hoc character, but he found it 'necessary for the purpose of making a quasi-static distribution of matter'. The value of the constant was unknown, except that in Einstein's model it was necessarily positive, since it expressed the mean density of matter in the universe. Moreover, in order for the equations to agree with planetary motions, it had to be very small: the general-relativistic equations $R_{\mu\nu} = 0$, valid for empty space, were known to agree with observations within the solar system and thus the cosmological term had to be exceedingly small not to spoil the agreement. How small he could not say, but de Sitter offered an estimate: 'Observations will never be able to prove that λ vanishes, only that λ is smaller than a given value. Today I would say that λ is *certainly* smaller than 10^{-45} cm^{-2} and is probably smaller than 10^{-50}. Maybe one day observations will also provide a specific value for λ, but up to now I have no knowledge of anything pointing to this.'[25]

It is helpful to think of the constant Λ as a term which introduces a cosmic repulsion proportional to the distance, negligible at small distances but increasingly important at very large distances. A particle of mass m will be repelled with a force $F \sim m\Lambda r$. In this picture, the evolution of the universe is determined by the competition between the repulsive Λ force and the attractive force of Newtonian gravitation. In Einstein's static universe, the two forces are in balance. The cosmological constant in Einstein's theory was not unlike the constant that appeared in Seeliger's Newtonian theory two decades earlier, but it had no classical counterpart in the form of a definite correction term to Newton's law. Although Einstein soon came to have second thoughts with regard to the cosmological constant, it played an important and natural role in the development of his theory of 1917. He saw it justified in particular by its connection to the mean density of matter in the closed universe, such as he explained in a letter to Besso of August 1918:

Either the world has a center point, has on the whole an infinitesimal density, and is empty at infinity, whither all thermal energy eventually dissipates as radiation. Or: All the points are on average equivalent, the mean density is the same throughout. Then a hypothetical constant λ is needed, which indicates at which mean density this matter can be at equilibrium. One definitely gets the feeling that the second possibility is the more satisfactory one, especially since it implies a finite magnitude for the world. Since the world just exists as a single specimen, it is essentially the same whether a constant is given the form of one belonging within the natural laws or the form of an 'integration constant.'[26]

The cosmological constant was introduced in order to maintain a static universe in accordance with observations. Would Einstein have discovered the dynamic solutions to his equations had he not introduced the constant? This might in principle have happened, but the lack of recognition of such solutions was not caused by the presence of the cosmological constant. After all, the first models of the expanding universe, those of Friedmann and Lemaître, included the constant. Although not necessarily ad hoc, the Λ term made the field equations a bit more complicated and a bit less appealing, at least in Einstein's view. It was such aesthetic considerations that at an early stage made him doubt if the cosmological constant could be justified. In 1919 he described the introduction of the constant as 'gravely detrimental to the formal beauty of the theory'.[27] However, at that time he could see no alternative, and it took another twelve years until he decided that the introduction of the cosmological constant had been a mistake, perhaps his 'greatest blunder'.[28]

The model of the universe derived by Einstein, and, he believed, the only one consonant with his equations, was homogeneously filled with dilute matter and thus could be ascribed a definite mass. He found the cosmological constant to be related to the density ρ and radius of curvature R by

$$\Lambda = \tfrac{1}{2}\kappa\rho = \frac{1}{R^2},$$

For the mass of the closed universe, he obtained

$$M = \pi^2\sqrt{\frac{32}{\kappa^3\rho}}.$$

The first of these relations reflects the key message of relativistic cosmology, that the density of matter determines the radius of curvature of the universe. As to the numerical values of the quantities, Einstein was understandably cautious. In his letter to Besso of March 1917, he erroneously suggested that $R \approx 10^7$ light years, on the basis of the much too high estimate of $\rho \approx 10^{-22}$ g/cm^3, and he believed that the most distant visible stars were merely 10 000 light years away from the Sun. In a letter to de Sitter of 12 March 1917 he repeated the suggestion, but wisely decided not to publish it. What mattered to Einstein was that he had succeeded in constructing a model of the universe with a constant positive curvature, and that this model was in full agreement with the theory of relativity and Mach's principle. 'From the standpoint of astronomy, of course, I have erected but a lofty castle in the air,' he admitted.[29]

3.1.4 *Solution A or solution B?*

Einstein originally believed that his static, matter-filled model was the only solution to the cosmological field equations. However, in his third report to the Royal Astronomical Society of 1917, de Sitter showed that there exists another solution, corresponding to an empty universe with $\Lambda = 3/R^2$ and spatially closed in spite of its lack of matter. As in the Einstein model, the pressure was taken to be zero.[30] De Sitter termed his new model solution B, to distinguish it from Einstein's solution A (but soon the two models became known under the names of de Sitter and Einstein, respectively). Compared with Einstein's model, de Sitter's was complex and difficult to conceptualize, in particular because it was unclear how to distinguish the properties of the model itself from those properties that merely reflected a particular coordinate representation of it. Although the de Sitter model would eventually be seen as representing an expanding universe, to de Sitter and his contemporaries it represented a static space–time.

When Einstein was confronted with de Sitter's alternative, he was forced to accept it as a mathematical solution to the field equations; yet he felt sure that it must be rejected on physical grounds. For one thing, it blatantly violated Mach's principle, which made Einstein write to de Sitter that 'your solution does not correspond to a physical possibility.' As he explained: 'In my opinion, it would be unsatisfactory if a world without matter were possible. Rather, the $g_{\mu\nu}$-field should be *fully determined by the matter and not be able to exist without the matter*.'[31] Even more importantly, Einstein argued that solution B contained a 'singularity' that signalled the breakdown of the theory at a surface, which to Einstein's mind was a clear indication that it was a toy model with no physical significance. Even after Felix Klein had shown that the alleged singularity was a result of the use of misleading coordinates, Einstein continued to consider solution B artificial and unphysical. As became clear in the 1920s, what Einstein took to be a singularity in de Sitter's solution was the presence of an event horizon, a distance from beyond which no light signal can reach the observer.

In his third paper to the *Monthly Notices*, de Sitter showed that if a particle was introduced at a distance r from the origin of a system of coordinates, it would appear to be moving away from the observer with an acceleration given by $\Lambda c^2 r/3$. He summarized the difference between the two models as follows:

> In A there is world-matter, with which the whole world is filled, and this can be in a state of equilibrium without any internal stresses or pressures if it is entirely homogeneous and at rest. In B there may, or may not, be matter, but if there is more than one material particle these cannot be at rest, and if the whole world were filled homogeneously with matter this could not be at rest without internal pressure or stress.

As to the cosmological constant, appearing in both of the systems A and B, he admitted that it 'is somewhat artificial, and detracts from the simplicity and elegance of the original theory of 1915.'[32] Interestingly, de Sitter's model indicated that, as a result of the metric, clocks would appear to run more slowly the farther away they were from the observer. Since frequencies are inverse time intervals, light would therefore be expected to be received with a smaller frequency, being more redshifted the larger the distance between source and observer. In de Sitter's theory, the redshift would increase quadratically with the distance. 'The lines in the spectra of very distant stars or nebulae must therefore be systematically displaced towards the red, giving rise to a spurious positive radial velocity,' he wrote.[33] Notice that de Sitter described the velocity as 'spurious': it was not a real velocity caused by the expansion of space, but an effect of the particular space–time metric he used. In spite of the redshift built into de Sitter's model, the model was thought of as static.

Keeping abreast with recent astronomical observations in spite of the difficulties caused by the war, de Sitter suggested that the predicted effect might be related to the measurements of radial nebular velocities reported by Slipher and others. This was the first suggestion that Einstein's theory might have connections to the observations of nebular redshifts. With a mean radial velocity of 600 km/s and an average distance of 10 parsecs (based on only three nebulae), he found $R \cong 3 \times 10^{11}$ AU. According to de sitter,

If... continued observation should confirm the fact that the spiral nebulae have systematically positive velocities, this would certainly be an indication to adopt the hypothesis B in preference to A. If it should turn out that no such systematic displacement of spectral lines towards the red exists, this could be interpreted either as showing A to be preferable to B, or as indicating a still larger value of R in the system B.[34]

At the end of his paper, de Sitter compared the two rival world models with available astronomical data. Adopting Kapteyn's estimate of a density of about 80 stars per 1000 cubic parsecs, he found that in Einstein's model the radius R would be about 10^{12} AU and the total mass of the universe M about 10^{12} Sun masses. Although he admitted that the estimates were highly uncertain, he found it remarkable that roughly the same figures could be obtained from different lines of reasoning.

Although de Sitter's model, being devoid of matter, might seem a very artificial candidate for the real world, it soon became a foundation for further theoretical work, among both astronomers and mathematicians. It was seen as particularly interesting because of its connection with the redshift observations of spiral nebulae, which by the early 1920s left little doubt that there was a systematic recession. The lack of content of matter in de Sitter's model certainly did not prevent researchers from investigating it as a possible model of the real universe. Although the universe is indeed filled with matter, it was known to be of very low density, and de Sitter suggested that the density might be so low that his model might apply as a zero-density approximation. Eddington appreciated at an early date the cosmological significance of the redshift measurements. In a letter of 1918 to Shapley, he wrote: 'de Sitter's hypothesis does not attract me very much, but he predicted this (spurious) systematic recession before it was discovered definitely; and if, as I gather, the more distant spirals show a greater recession that is a further point in its favour.'[35]

Whatever the credibility of solutions A and B as candidates for the real structure of the universe, from about 1920 there developed a minor industry based on the two models. It was predominantly a mathematical industry, with mathematically minded physicists and astronomers analysing the properties of the two solutions and proposing various modifications of them. The mathematical appeal of the field is underlined by the fact that it attracted the attention of distinguished mathematicians such as Felix Klein, Tullio Levi-Civitá, and Hermann Weyl. The basic aim of these early investigations was to determine which of the two relativistic models of the static universe was the most satisfactory from the point of view of fundamental physics; astronomical observations were still too scarce and uncertain to enter as a decisive criterion.

The question of the relationship between cosmology and the observed redshifts remained unresolved for a decade or so, for other reasons, because it was difficult to distinguish a cosmological redshift (the de Sitter effect) from gravitational redshifts and the Doppler shifts caused by relative motion.[36] The German astronomer Carl Wilhelm Wirtz discussed in 1922 the possibility of a velocity–distance relation, and in 1924 he related the radial velocities of the spirals to de Sitter's cosmology. Since he did not know the distances to the spiral nebulae, he used, as a substitute, their apparent diameters and found that the radial velocities decreased with the logarithm of the distances derived from the diameters. Wirtz concluded that his relation was consistent with de Sitter's world model, but his work was not further developed. Whereas de Sitter had obtained a quadratic relation between redshift and the distance r, Weyl calculated in 1923 in his celebrated *Raum-Zeit-Materie* a result that for relatively small distances amounted to a linear relation, the first of its kind.[37]

The following year Ludwik Silberstein, a Polish-born physicist then staying in England, argued for a relation of the form

$$z = \frac{\Delta\lambda}{\lambda} \cong \pm\frac{r}{R},$$

referring to blueshifts as well as redshifts. Silberstein compared redshift data with estimates of the distances to globular clusters and obtained in this way a value for the radius of curvature R of about 6×10^{12} AU. However, most astronomers agreed with the Swedish astronomer Knut Lundmark that the available data did not support a linear relationship of the kind suggested by Silberstein. Not only was Silberstein under suspicion for having selected only data which agreed with his prediction, but also his value for R conflicted with the island universe theory that Lundmark preferred. As a consequence, his theory was strongly criticized by several leading astronomers, who tended to ridicule the 'Silberstein effect'. Partly as a result of the hostile reaction to Silberstein's prediction, redshift–distance relations were for a period regarded with scepticism in the astronomical community.

The cosmological models of Einstein and de Sitter were not well known among observational astronomers in the 1920s, but de Sitter's solution B did attract some attention because of its connection to the redshift observations. In this respect an important paper of 1926 by Edwin Hubble on the classification of nebulae stands out as significant, especially in regard to later events. By assuming that all nebulae had the same absolute luminosity and a mean mass of 2.6×10^8 solar masses, Hubble obtained an average mass density of the

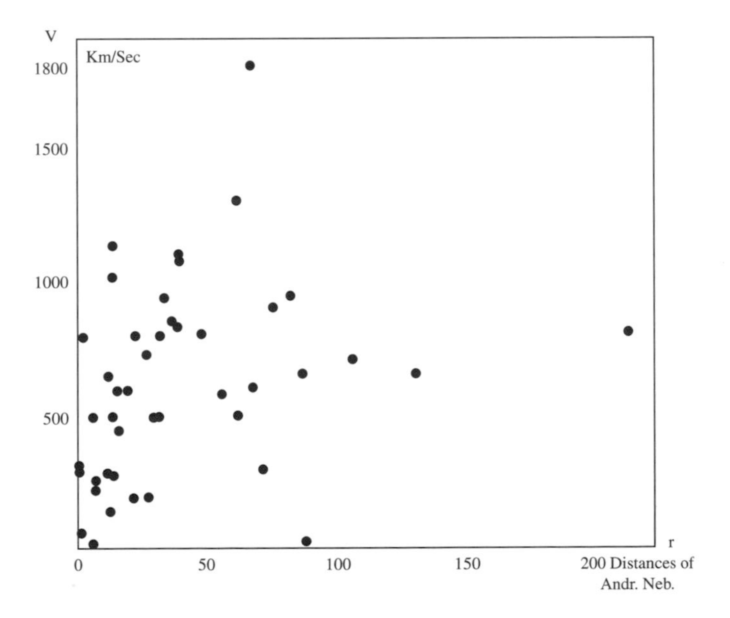

Fig. 3.2 Knut Lundmark investigated in 1924 the possible existence of a redshift–distance relation among spiral nebulae. He suggested that there was some kind of relation, although 'not a very definite one'. K. Lundmark, 'The determination of the curvature of space–time in de Sitter's world', *Monthly Notices of the Royal Astronomical Society* **84** (1924), 747–770.

universe of 1.5×10^{-31} g/cm^3, much less than previous estimates. He ended his paper by inserting his new density value into Einstein's expressions for the radius and mass of the universe, ending up with the values $R = 2.7 \times 10^{10}$ parsecs and $M = 9 \times 10^{23}$ solar masses = 3.5×10^{15} normal nebular masses. Hubble ended his paper as follows:

> The distance to which the 100-inch reflector should detect the normal nebula was found to be of the order 4.4×10^7 parsecs, or about 1/600 the radius of curvature. Unusually bright nebulae, such as M 31, could be photographed at several times this distance, and with reasonable increases in the speed of plates and size of telescopes it may become possible to observe an appreciable fraction of the Einstein universe.[38]

Silberstein, a seasoned polemicist, sharply criticized Hubble's conclusions and came up with his own result for the size of the universe, based on analysis of globular clusters, cepheids, and stars. His universe was curiously small, having a curvature radius of merely 1.5 Mpc (megaparsecs).[39]

The idea of treating the entire world (or an idealization of it) by means of Einstein's cosmological field equations constituted a revolution in the age-old conception of the universe. The world or universe had traditionally been thought of as those parts within the limits of observation. It now became everything, the totality of events in space and time, and—most importantly—space–time itself. To the majority of astronomers, and of course to most laypersons, Einstein and de Sitter's reconceptualization of the universe was either unknown or, if known, unintelligible and objectionable. The eminent French mathematician Emile Borel published in 1922 a book in which he gave a popular exposition of the theory of relativity. In the final chapter on cosmology—'the most interesting [consequence of relativity theory] from the point of view of philosophy'—he asked rhetorically about what use these new cosmological speculations could possibly be:

> It may seem rash indeed to draw conclusions valid for the whole universe from what we can see from the small corner to which we are confined. Who knows that the whole visible universe is not like a drop of water at the surface of the earth? Inhabitants of that drop of water, as small relative to it as we are relative to the Milky Way, could not possibly imagine that beside the drop of water there might be a piece of iron or a living tissue, in which the properties of matter are entirely different.[40]

Borel did not subscribe to the objection, but it is a natural one that cannot easily be brushed aside. As we have seen, similar objections against the possibility of a science of the universe had been raised in the nineteenth century, and the argument would, in different guises, continue to play a role in the cosmological discussion of the twentieth century. However, Einstein, and those who followed him, decided that if cosmology were to progress, the objection had to be ignored.

Although Einstein's theory of general relativity was adopted by most experts, it was not the only non-Newtonian theory of gravitation in the 1920s. Alfred North Whitehead, the mathematician and philosopher, presented in 1922 an alternative theory of gravitation with cosmological implications that were, he claimed, directly deducible from his natural philosophy. His non-covariant, action-at-a-distance theory differed conceptually and mathematically from Einstein's, yet, as Eddington showed, it led to the very same predictions.[41] Whitehead's gravitation theory was taken up by a few physicists, but soon fell into oblivion. Only after the Second World War was it further developed and applied to cosmological problems by John Synge, C. B. Rayner, and others.[42] However, mainstream relativists and cosmologists ignored it. The same was the case with yet another new theory of gravitation, proposed by the American mathematician Garrett D. Birkhoff in 1942. Interest in Birkhoff's theory remained limited to a small group of physicists and mathematicians.

Whitehead's principal work, *Process and Reality: An Essay in Cosmology*, appeared in 1929. It was an ambitious attempt to understand the world in its totality, a metaphysical system rather than a cosmological theory in the tradition of Einstein, de Sitter, and Eddington. Nonetheless, it included some features that later attracted attention, in particular Whitehead's evolutionary and organistic conception of the universe, his claim that our universe was ephemeral and would be replaced by another one with different laws of nature and different dimensions of the space–time continuum.[43]

3.2 The expanding universe

As far as observations are concerned, it is generally recognized that modern cosmology rests on two momentous discoveries, one being that of the expansion of the universe in 1929 and the other that of the cosmic microwave background radiation in 1965. Neither of these cosmic phenomena was discovered in any direct observational sense: they were turned into discoveries in processes where theory was no less important than observation. The expanding universe was predicted in two stages, first as one possibility among many (by Friedmann in 1922) and then as the solution favoured by measurements (by Lemaître in 1927). By 1927 there was a definite prediction of a universe expanding in such a way that the galactic redshifts varied proportionally to their distances, yet Hubble's subsequent observation-based establishment of the relationship was wholly independent of the prediction. From today's vantage point it may be hard to understand why the expansion of the universe was recognized only in 1930, at a time when the 'obvious' solution had been there for several years. But, from the point of view of cosmologists in the 1920s, the expanding universe was far from obvious. It was a concept outside their mental framework, something not to be considered, or, if it was considered, to be resisted.

3.2.1 *Non-static world models*

During the course of their work to understand and elaborate the two relativistic world models, some physicists and astronomers proposed solutions that combined features of Einstein's model A and de Sitter's model B. Towards the end of the 1920s there was a tendency to conclude that neither of the two models could represent the real universe, yet finding a compromise turned out to be frustratingly difficult. Eddington wavered in his sympathies. On the one hand, he considered it a point in favour of de Sitter's model that it promised an explanation of the redshifts of the spiral nebulae. As he wrote in 1923, 'It is sometimes urged against De Sitter's world that it becomes non-statical as soon as any matter is inserted in it. But this property is perhaps rather in favour of De Sitter's theory than against it.'[44] On the other hand, the Einstein model had the advantage that it was filled with matter and therefore compatible with the occurrence of very large pure numbers, such as the ratio of the electromagnetic radius of an electron (e^2/mc^2) to its gravitational radius (κm). This number greatly appealed to Eddington, who quoted it as 3×10^{42}. He suggested further that the number might be related quadratically to the number of particles (electrons and protons) in the universe, yielding a 'cosmical number' of about 10^{85} particles. This was an early example of Eddington's expression of interest in connecting numbers from the microworld and the macroworld, a theme that would dominate much of his later thinking about cosmology and fundamental physics.[45]

In a formal sense, a non-static world model was included in an article by Cornelius Lanczos, a Hungarian–German physicist, published in 1922. By an ingenious change of coordinates, Lanczos found a model in which the radius of curvature varied hyperbolically with time. However, Lanczos's main interest was mathematical rather than physical, and he did not explicitly attempt to apply his model to the real world. In this respect, his paper was similar to that of Friedmann from the same year (see below), except that whereas Friedmann's was largely ignored, Lanczos's attracted considerable attention from the small community of relativity experts. It was discussed and criticized by Weyl in particular.

In part inspired by Eddington, the Belgian physicist Georges Lemaître (of whom more later) introduced in 1925 division of space and time other than that used by de Sitter and was in this way able to derive a model in which 'the radius of space is constant at any place, but is variable with time'.[46] What Lemaître and a few other relativists did in the 1920s was not really to abandon the static universe in a physical sense, but rather to transform de Sitter's line element in such a way that one or more of the components $g_{\mu\nu}$ depended on the time coordinate. In this case, the metric could be written in a form where space was Euclidean, namely as

$$ds^2 = c^2\,dt^2 - F(t)[dx^2 + dy^2 + dz^2],$$

where $F(t)$ is a function of the time parameter. In his paper of 1925, Lemaître found for $F(t)$ the expression $\exp(2ct\sqrt{\Lambda/3})$ and he suggested a Doppler interpretation of the receding motion of the spiral nebulae. He judged that the non-static nature of the line element 'may probably be accepted', whereas the lack of space curvature was 'completely inadmissible'. Lemaître's reason for dismissing the idea that space was Euclidean seems to have been his conviction that the amount of matter in the universe must necessarily be finite. To a modern reader, Lemaître's line element looks like an exponentially expanding universe, but neither Lemaître nor others adopted this interpretation at the time. Although Lemaître did not suggest an expanding universe in 1925, his paper anticipated in some respects his masterpiece of two years later, in which the expansion of the universe did appear explicitly.

Unaware of Lemaître's work and following a different approach, in 1928 the American mathematical physicist Howard Percy Robertson duplicated Lemaître's result. Robertson derived for the redshift a formula which, if interpreted in terms of a Doppler effect, corresponded to a linear relation between the recessional velocity of the light source and its distance from the observer. One should, he wrote, 'expect a correlation $v \cong cl/R$ between assigned velocity v, distance l, and radius of the observable world R'.[47] When he compared his prediction with observational data due to Slipher and Hubble, Robertson found the relation to be roughly satisfied. For the world radius R, he obtained 2×10^{27} cm, in approximate agreement with Hubble's value of 1926 based on the Einstein model.

3.2.2 *An unnoticed revolution*

In retrospect, it may appear surprising that the dynamic solutions to the cosmological field equations were only discovered in 1922; and it is even more surprising that, when this happened, they attracted almost no attention. The Russian physicist Alexander Friedmann started his scientific career in pre-Communist Russia and did research in theoretical meteorology until the First World War and the October Revolution put a stop to scientific activity.[48] After the end of the civil war, he returned to St Petersburg, the city where he was born and which in 1924 would become Leningrad (until it resumed its old name 67 years later). Here he founded an important school of theoretical physics and began focusing his work on problems of general relativity.

In a paper in the *Zeitschrift für Physik* of 1922, Friedmann offered a complete and systematic analysis of the solutions of Einstein's cosmological equations that went beyond earlier analyses, as it also included non-static solutions. He first showed that the solutions of Einstein and de Sitter exhausted all the possibilities for stationary world models, and then demonstrated 'the possibility of a world in which the curvature of space is independent of the three spatial coordinates but does depend on time.'[49] For closed models, he rewrote Einstein's field equations as a simple pair of differential equations for $R(t)$, the scale factor measuring the rate of change of distances in the universe. By integration, he found a class of solutions that included homogeneously expanding world models, what he called 'monotonic worlds of the first class'. For such models, 'Since R cannot be negative, there must be, as one decreases the time, a time when R vanishes . . . [a] beginning of the world.' And, in a footnote, 'The time since the creation of the world is the time which has passed from the moment at which space was [concentrated at] a point ($R = 0$) to the present state ($R = R_0$).' Friedmann also considered the case where the cosmological constant had such a value that the equations described a cyclical or periodic world. This he interpreted as either a finite-time universe or a universe oscillating forever between $R = 0$ and a maximum value. He ended his paper with a particular case for illustration: 'If we set $\lambda = 0$ and $M = 5 \times 10^{21}$ solar masses we get a world period of about ten billion years.'

In a companion paper, published in 1924, Friedmann introduced hyperbolic models of constant negative curvature. Such a world, he showed, can only be stationary if the density of matter is either zero or negative, whereas a positive density corresponds to a non-stationary world. With respect to the question of the spatial finiteness of the world, he realized that this could not be answered on the basis of the metric alone. The same point was made in a semi-popular book of 1923, *The World as Space and Time*, in which he wrote: 'Thus, the world's metric alone does not enable us to solve the problem of the finiteness of the universe. To solve it, we need additional theoretical and experimental investigations.... *From a constant and positive curvature of the universe it follows by no means that our world is finite*.'[50] As the first publication ever to do so, the book included a discussion of expanding, contracting, and oscillating world models. It was published in Russian and remained for a long time unknown outside the country.

Although Friedmann realized the hypothetical nature of his models, he could not resist the temptation 'to calculate, out of curiosity, the time which has passed since the moment when the universe was created out of a point to its present stage'. Without revealing the basis for his result, he stated it to be 'tens of billions of ordinary years'. Although he did not indicate any preference for a particular world model, he seems to have been fascinated by the possibility of a cyclical universe. 'The universe contracts into a point (into nothing) and then increases its radius from the point up to a certain value, then again diminishes its radius of curvature, transforms itself into a point, etc.' He added that 'all this should at present be considered as curious facts which cannot be reliably supported by the inadequate astronomical material'.[51]

Friedmann discovered mathematically the possibility of dynamic world models, but he did not highlight expanding solutions or argue that our universe was in fact in a state of expansion. His emphasis was clearly on the mathematical aspects, whereas he showed little interest in physics and astronomical data. In neither of the two papers did physical terms such as 'nebula', 'energy', or 'radiation' occur, and there was no reference to the redshifts of spiral nebulae, although these must have been known to him.[52] It is hard to avoid the impression that Friedmann did not recognize the physical importance of his calculations and to a large degree considered them to be nothing but a mathematical game. Thus, the notion of the age of the universe that appears in his papers should not be taken in a realistic sense and was rather seen by Friedmann himself as a mathematical curiosity. Nonetheless, his contributions were of momentous importance, as they broke with the ingrained belief in a static world. But this was far from recognized at the time, when his papers failed to attract the attention later scientists would think they deserved.

In the autumn of 1929, Robertson deduced, in a more general and rigorous way than Friedmann had done, all the line elements of a homogeneous, isotropic universe.[53] He had evidently studied both of Friedmann's papers, but he did not fully recognize the dynamic nature of the universe hidden in his and Friedmann's equations. Robertson's paper included the first version of the Robertson–Walker metric, the most general expression for the line element valid for all homogeneous and isotropic world models (this metric was derived again in a very different way by Arthur G. Walker in 1935). Prior to Robertson's work, Friedmann's papers went unnoticed, in spite of being published in German in a prestigious physics journal. What is more, Einstein responded to the 1922 paper with a note in the *Zeitschrift für Physik* in which he claimed that Friedmann had made a mistake and that his equations were not compatible with the cosmological field equations. Only one and a half years later did he realize that it was he, not Friedmann, who had made a mistake. 'I am convinced that Mr. Friedmann's results are both correct and clarifying,' Einstein now wrote. 'They show that in addition to the static solutions there are time varying solutions with a spatially symmetric structure.'[54]

However, Einstein's concession did not imply that he accepted an evolutionary universe as a candidate for the real world. That Friedmann's work did not change Einstein's attitude (or anyone else's) is illustrated by an article on 'Space–time' he wrote for the 1929 edition of the *Encyclopedia Britannica*. 'Nothing certain is known of what the properties of the space–time continuum may be as a whole,' Einstein stated. 'Through the general theory of relativity, however, the view that the continuum is infinite in its time-like extent but finite in its space-like extent has gained in probability.' It looked as if Friedmann might just as well not have written his later papers that were so celebrated. One reason for the lack of impact was undoubtedly that Friedmann died prematurely in 1925. Another and probably more important reason was the mathematical character of his papers and their lack of reference to astronomical data.

3.2.3 *Lemaître's expanding model*

Georges Lemaître served in the Belgian army throughout the First World War and subsequently embarked on a remarkable double career, as a theoretical physicist and simultaneously as a priest in the Catholic church.[55] In the autumn of 1923, he was ordained a Catholic priest and immediately thereafter went to Cambridge to spend a year as a postgraduate student with Eddington, under whose influence he specialized in the general theory of relativity. In 1924–25 Lemaître continued his postgraduate studies in the United States, at Harvard University (under Shapley) and Massachusetts Institute of Technology (MIT). While in the United States he became increasingly interested in cosmology and suggested the previously mentioned modification of de Sitter's cosmological theory, at the time still conceiving it as a static model. Having received his PhD degree from MIT, he returned to Belgium, where he was appointed to a professorship at the Catholic University of Louvain. In the same year, 1927, he made his breakthrough in cosmology by arguing that the universe was expanding in accordance with the laws of general relativity.

From a formal point of view, Lemaître did little more than Friedmann had done in his ill-fated paper of 1922, which Lemaître did not know at the time and the results of which he thus unknowingly duplicated. It was only in 1927, when he had a chance to present his theory to Einstein, that he realized that Friedmann had anticipated a major part of his work. Incidentally, Einstein's response to the idea of an expanding universe was essentially the same in 1927 as five years earlier: he found no errors in the mathematics of Lemaître's theory, yet he was convinced that 'from the physical point of view it was "tout à fait abominable"'.[56] It is unknown precisely how Lemaître arrived at his theory of the expanding universe, but he clearly relied on his discussion of 1925 of de Sitter's model. It was while searching for a solution that combined the advantages of this model and Einstein's solution A that he was led to 'consider an Einstein universe where the radius of space (or of the universe) varies in an arbitrary way'.[57] With a time-dependent space curvature $R(t)$, he found the same differential equations as Friedmann, except that Lemaître included radiation pressure (p). With the velocity of light taken to be unity, the equations read

$$3\left[\frac{R'}{R}\right]^2 + \frac{3}{R^2} = \Lambda + \kappa\rho$$

and

$$2\frac{R''}{R} + \left[\frac{R'}{R}\right]^2 + \frac{1}{R^2} = \Lambda - \kappa\rho,$$

where $R' = \mathrm{d}R/\mathrm{d}t$ and $R'' = \mathrm{d}^2R/\mathrm{d}t^2$. Lemaître also introduced thermodynamical arguments by deriving energy conservation in the expanding universe in the form

$$dE + pdV = d(\rho V) + pdV = 0,$$

where V is the volume of the universe, $\pi^2 R^3$. The model favoured by Lemaître was a closed universe expanding from an Einstein state of radius $R_0 = \Lambda^{-1/2}$, which he estimated from astronomical data to be about 270 Mpc. As the expansion continued, the mass density would gradually diminish and eventually approach that of the empty de Sitter state.

Although from a mathematical point of view Lemaître's paper was very similar to Friedmann's, otherwise it differed strikingly from it. He was not interested in any kind of variation $R(t)$ allowed by the equations, but only in the expanding solution that seemed to correspond to the redshift data. The importance of these data was reflected in the title of the paper, 'A Homogeneous Universe of Constant Mass and Increasing Radius Accounting for

The Radial Velocity of Extra-Galactic Nebulae.'[58] Lemaître was the first to introduce the crucial notion that 'The receding velocities of extra-galactic nebulae are a cosmical effect of the expansion of the universe.' That is, he realized that the redshifts were caused not by galaxies moving *through* space, but by galaxies being carried *with* the expanding space. If light was emitted when the radius of the universe is R_1 and received when it had increased to R_2, the 'apparent Doppler effect' would be

$$\frac{\Delta\lambda}{\lambda} = \frac{R_2}{R_1} - 1.$$

Lemaître found the approximate relationship between recession velocity and distance to be

$$v = (R'/R)cr = kr.$$

Relying on data from Gustaf Strömberg (for radial velocities) and Hubble (for apparent magnitudes, and hence distances), he obtained for the expansion constant a value of 625 km/s/Mpc. This is what later would be known as Hubble's constant, but Lemaître had no empirical evidence for the linear relationship and therefore used the mean distances to obtain the constant. In the conclusion of his 1927 paper, he noted that 'the largest part of the universe is forever out of our reach. The range of the 100-inch Mount Wilson telescope is estimated by Hubble to be 5×10^7 parsecs, or about $R/200$. The corresponding Doppler effect is 3,000 km/sec. For a distance of $0.087R$ it is equal to unity, and the whole visible spectrum is displaced into the infra-red.' Lemaître underlined the physical nature of his model by tentatively suggesting a cause for the expansion of the universe, namely the pressure of radiation.

Lemaître's prediction of an expanding universe made no more impact than did Friedmann's work. On the contrary, his paper seems to have been almost completely unknown and to have received no citations from other scientists until 1930. It was, like Friedmann's article of 1922, mentioned in the leading abstract journal *Astronomischer Jahresbericht*, but by title only (Friedmann's was listed under 'Relativity Theory', and Lemaître's under 'Nebulae'). While Friedmann's 1922 paper received a review in the American-based *Physics Abstract*, physicists would look in vain for a review of Lemaître's work. In a survey of cosmology of early 1929, Lemaître briefly referred to his work two years earlier (and also to Friedmann's theory), but this paper did not attract any attention either. There is little doubt that part of the reason why Lemaître's expanding universe was met with silence was his decision to publish it in the relatively obscure *Annales Scientifique Bruxelles*. At any rate, after a period of neglect of nearly three years, it was rediscovered in 1930 and then recognized as a crucial contribution to theoretical cosmology.

3.2.4 *The Hubble law*

By the late 1920s, Edwin Hubble had turned to the problem of the redshifts of the extragalactic nebulae (or galaxies), that is, why they 'shun us like a plague', as Eddington expressed it.[59] Slipher's measurements of radial velocities had effectively come to an end as his 24-inch refractor at the Lowell Observatory was unable to penetrate further into space and record the spectra of the numerous fainter nebulae. At a meeting of the International Astronomical Union in the Netherlands in 1928, Hubble discussed the matter with other experts and decided to use the 100-inch telescope at Mount Wilson to solve the puzzle of the redshifts and their relation to distances. In a paper submitted to the *Proceedings of the National Academy of Sciences* in January 1929 and appearing in the 15 March issue, he reported data which substantiated the linear velocity–redshift relation that some theorists had anticipated. Hubble did not specifically state the relationship as

$$v = Hr,$$

but wrote that 'The results establish a roughly linear relation between velocities and distances among nebulae for which velocities have been previously published, and the relation appears to dominate the distribution of velocities.'[60]

Most of the redshifts were taken from Slipher's work, and new observations were made by Milton Humason, Hubble's collaborator and a self-trained expert in galactic spectroscopy. Humason had found a galaxy with the record-high recessional velocity of 3779 km/s, and Hubble used this value to check the linearity in his sample, ranging from −12.7 to −17.7 in absolute magnitude. The result was a confirmation of the linear relationship between redshift and distance. Hubble's data consisted of radial velocities of 46 galaxies, of which he believed to have fairly accurate distances for 24. The most distant of these galaxies was 2 megaparsecs away and receded from the Earth with a velocity of about 1000 km/s. In a now famous diagram, he plotted the redshifts (velocities) of the receding galaxies against their distances, obtaining what he judged to be a reasonably linear correlation.

The linear Hubble relation was rapidly accepted by the majority of astronomers. About the only one who seriously objected to Hubble's claim was Shapley, who found the data insufficient and argued that it was premature to conclude either for or against a linear law. However, new measurements made at Mount Wilson by Humason soon dispelled whatever uncertainty there might have been. In a paper of 1931, Hubble and Humason greatly extended the data base with 40 more galaxies, now reaching out to a distance

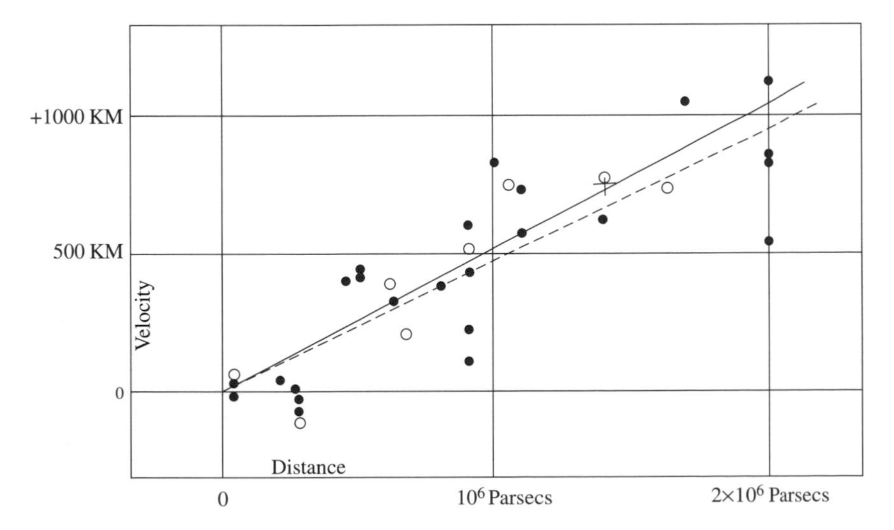

Fig. 3.3 Hubble's velocity–distance diagram of 1929. The solid and broken lines represent two different ways of correcting for the motion of the Sun. From Hubble 1929.

of 32 megaparsecs, and proved without doubt the linear relationship, which eventually came to be known as the Hubble law. For the recession constant, they found a value of 558 km/s/Mpc, whereas the value of 1929 had been about 500 km/s/Mpc. Although aware of the theoretical significance of his work, Hubble was reluctant to go much beyond observations. At the end of his 1929 paper, he wrote, 'The outstanding feature . . . is the possibility that the velocity–distance relation may represent the de Sitter effect, and hence that numerical data may be introduced into discussions of the general curvature of space.' However, he interpreted the spectral shifts as a compound phenomenon caused by two different effects in de Sitter's model:

In the de Sitter cosmology, displacements of the spectra arise from two sources, an apparent slowing down of atomic vibrations and a general tendency of particles to scatter. The latter involves an acceleration and hence introduces the element of time. The relative importance of these two effects should determine the form of the relation between distances and observed velocities; and in this connection it may be emphasized that the linear relation found in the present discussion is a first approximation representing a restricted range in distance.[61]

That Hubble did not interpret the redshifts as simply Doppler shifts caused by the recession of galaxies is also indicated by a letter to Shapley of May 1929, where he suggested that the redshifts might be gravitational.

Hubble discovered the linear law connecting redshifts and distances, but this does not mean that he also discovered the expansion of the universe.[62] Nowhere in his paper did he conclude that the galaxies were actually receding from us or otherwise suggest that the universe was expanding. He did present the law in terms of velocities, but he referred to 'apparent velocities', that is, redshifts that could be conveniently transformed to velocities by means of the Doppler formula but for which other interpretations might be equally valid. 'We [Hubble and Humason] use the term "apparent" in order to emphasize the empirical features of the correlation,' he wrote to de Sitter in 1931. 'The interpretation, we feel, should be left to you and the very few others who are competent to discuss the matter with authority.'[63] Humason followed Hubble in his cautious attitude and wrote that the observations could be explained 'both by the apparent slowing-down of light vibrations with distance and by a real tendency of material bodies to scatter in space.'[64] Similarly, in their 1931 paper, Hubble and Humason noted that the 'present contribution concerns a correlation of empirical data of observation. The writers are constrained to describe the "apparent velocity-displacements" without venturing on the interpretation and its cosmologic significance.'[65]

Hubble's celebrated paper appeared in the spring of 1929 but was not immediately seen as a revolutionary contribution to cosmology, i.e. it was not seen as an observational proof that the universe was expanding. For example, when Robertson completed his paper on relativistic cosmology in the autumn of 1929, he was undoubtedly aware of Hubble's work, but he did not find it relevant to mention it. Similarly, in his *The Size of the Universe*, prefaced November 1929, Silberstein dealt at length with Hubble's 1926 paper but did not mention his more recent work. Richard Tolman, in a paper published in May 1929, did refer to 'The correlation between distance and apparent radial velocity for the extra-galactic nebulae obtained by Hubble', but without concluding that it demonstrated an expanding universe.[66] It took more than observations to effect a change in the world view from a static to an expanding universe. This change only occurred in the early part of 1930, when it was realized that both theory and observation indicated that there was something seriously wrong with the static universe.

At a meeting of the Royal Astronomical Society on 10 January 1930, Eddington argued that since neither of the solutions A and B had proved adequate, interest should focus on non-static solutions. A similar conclusion was obtained by de Sitter, who was present at the meeting, and also by Robertson and Tolman in the United States. When Lemaître, Eddington's former student, read about the meeting, he immediately wrote to Eddington and reminded him about his paper of 1927, a reprint of which he had sent to him but which Eddington had forgotten about:

> I just read the February n° of the Observatory and your suggestion of investigating nonstatical intermediary solutions between those of Einstein and de Sitter. I made these investigations two years ago [sic]. I consider a universe of curvature constant in space but increasing with time. And I emphasize the existence of solution in which the motion of the nebulae is always receding one from time minus infinity to plus infinity.[67]

The oversight was remedied by a letter to *Nature* of 7 June 1930, in which Eddington drew attention to Lemaître's work. It was, he wrote to de Sitter, a 'brilliant solution'. De Sitter agreed.

Having digested Lemaître's paper, Eddington reanalysed in the spring of 1930 Einstein's matter-filled solution, and he now conluded that it was only static because of the specific value assigned to the cosmological constant; any small perturbation would cause the Einstein universe either to contract or to expand. From Lemaître's equations with $p=0$, it follows that

$$3R'' = R(\Lambda - \kappa\rho/2).$$

If $\Lambda=\kappa\rho/2$, the solution corresponds to the static Einstein universe of 1917. But if there is a slight disturbance which causes ρ to drop below $2\Lambda/\kappa$, say by particles disappearing by annihilation, the universe will start to expand; conversely, if there is an excess of mass ($\rho > 2\Lambda/\kappa$), it will contract. 'The initial small disturbance can happen without supernatural interference,' Eddington stressed; for example, it could happen by some gravitational instability. Once started, the universe would continue to expand at an increasing rate. Comparing his results with astronomical data, he concluded: 'The radius of space was originally about 1200 million light-years . . . [and] its present rate of expansion is 1 per cent in about 20 million years.'[68] His model and size of the initial universe were thus similar to Lemaître's, for which reason this model entered the literature as the Lemaître–Eddington (or sometimes Eddington–Lemaître) model.

In a letter to Shapley of April 1930, de Sitter abandoned his solution B and described Lemaître's theory of the expanding universe as 'the true solution, or at least a possible solution, which must be somewhere near the truth'.[69] Having studied Lemaître's work, de Sitter presented in June his own version of the expanding universe, which included a recession constant of 490 km/s/Mpc, not far from Hubble's value. Also, Einstein, who had previously dismissed the theories of Friedmann and Lemaître, accepted the dynamic solutions and realized that in an expanding universe the cosmological constant was no longer necessary. In a paper of 1931, in which he advocated an oscillatory model of the universe, he noted that the instability of his solution A followed from Friedmann's equations. He now definitely abandoned the cosmological constant, never to return to it.[70]

By 1933, the theory of the expanding universe was accepted by a majority of astronomers and subjected to detailed reviews in works by Otto Heckmann, Robertson, and Tolman.[71] It was also disseminated to the public through a number of popular works, such as James Jeans's *The Mysterious Universe* (1930), James Crowther's *An Outline of the Universe* (1931), de Sitter's *Kosmos* (1932), and Eddington's *The Expanding Universe* (1933). The latter book was based on a public lecture that Eddington gave at the meeting of the International Astronomical Union in Cambridge, Massachusetts. Eddington opened his lecture by pointing out how truly international were the recent developments in cosmology: 'This is an International Conference and I have chosen an international subject. I shall speak of the theoretical work of Einstein of Germany, de Sitter of Holland, Lemaître of Belgium. For observational data I turn to the Americans, Slipher, Hubble, Humason, recalling however that the vitally important datum of distance is found by a method which we owe to Hertzsprung of Denmark.... My subject disperses the galaxies, but it unites the earth. May no "cosmical repulsion" intervene to sunder us!'[72]

Although by the early 1930s the expansion of the universe had become a reality, it was far from clear why it expanded or why it was an expansion and not a contraction. The Friedmann–Lemaître equations are symmetric with respect to the direction of time and therefore describe also a contracting universe different from the one we live in. In England, the problem was studied by William McCrea and George McVittie, who investigated

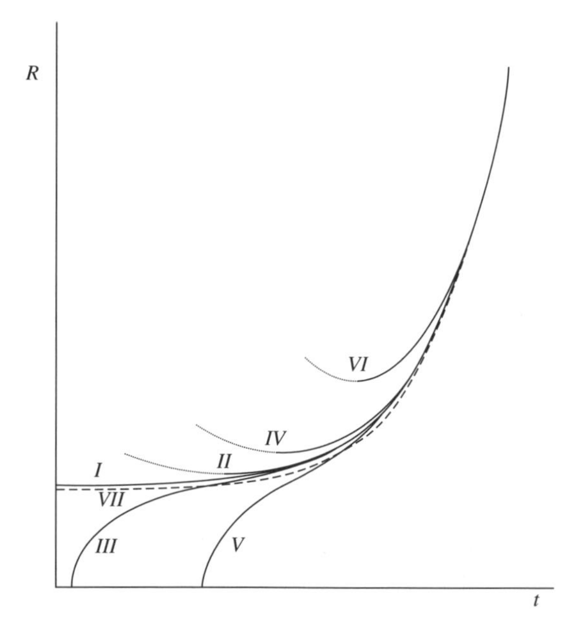

Fig. 3.4 After de Sitter had become aware of the Friedmann–Lemaître equations in 1930, he derived world models corresponding to physically acceptable solutions. In his textbook of 1934, Richard Tolman included de Sitter's graphical representations (Tolman 1934, p. 411). Curves I and VII, which include the Lemaître–Eddington model, expand from an original static state, whereas II, IV, and VI first contract to a minimum radius and then expand to the empty de Sitter state. Curves III and V are models expanding from a singular state, such as Lemaître's 1931 Big Bang model. Oscillating models are not included.

the effect of condensations in the initial state of the universe. The young Leningrad physicist Matvei Bronstein believed that the problem could not be solved by referring to a time-asymmetric initial situation, but that it required a modification of the cosmological equations. His radical suggestion was that the asymmetry of cosmic history was due to a time-dependent cosmological constant acting as an arrow of time. The price to pay was a violation of energy conservation on a cosmic scale, as the energy-conserving equation $dE + p\,dV = 0$ would no longer hold. Bronstein further suggested that the cosmological constant might represent some form of energy, and that there might exist an energy transfer between ordinary matter and the matter or energy associated with the constant. Without knowing of Bronstein's paper, in 1934 Lemaître argued that the cosmological constant might be understood as a vacuum energy density given by $\rho = \Lambda c^2/4\pi G$.[73] Neither Bronstein's nor Lemaître's paper attracted attention in the 1930s, but they dealt with a topic that would become of great importance in cosmology many years later.

3.3 Towards a finite-age universe

It is often assumed that the idea of the expanding universe entails the Big Bang universe, that is, the notion of a catastrophic beginning of the world a definite time ago. That is not the case, however, neither from a logical point of view nor from a historical perspective. The two notions are distinctive. The Friedmann (or Friedmann–Lemaître) equations can be seen as the first step towards the Big Bang universe, but this step was neither necessary nor sufficient. As the historial record documents, there were ideas about an explosive origin of the universe completely independent of the cosmological field equations; furthermore, the recognition that the universe was expanding did not lead automatically to the conclusion that it had an origin in time. On the contrary, when the idea of a finite-age universe was eventually proposed—and that occurred only two years after Hubble's discovery—most astronomers and physicists resisted it. All the same, during the 1930s, the 'exploding' models of Lemaître and Edward Milne were not completely ignored, and at the end of the decade they were seriously discussed by at least a few physicists and astronomers.

Although the high road of cosmological history goes from Hubble via Lemaître to Gamow, and from there to the the modern theory of the Big Bang, this was not the road of history. As usual, this road was more messy and less straightforward. Today's cosmologists (and, unfortunately, many historians of cosmology) naturally identify Big Bang cosmology with the general theory of relativity, but from a historical point of view such an identification is not warranted. In fact, in the 1930s many observers would associate theories of the Big Bang type with a cosmological theory entirely different from that of Einstein and Lemaître, namely with the one proposed by Milne in England. Generally speaking, the situation in theoretical cosmology during the 1930s was confusing. The situation in observational cosmology was no less so. To put it briefly, the 1930s were a decade of uncertainty and—in retrospect—a preparation for what happened in the next two decades.

3.3.1 *The beginnings of physical cosmology*

During the first three decades of the twentieth century, the notion of a universe of finite age was rarely considered and never seriously advocated. Cosmologists were not concerned with the temporal extension of the world, except in that they tended to take it for granted that the world had always existed, whatever its spatial structure. If astronomers in a speculative mood looked far back in time, they generally conjectured that 'at the beginning of time' the whole of space was uniformly filled with a tenuous gas. They did not think of space itself as evolving from some initial state.

In the few cases where the idea of a finite-age universe surfaced, it was either dismissed or entertained only in a vague, uncommitted way. It was clearly an unwelcome idea. The leading geophysicist Arthur Holmes discussed in 1913 the entropic paradox, realizing that it might be solved if it was assumed that the universe had a definite beginning; however, having mentioned the possibility, he dismissed it and instead affirmed the eternity of the universe.[74] As another example, consider the article on 'Cosmogony' which appeared in the ten-volume *Handwörterbuch der Naturwissenschaften* in 1914. The author, Otto Knopf from the Jena Observatory, discussed the same problems as Holmes, and he shared his uneasiness about the notion of an origin of the universe. Fortunately, he wrote, this was not an inevitable consequence, for 'According to the present state of science, one must accept a periodical change in creative and destructive individual processes, and that the whole [universe] is unchangeable.'[75]

That the notion of a universe with a definite beginning in time was hard to accept, indeed nearly unthinkable, is also illustrated by Weyl's response to the cosmological model that Lanczos had suggested in 1922. In this model there appeared an initial space–time singularity, a feature Weyl found to be an unacceptable blemish. His own version, he pointed out, 'has the great advantage [over Lanczos's] of not introducing a singular initial moment, of conserving the homogeneousness of time'.[76] As mentioned, Friedmann explicitly introduced the idea of a created world in his paper of 1922, but without clearly interpreting it as a possible physical reality. Again, in a book of 1924 the British astrophysicist Herbert Dingle made the innovative suggestion that the galactic redshifts measured by Slipher reflected the history of the universe, that the observed recession might be 'the legacy of a huge disruption, in the childhood of matter, of a single parent mass'. In that case, 'the apparent repulsion would be merely the effect of inertia, and might or might not in time be overcome by the persistent influence of gravitation'. Might it not be, he continued, that 'we exist at a special point of time in a Universe which had a beginning in time'? But Dingle raised the possibility only to dismiss it as an unwarranted speculation. He did not believe in a universe of a finite age, and concluded that there was no evidence for 'a beginning of things'.[77] This attitude was shared by most of his colleagues.

As Dingle vaguely anticipated an evolutionary universe of finite age, so did Jeans. In a lecture given in Bristol in the autumn of 1928, Jeans surveyed some of the more speculative aspects of cosmology, including the heat death and the beginning of the universe. He suggested that present matter had not existed forever, that it 'must have begun to exist at some time not infinitely remote, and this leads us to contemplate a definite event, or series of events, or continuous process, of creation of matter'. Jeans imagined that originally atoms, made up of electrons and protons, were created by high-energy photons. 'If we want a concrete picture, we may think of the finger of God agitating the ether,' he famously wrote.[78] Although Jeans thus discussed the creation of matter and the beginning of the universe, it is characteristic that he did so in a popular lecture and without committing himself to a particular view. It would take another three years until a universe with a beginning in time was introduced into cosmology.

Before proceeding to this breakthrough, it should be noted that although mathematical cosmology in the 1920s did not include the physics of matter and radiation, a few scientists did attempt to adopt a more physical perspective in the study of the universe.[79] Most of these attempts were not of a truly cosmological nature, but in retrospect they can be seen as early examples of physical cosmology, a branch of science that would reach a mature state only after the Second World War. For example, as early as 1922 Richard Tolman applied chemical equilibrium theory to investigate the relative abundances of hydrogen and helium in the universe, a problem that was also taken up six years later by Seitaro Suzuki, a Japanese physicist. Suzuki reached the conclusion that the observed cosmic ratio of helium and hydrogen—at the time known only very roughly—could be explained on the assumption of a very hot early universe. However, the remarkable assumption of an early hot state failed to make any impact among European and American cosmologists.

The application of the laws of thermodynamics to the entire universe goes back to the nineteenth century, and in the 1920s a few physicists studied cosmological models from the perspective of thermodynamics. Wilhelm Lenz, a German physicist, was probably the first to apply thermodynamics to a definite cosmological model. In 1926 he studied the equilibrium of matter and radiation in an Einstein universe and found that the temperature of the black-body radiation would depend on the radius of the universe as

$$T^2 = \frac{1}{R}\sqrt{\frac{2c^2}{\kappa a}} \cong \frac{10^{31}}{R},$$

where a is the constant in the Stefan–Boltzmann law, and T is measured in absolute degrees and R in cm. Arbitrarily assuming the radiation temperature to be about 1 K, he was led to suggest a world radius of the order 10^{31} cm or 10^5 Mpc. Two years later, Tolman improved Lenz's calculations and derived expressions for the energy and entropy of the same kind of world model. Lenz also considered the vacuum energy, which appears here for the first time in a cosmological context. He stated that if the zero-point energy was included, one would obtain for the vacuum energy an equivalent amount of matter so great that it would produce a universe with a curvature radius smaller than the distance to the Moon![80] He consequently concluded that the vacuum energy could not act gravitationally.

Also, the Swiss–American astronomer Fritz Zwicky studied the Einstein universe as a thermodynamical equilibrium system. After the expanding universe had been recognized, Tolman studied expanding models within the framework of relativistic thermodynamics. In a paper of May 1931, he discussed the classical entropic paradox and mentioned some of the solutions that had been proposed. One of these was Boltzmann's idea of fluctuations in entropy; another was 'that the universe was indeed created at a finite time in the past with sufficient available energy so that the entropy has not yet reached its maximum value'.[81] This was precisely what Lemaître suggested at the same time, but Tolman (who probably did not yet know of Lemaître's hypothesis) refused to take take the idea seriously. It was, he claimed, an ad hoc hypothesis with no scientific merit.

The idea that the pattern of element abundances reflects nuclear processes in the early universe, and may therefore be used to reconstruct this phase in the history of the cosmos, was to play an important role in early Big Bang cosmology. Among the pioneers of this kind of reasoning—a continuation of the astrochemistry of the Victorian era—was the American chemist William Harkins, who as early as 1917 argued that the observed abundances of the various elements were the result of nuclear evolution processes starting from hydrogen. Basing his work on analyses of meteorites in particular, Harkins found that the first seven elements in order of cosmic abundance all had even atomic numbers and made up almost 99% of the matter in the universe. This was at an early date in the history of nuclear physics, only a few years after the notion of the atomic number (the positive charge of the atomic nucleus) had been introduced as an ordering number for chemical elements. Several other chemists shared with Harkins an interest in the chemistry of the stars. One of them was the brilliant American physical chemist Gilbert Lewis, who in 1922 argued that as astronomy could learn from chemistry, so chemistry could learn from astronomy:

While the laboratory affords means of investigating only a minute range of conditions under which chemical reactions occur, experiments of enormous significance are being carried out in the great laboratories of the stars. It is true, the chemist can synthesize the particular substances which he wishes to investigate and can expose them at will to the various agencies which are at his command; we cannot plan the processes occurring in the stars, but their variety is so great and our methods of investigation have become so refined that we are furnished an almost unbounded field of investigation.[82]

Eddington, too, realized that the content of matter in the universe might reflect its history. Like a few others before him, he called attention to the existence of radioactive minerals on the earth, suggesting that it indicated 'a mechanism running down which must at some time have been wound up'. According to Eddington, 'it would seem clear that the winding-up process must have occurred under physical conditions vastly different from those in which we now observe only a running-down'. This may look like an anticipation of the Big Bang scenario, except that Eddington did not refer to the early phase of the universe but to 'the general brewing of material which occurs under the intense heat in the interior of the stars'.[83]

In a brief paper of 1931, Ladislaus Farkas and Paul Harteck, two German physical chemists, applied equilibrium theory to a primeval mixture of nuclei, protons, and electrons at a temperature of two billion degrees. They calculated the relative abundances for elements up to sodium, claiming that their calculations agreed roughly with experimental data. Their work would later inspire Weizsäcker to develop a cosmological theory of element formation and stellar energy generation. However, Harold Urey and Charles Bradley argued in the same year (1931) that the relative abundances of terrestrial elements could not be reconciled with the equilibrium hypothesis, whatever the temperature of the equilibrium mixture. None of the chemists engaged in this kind of work adopted a truly cosmological perspective or sought to relate their work to cosmological models in the relativistic tradition. Such a connection still lay in the future, though not by many years.

3.3.2 *The primeval-atom hypothesis*

The expanding model that Lemaître suggested in 1927 became known as the Lemaître–Eddington model because it was adopted and further developed by Eddington after he converted in 1930 to the new idea of the expanding universe. Whereas Eddington remained throughout his life faithful to an expanding, yet eternally existing world, his former Belgian postdoc moved on, proposing in May 1931 the first version of what would eventually be called the Big Bang universe.

Lemaître's main inspiration did not come from either astronomy or relativity (nor from theology), but from recent developments within quantum theory coupled with considerations about the presence of radioactive substances. The half-lives of radioactive elements such as uranium and thorium were known to be very long and of the same order of magnitude as the Hubble time, that is, the inverse of the Hubble constant (the half-life of U-238 is 4.5 billion years, and that of Th-232 is 14 billion years). Was this just a coincidence? Did it not indicate that our present world might be looked upon as the nearly burned-out result of a previous radioactive universe? According to Lemaître, the universe had originated in a giant radioactive flash and hence could be ascribed a definite age. To formulate this idea of a cosmic beginning in a satisfactory way, he had to avoid Kant's first antinomy, essentially an argument based on the impossibility of giving a causal–deterministic account of the beginning of the universe. At this stage quantum mechanics entered Lemaître's line of reasoning, for he knew that quantum-mechanical processes such as radioactivity are non-causal and indeterministic and might therefore be used to circumvent the logical problems discussed by Kant.

In a popular address of early 1931, Eddington dealt with one of his favourite themes, the heat death and the role of entropy as an arrow of time. He considered briefly what the state of the world would have been like if time were traced backwards to a state of minimum entropy. Would this state correspond to the beginning of the world? Not according to Eddington, who conceded that the state of maximum possible organization might be called a 'beginning' but denied that it could be a physical reality, a subject of scientific reasoning. 'Philosophically, the notion of a beginning of the present order of Nature is repugnant to me,' he stated.[84] Lemaître begged to disagree. In his brief paper of 9 May, he expressed his heretical view, that he was 'inclined to think that the present state of quantum theory suggests a beginning of the world very different from the present order of Nature'. He suggested that 'we could conceive the beginning of the universe in the form of a unique atom, the atomic weight of which is the total mass of the universe . . . [and which] would divide in smaller and smaller atoms by a kind of super-radioactive process'. Quantum-mechanical indeterminacy helped him explain how the present world in all its colourful diversity could be the result of a single, undifferentiated quantum:

> Clearly the initial quantum could not conceal in itself the whole cause of evolution; but, according to the principle of indeterminacy, that is not necessary. Our world is now understood to be a world where something really happens; the whole story of the world need not have been written down in the first quantum like the song on the disc of a phonograph. The whole matter of the world must have been present at the beginning, but the story it has to tell may be written step by step.[85]

Lemaître's hypothesis, as it appeared in *Nature*, was more a piece of cosmic poetry than a scientific theory. But he soon presented a better argued and more elaborate version, first in an address to the British Association for the Advancement of Science in October 1931. Although on this occasion he did not present a quantitative theory, he did clarify his idea of what he now called a 'fireworks theory' of cosmic evolution. He argued that the cosmic radiation was the remnants of the disintegration of the primeval superatom—'ashes and smoke of bright but very rapid fireworks'. At about the same time, he developed his ideas into a definite model of the universe, the first example ever of a relativistic Big Bang universe. In a paper in *Revue des Questions Scientifiques*, he described his world model as follows:

> The first stages of the expansion consisted of a rapid expansion determined by the mass of the initial atom, almost equal to the present mass of the universe. . . . The initial expansion was able to permit the radius [of space] to exceed the value of the equilibrium radius. The expansion thus took place in three phases: a first period of rapid expansion in which the atom-universe was broken down into atomic stars, a period of

slowing-down, followed by a third period of accelerated expansion. It is doubtless in this third period that we find ourselves today, and the acceleration of space which followed the period of slow expansion could well be responsible for the separation of the stars into extra-galactic nebulae.[86]

This is what is called the Lemaître model of the universe, a model which has some similarities to the Lemaître–Eddington model (both of them are ever-expanding and spatially finite) but differs from it by being of finite age. In addition to the explosive beginning, what characterized Lemaître's model was the second phase of slowing-down or 'stagnation', a phenomenon made possible by the assumption of a positive cosmological constant. Lemaître had first introduced stagnation within the context of the Lemaître–Eddington model in an attempt to explain how the expansion from the Einstein state was caused by condensation processes. There was a double bonus from introducing the stagnation phase: for one thing, it stretched the timescale and thus provided a solution to the age paradox (see below); for another thing, it made it easier to explain how galaxies were formed in the early universe.

Lemaître was a great advocate of the cosmological constant. Not only did his favoured model of the universe rely on it, but he also found it methodologically convenient to operate with the constant because it would provide relativistic cosmology with an extended empirical content. In sharp contrast to Einstein's attitude, Lemaître referred to the cosmological

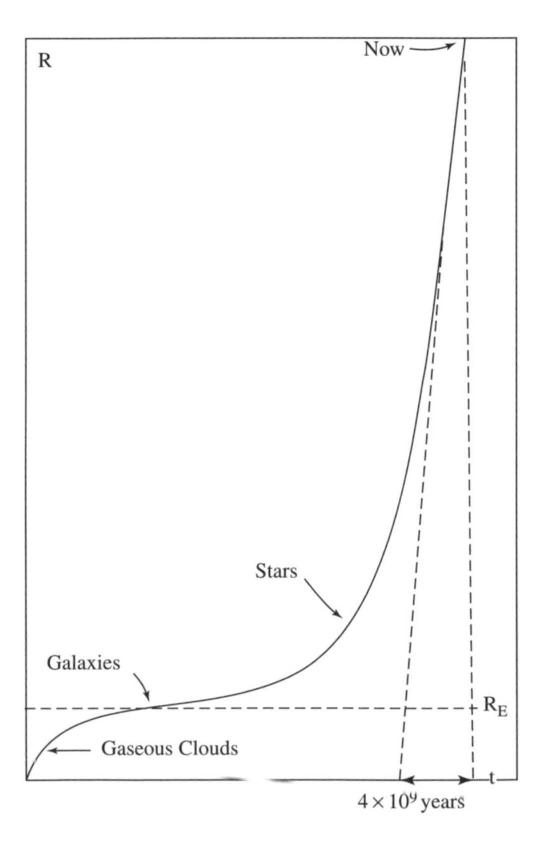

Fig. 3.5 Lemaître's finite-age world model, as represented in a paper of 1958. The present radius was assumed to be ten times the Einstein equilibrium radius. From Godart and Heller 1985, © 1985 by the Pachart Foundation dba Pachart Publishing House, and reprinted by permission.

constant by positive names such as 'happy accident', 'logical convenience', and 'theoretical necessity'. He tried several times to convince Einstein about the necessity of keeping a non-zero cosmological constant, but in vain. All Einstein would admit was that the quantity was helpful in solving the age paradox, and this was not enough to change his view.[87]

From a mathematical point of view the Lemaître model belongs to the class of Big Bang models, in the sense that it includes $R = 0$ for $t = 0$. However, Lemaître adopted a physical point of view which caused him to resist the idea of an initial singularity in the strict meaning of the term. According to his thinking, at $t = 0$ the universe already 'existed' in the shape of the material primeval atom that contained within it the entire mass of the universe, and the radius of which he estimated to be a few astronomical units. The matter density would correspond to that of an atomic nucleus, roughly 10^{15} g/cm^3, the highest density physicists could imagine. Lemaître's primeval atom was simple in an absolute, almost metaphysical sense. It was inaccessible to scientific inquiry, devoid of physical properties, and hence non-existent from a physical point of view. While he originally spoke of the original state as a unique atom, a kind of super-transuranic element, he later likened it to a gigantic 'isotope of the neutron'.

As to the cosmic singularity formally turning up at $t = 0$, he denied that it could have any physical meaning. When the movie of cosmic development was run backwards in time, at some stage physical conditions would surely prevent the unwelcome singularity. This is what Lemaître argued in an important paper of 1933, which is also noteworthy because it included the first calculation of an inhomogeneous world model with properties corresponding to later ideas of 'bubble universes' (the model is today often known as the Tolman–Bondi model). Lemaître insisted that 'matter has to find a way to avoid the annihilation of its volume' and concluded that 'only the subatomic nuclear forces seem capable of stopping the contraction of the universe when the radius of the universe is reduced to the dimension of the solar system'. He briefly considered cyclic solutions—and endless series of contractions and expansions—but only to conclude that they were ruled out observationally. Yet he shared Friedmann's fascination with such solutions, which 'have an indisputable poetic charm and make one think of the phoenix of the legend'.[88]

Lemaître realized that his Big Bang scenario was hypothetical and that sceptics might consider it nothing but 'a brilliantly clever *jeu d'esprit*', as it was characterized by Ernest Barnes, the mathematically trained bishop of Birmingham.[89] Was the model physically testable? If the universe had really once been in a highly compact, hot, and radioactive state, would it not have left some traces that could still be subjected to analysis? Lemaître believed that there were indeed such traces from the far past, and that these were to be found in the cosmic radiation. If the cosmic rays were remnants of the original explosion, they would have to consist of high-velocity charged particles (and not of photons, as Robert Millikan and others believed). With the expansion of the universe the energy of the radiation would decrease, which meant that this energy must have been enormous shortly after the Big Bang. In collaboration with Manuel Vallarta, a Mexican physicist, Lemaître developed his idea in a series of calculations of the energies and trajectories of charged particles in the Earth's magnetic field. He believed that there was experimental support for the view of a super-radioactive origin of cosmic rays, but failed to convince the majority of physicists. It was a wrong hypothesis, but it was a brave attempt to argue physically for the origin of the universe.

3.3.3 *Responses to the exploding universe*

Given that Lemaître's idea of a beginning of the universe was so radical and novel, it is not surprising that it was received with scepticism by his fellow cosmologists. In so far as it was considered as a mathematical model only, as one possible solution to the Friedmann equations, it was not seen as particularly problematic. However, viewed as a physical model, it was a different matter. In fact, as early as June 1930 de Sitter had included the model in a survey of dynamic world models, but he considered it to be just a mathematical solution of no particular physical importance. During the first several years after 1931, most experts either ignored or rejected the primeval-atom hypothesis. If there was a paradigm shift, from an eternal to a created universe, it occurred only in the 1960s.

Einstein was one of the first physicists to accept Lemaître's Big Bang or something like it, although not without hesitation. The cosmological singularity was a problem, but 'one can try to escape this difficulty by pointing out that the inhomogeneity of the distribution of stellar material makes our approximate treatment illusory'. He further noticed that 'indeed hardly any theory that explains Hubble's enormous shift in the spectral lines as a Doppler effect can avoid this difficulty in a comfortable way'.[90] Almost all relativistic models considered as candidates for the real universe in the interwar years were closed. It was as if cosmologists wanted to repress the idea of infinite space associated with open models, an idea which Barnes in 1931 called 'a scandal of human thought' and which Lemaître in 1950 labelled 'the nightmare of infinite space'. The only open model widely noticed in the 1930s was the Einstein–de Sitter model.

In 1932 Einstein collaborated with de Sitter in suggesting a model of the universe in which there was no space curvature, no pressure, and no cosmological constant. From these assumptions of a parsimonious universe, it follows from the Friedmann equations that the matter density is

$$\rho = \frac{3H^2}{8\pi G} = \frac{3}{8\pi G T^2} = \rho_c \Omega,$$

where $T = 1/H$ is the Hubble time. With $H = 500$ km/s/Mpc, this gives a density of 4×10^{-28} g/cm^3, which 'may perhaps be on the high side, [but] it certainly is of the correct order of magnitude'.[91] The Einstein–de Sitter density is 'critical' in the sense that the gravitational attraction is precisely balanced by the expansion. If (for $\Lambda = 0$) there is more matter in the universe, gravitation will take over and the expansion will be followed by a contraction; space in a subcritical universe will be negatively curved and expand forever. In more recent literature the density is often given by the parameter $\Omega \equiv \rho/\rho_c$, which has the value 1 in the Einstein–de Sitter model. Also, the Hubble constant is often written as $H = 100h$ km/s/Mpc, where h is a pure number (not to be confused with Planck's constant). Converted into billions of years, the Hubble time can be written $T = 9.8h^{-1}$, and for the critical density we have $\rho_c \equiv (3/8\pi G)10^4 h^2 = 1.9h^2 \times 10^{-29}$ g/cm^3. The density parameter Ω then varies inversely with the square of h.

It is easy to show that, according to the Einstein–de Sitter model, the scale factor increases as $R(t) \sim t^{2/3}$, meaning that $R = 0$ for $t = 0$, and the age of the universe is given by $2T/3$. Remarkably, Einstein and de Sitter did not write down the variation of $R(t)$, and neither did they note that it implies an abrupt beginning of the world such as suggested by Lemaître (to whom they did not refer). De Sitter, who died in 1934, never felt at home with the Big Bang theory, whereas Einstein seems to have accepted it at an early date. In a letter to Eddington of 1933, Lemaître reported about his fireworks theory that, 'I had the great pleasure to find professor Einstein very enthusiastic about it' (he also told that Einstein had 'great prejudice' against the cosmological constant).[92]

There were, basically, two reasons why Lemaître's theory was received so reservedly. One was in part observational, and in part emotional; the other was related to the timescale problem shared by most Big Bang models (although not, ironically, by Lemaître's). A universe with a beginning in time was considered very strange, especially as long as the hypothesis was not supported by observational evidence. To entertain it seriously, strong reasons were needed, and few physicists and astronomers could find such reasons (they did not consider Lemaître's claimed support from the cosmic radiation to be convincing). Because of the absence of hard data, much of the discussion was kept in an emotional language, indicating that it was as much a matter of taste as of scientific evidence. To John Plaskett, a Canadian observational astronomer, Lemaître's hypothesis was 'the wildest speculation of all', nothing less than 'an example of speculation run mad without a shred of evidence to support it'.[93] This was a rather extreme judgement, but scepticism was widespread also among experts in relativistic cosmology. At the 1931 meeting of the British Association, de Sitter stressed that we had direct knowledge only of the part of the universe that we could observe ('our neighbourhood'), and that assertions about other portions were, strictly speaking, scientifically meaningless. On the other hand, he admitted that such assertions, necessarily based on extrapolations and 'philosophical taste', were necessary in cosmology.

Robertson found the Big Bang solution to be contrived and unappealing and much preferred the Lemaître–Eddington model because it avoided a catastrophic behaviour in the past. A similar attitude was expressed by Tolman, who warned against the danger of dogmatism and what he called 'the evils of autistic and wishfulfilling thinking'. Among such prejudices he included the belief that the universe was created in the past. 'The discovery of models, which start expansion from a singular state of zero volume, must not be confused with a proof that the actual universe was created at a finite time in the past.'[94] Eddington never believed that the universe had a beginning, a concept he found philosophically monstruous. As he wrote in 1933, 'The beginning seems to present insuperable difficulties unless we agree to look on it as frankly supernatural.'[95] Eddington continued to defend the Lemaître–Eddington model, which was also the subject of his last paper to the Royal Astronomical Society, published in 1944. In this paper, which was based on his unorthodox attempt to integrate quantum mechanics and cosmology, he derived for the original Einstein state a radius of about 300 Mpc, which by now would have increased to about 1500 Mpc. Eddington's result was theoretically based and did not depend on observations.

A few years after having firmly established the empirical redshift–distance relation, Hubble initiated an ambitious research programme at Mount Wilson with the aim of determining from this relation, supplied with counts of galaxies, the structure of the universe. His principal aim was not to establish which kind of relativistic model agreed best with the data, but to determine if the observed redshifts were cosmological in nature, that is, due to an expansion of the universe in accordance with the Friedmann equations. He believed that a critical test was possible, since rapidly receding galaxies should appear fainter than stationary galaxies at the same distance. However, the result of his efforts was disappointing.

In a major work of 1935, written jointly with Tolman, the two cosmologists admitted that they were unable to decide in favour of any particular model; instead, they chose to emphasize the uncertainty of the data. 'It might be possible', they tentatively concluded, 'to explain the results on the basis of either a static homogeneous model with some unknown cause for the red-shift or an expanding homogeneous model with the introduction of effects from spatial curvature which seems unexpectedly large but may not be impossible.'[96]

In a study of 1936, described in his monograph *The Realm of the Nebulae*, Hubble reported counts of faint galaxies up to apparent magnitude $m=21$. If space is flat and the galaxies are distributed uniformly, the number with a magnitude smaller than or equal to m is given by

$$\log N(m) = 0.6m + \text{constant}.$$

Hubble found that this expression did not hold for very faint galaxies, where a slope of 0.5 agreed better with the data (see Fig. 3.6). He concluded that the deviation indicated effects of redshift on the apparent magnitudes, but had to admit that his data provided no unequivocal information about the structure of space. All what he could conclude was that *if* the observations had to fit an evolving universe, this would be 'a curiously small-scale universe' with a disturbingly high density, namely, $R=145$ Mpc and $\rho \approx 10^{-26}$ g/cm^3. The alternative was that the redshifts were not caused by the expansion of space, but were the result of some other, unidentified mechanism. That Hubble was not convinced about the

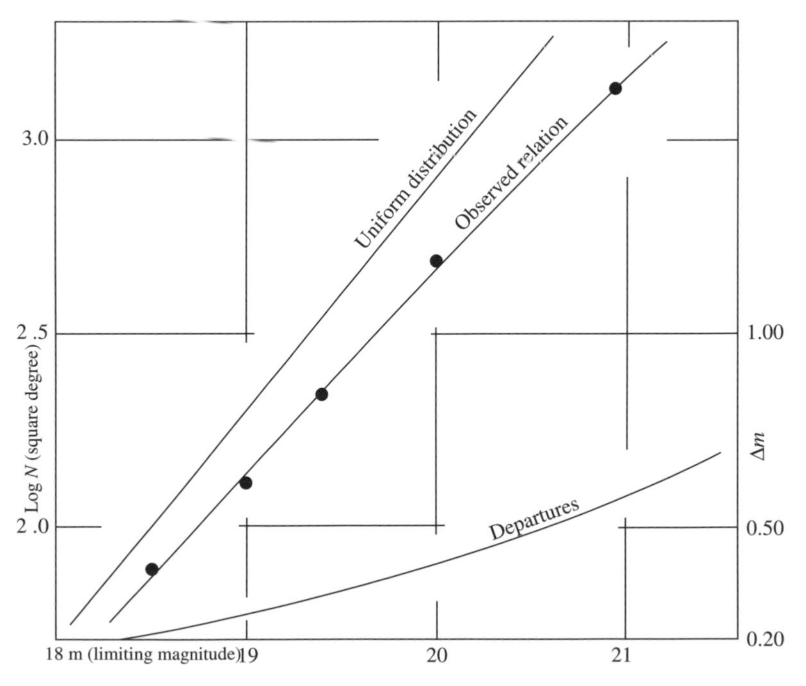

Fig. 3.6 Hubble's determination of a log N–m relation. He interpreted the departures of the observed relation from the relation to be expected from a uniform distribution of nebulae as effects of redshifts. From Hubble 1936, p. 186.

expansion of the universe is indicated by a letter Nicholas Mayall wrote to him in 1937. 'It is perhaps unnecessary to mention how pleased are some of the people here [Lick Observatory] to note the way that your interpretation of the nature of the redshift casts doubt upon the validity of theories of the expanding universe.'[97]

Throughout his life, Hubble avoided committing himself to the expanding universe, in spite of a philosophical preference for such a universe governed by the laws of general relativity. In a review of 1942, he again emphasized the lack of sufficient and reliable data and the dilemma implied by current observations. To accommodate observations, he saw no other choice than to conclude either that the universe expanded in agreement with relativistic cosmology, or that the redshifts were not recessional but due to some hitherto unknown mechanism. The conservative Hubble found the second alternative unattractive and the first implausible, as it implied a 'strangely small' high-density universe. Its volume would be merely four times the observational region of space and its mass far greater than what could be ascribed to visible matter alone.[98] Faced with this situation, he preferred to suspend judgement until better data were obtained.

> Thus the explorations of space end on a note of uncertainty.... With increasing distance, our knowledge fades, and fades rapidly. Eventually, we reach the dim boundary—the utmost limits of our telescopes. There, we measure shadows, and we search among ghostly errors of measurement for landmarks that are scarcely more substantial. The search will continue. Not until the empirical resources are exhausted, need we pass on to the dreamy realms of speculation.[99]

Thus spoke the observational cosmologist. In his Darwin lecture of 1953, given shortly before his death and at a time when the Big Bang theory had been much developed, if not yet generally accepted, he maintained this cautious view concerning the reality of the expansion.

Hubble was not alone in using observational data in an attempt to determine the structure of the universe. His methods and assumptions were criticized in particular by Eddington and George McVittie in England and Otto Heckmann in Germany, who in the late 1930s introduced more sophisticated methods of analysis. Yet, in spite of their methodological improvements, these people arrived at results that did not differ significantly from those obtained by Hubble.[100] For example, Heckmann reported as his best offer a spherically closed world model with a radius of about 160 Mpc and a density of about 5.4×10^{-26} g/cm^3, although he recognized that the result was very sensitive to small observational errors. In Hubble's cosmological programme—or in the work done by McVittie, Heckmann, and others—models with a beginning in time played no significant role. Hubble was aware of the possibility, though, and in a footnote to his and Tolman's 1935 paper he referred to Lemaître's model with a cosmological constant. In the Rhodes Memorial Lectures of 1936, he concluded that if the universe was expanding, then a Lemaître-like universe with $\Lambda \cong 4.5 \times 10^{-18}$ (light years)$^{-2}$, or 5×10^{-50} m^{-2} was the only possibility. However, he in no way found the possibility attractive, and concluded that the Lemaître model, although it could not be ruled out, was 'rather dubious.'[101]

3.3.4 *The age paradox and element formation*

One important reason for the resistance with which models of the Big Bang type were met in the 1930s and 1940s was the age paradox, or timescale difficulty.[102] For logical and semantic reasons, the age of the universe must be greater than (or, in principle, equal to) the age of any of its constituents, such as planets, stars, and nebulae. In other words, a cosmological model which violates this criterion must be seriously wrong. The problem stands out most clearly in the case of Friedmann models with a zero cosmological constant, where the age of the universe is of the same order as, but smaller than, the Hubble time T (recall that the age of the flat Einstein–de Sitter universe is $\frac{2}{3}T$). Hubble's value of $H \cong 500$ km/s/Mpc corresponds to $T \cong 1.8$ billion years, much smaller than the ages astronomers ascribed to stars and galaxies. According to Jeans's calculations, which enjoyed wide support in the early 1930s, it took no less than 10^{13} years for nebulae to form stars, and so the age of the universe must be much greater than the billion years or so that most exploding models prescribed.

Jeans's long timescale began to lose authority in the mid 1930s, but the timescale difficulty remained. At the end of the decade, most astronomers agreed that a more sensible age for the stellar universe was 3–5 billion years, still a value in glaring contradiction to the age inferred from cosmological models. What is more, the age of the Earth was known with reasonable accuracy from radioactive dating methods, and even this number came out too high. The accepted value for the age of the Earth in the 1930s was between two and three billion years, about double the age of the Einstein–de Sitter universe. Clearly, there was something very wrong.

This problem was considered a most serious one during the early phase of Big Bang cosmology. De Sitter was deeply worried about it, and to Hubble it was a further reason not to embrace expanding models with a definite span of time. Tolman, too, was worried, but optimistically believed that the timescale difficulty might not be real. After all, cosmological models were idealized to such an extent that the very concept of age might lose its meaning. Perhaps the discrepancy was an artefact resulting from the use of homogeneous models and illegitimate extrapolations to the assumed singular state? Such considerations turned up in his textbook of 1934, and in his last paper, published posthumously in 1949, Tolman returned to the subject. 'I see at present no evidence against the assumption that the material universe has always existed,' he wrote.[103] Although he could not resolve the age paradox, he thought that modified relativistic models might well lead to an answer. For example, if the assumption of homogeneity was abandoned, it was possible to construct a model universe with an age of 3.64 billion years, still at the low end but a step in the right direction. The point to notice is that although physicists and astronomers recognized the difficulty, most of them did not see it as insurmountable or something that seriously questioned the credibility of relativistic cosmology.

If no other solution was in sight, one could always appeal to the cosmological constant. For models with $\Lambda > 0$ and positive acceleration, such as Lemaître's, the Hubble time is shorter than the age of the universe, and then the timescale problem can be avoided. Indeed, the main reason why models with a cosmological constant were kept alive in post-Second-World-War cosmology was that they avoided the age paradox. In his presentation at the 1958 Solvay congress, Lemaître concluded that the age of the universe was somewhere between twenty and sixty billion years. As we have seen, Einstein would have nothing to do with the cosmological constant (nor with inhomogeneous models), and for this reason he could not adopt the somewhat cavalier attitude of other cosmologists. He recognized that a comparison between the age of the universe and the reliably determined age of the Earth resulted in a genuine dilemma, a view he held as early as 1931. In 1945 he even indicated that if the dilemma could not be resolved, he would be forced to abandon the relativistic theory of the universe.[104]

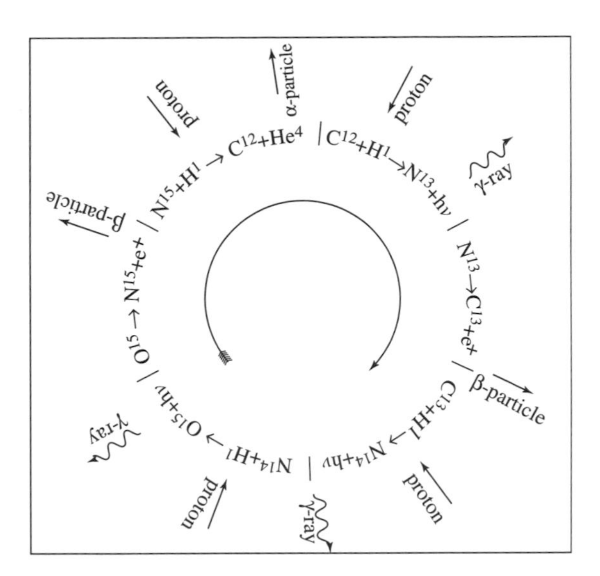

Fig. 3.7 Gamow's drawing of Bethe's CN cycle. Gamow accepted Weizsäcker's cosmic scenario, which he took to indicate that 'such extreme conditions could not be found even in the central regions of the hottest stars, [but] we are forced to look for them in the early superdense and superhot stages of the universe'. 'The alchemy of the sun', from *The Birth and the Death of the Sun* by George Gamow, copyright 1946 by George Gamow (first published in 1940). Used by permission of Viking Penguin, a division of Penguin Group (USA) Inc.

The early tradition in nuclear astrophysics reached a provisional climax in 1938–39, when Hans Bethe in the United States and Carl Friedrich von Weizsäcker in Germany independently proposed how stellar energy was produced in nuclear reactions. Bethe, in part collaborating with Charles Critchfield, suggested two mechanisms by means of which four protons fuse into a helium-4 nucleus, one called the pp cycle and the other the CN cycle. Bethe's very successful theory (for which he was awarded the Nobel Prize in 1967) was a breakthrough in nuclear astrophysics but of no direct cosmological consequence, since it was not concerned with the synthesis of heavier elements or with the early, prestellar universe.

Contrary to Bethe, Weizsäcker wanted in his theory of 1938 to account for the formation of the heavier elements. He argued that it was impossible to build up these elements in the interior of the stars and was consequently led to consider that they were formed cosmologically, in an early hot state of the universe. 'It is quite possible', he wrote, 'that the formation of the elements took place before the origins of the stars, in a state of the universe significantly different from today's.' But how could the physical conditions of this early state be known? Weizsäcker realized that direct empirical evidence was out of the question and suggested it was necessary 'to draw from the frequency of distribution of the elements conclusions about an earlier state of the universe in which this distribution might have originated'.[105] This research programme is what is known as 'nuclear archaeology', namely, the attempt to reconstruct the history of the universe by means of hypothetical cosmic or stellar nuclear processes, and to test these by examining the resulting pattern of element abundances.

Weizsäcker suggested tentatively that in the earliest state of the universe the density was about that of an atomic nucleus and the temperature about 100 billion degrees.

He imagined a great primeval aggregation of hydrogen collapsing under the influence of gravitation to form the extreme conditions under which element formation would take place in an explosive act. 'How large should one imagine the first aggregations to have been? Theory sets no upper limit, and our fancy has the freedom to imagine not only the Milky Way system but also the entire universe as known to us combined in it.' Weiszäcker saw his speculation as justified because it promised a physical explanation for the recession of the galaxies:

> The energy released in nuclear reactions is about 1% of the rest energy of matter and imparts to the nuclei on the average a velocity of the order of magnitude a tenth of the velocity of light. At approximately this speed the fragments of the [primordial] star should fly apart. If we ask where today speeds of this order of magnitude may be observed, we find them only in the recessional motion of the spiral nebulae. Therefore, we ought at least to reckon with the possibility that this motion has its cause in a primeval catastrophe of the sort considered above.[106]

Weizsäcker's picture had a good deal in common with Lemaitre's hypothesis of a primeval atom, but it belonged to a different tradition as it did not refer to relativistic models and had nothing to say about the geometry of the universe. His entire argument rested on cosmic-abundance data for chemical elements, a new field of research, which to a large degree relied on work done by the Norwegian mineralogist and geochemist Victor Goldschmidt.

Although his main interest was in the chemical composition of the Earth, Goldschmidt believed that only through comparison with the universe at large would it be possible to understand the geochemical evolution of the Earth. As he expressed it in a lecture of 1944,

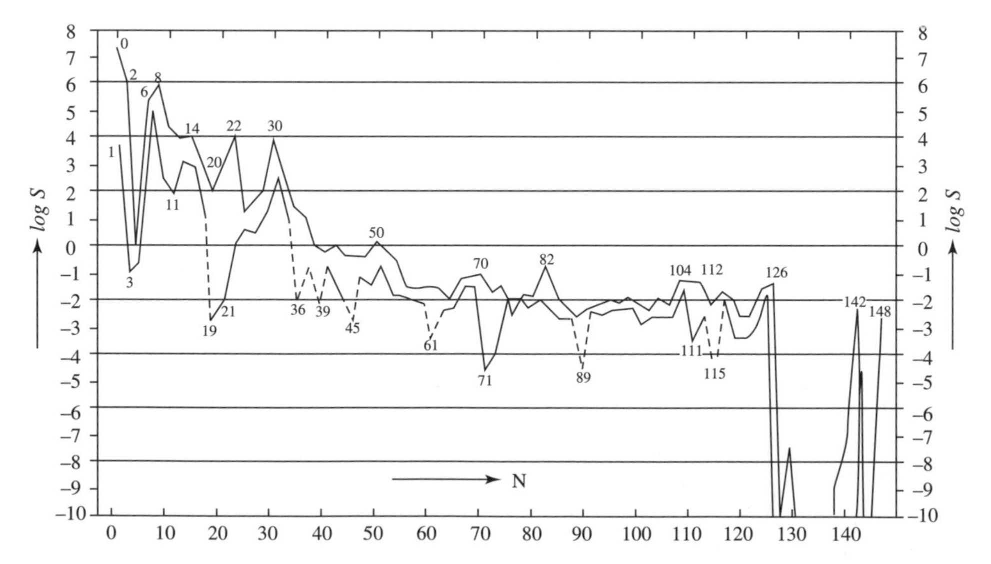

Fig. 3.8 Goldschmidt's 1938 summary data for the cosmic abundances of the elements as a function of their neutron number.

geochemistry was closely connected with 'astrophysics and nuclear physics, leading up to the final problem of the origin and evolution of matter itself'.[107] Goldschmidt published in 1938 an important compilation of tables of element distributions based on solar, stellar, and meteoritic data.[108] These tables resulted in the first diagrams showing not only the variation of the cosmic abundances of elements with the atomic number, but also the variation with the mass and neutron numbers. The Norwegian professor of mineralogy suggested that the data might be explained astrophysically or cosmophysically, possibly by the new theory proposed by Weizsäcker. However, Goldschmidt argued that Weizsäcker's theory was unable to explain the sudden decrease in abundance following iron and therefore suggested a revision of the theory.

3.4 Alternative cosmologies

During the period between the two world wars, cosmology was a very small research area, with no professional community and no clear scientific identity. It was a science seeking a paradigm—and there were those who questioned whether it was a science at all. In such a situation of dissension, it is difficult to speak of mainstream cosmology, except that a majority of those working with cosmological models accepted general relativity as the theoretical framework for their studies. However, a large minority disagreed and chose to disregard the still immature paradigm-candidate based on the works of Friedmann, Lemaître, Robertson, and others. Many cosmologists denied that the universe was expanding and that space was curved, and they looked for other ways to explain Hubble's observations. Again, acceptance of the expansion of the universe did not necessarily mean acceptance of relativistic cosmology. From the mid 1930s, A. E. Milne developed a strong alternative which attracted much interest among mathematically inclined astronomers. Milne's alternative and also the unorthodox theories proposed by Dirac and Jordan were 'modern' in the sense that they incorporated not only the expanding universe but also the idea of a sudden beginning, an idea still resisted by most cosmologists of the relativistic school.

3.4.1 *Against cosmic expansion*

After 1929, any plausible cosmological theory would have to account for Hubble's linear relationship between the redshifts and distances of the galaxies. Relativists explained the Hubble law by the expansion of the universe, but it could easily be explained differently, without assuming the observed redshifts to be caused by receding galaxies.[109] For example, one could assume—purely ad hoc—that the velocity of light decreased slowly with time, as did the Japanese physicist Tokio Takeuchi. By adopting a suitable variation of the velocity, he derived in 1931 a galactic redshift proportional to the distance. From a review of determinations of the velocity of light since 1849, it could even be argued that there was good historical evidence for a linear slowing-down of the velocity, corresponding to a decrease in the velocity of light during its journey from a distant galaxy to the Earth.[110] The hypothesis was not taken very seriously, but it caused Eddington to lament that 'The speculation of various writers that the velocity of light has changed slowly in the long periods of cosmological time... has seriously distracted the sane development of cosmological theory.' According to Eddington and most other physicists, 'The speculation is nonsense because a change of the velocity of light is self-contradictory.'[111] As we shall see in Section 5.3, this was not the final verdict on the subject of a possible varying speed of light.

The first non-Doppler explanation of the redshift–distance relation came as early as 1929, when Zwicky proposed that a gravitational analogue of the Compton effect might be responsible. By assuming light quanta to transfer energy to matter in intergalactic space, he arrived at a relation similar to Hubble's and with a redshift of the right order of magnitude. Zwicky found that his 'gravitational drag' explanation was in qualitative accordance with all known observational facts. In 1935 he summarized his objections against the standard interpretation, cautioning us 'not to interpret too dogmatically the observed redshifts as caused by an actual expansion'. As he pointed out, the theory of the expanding universe was unable to explain the very large dispersion in redshifts of galaxies belonging to clusters such as the Coma cluster (with a redshift of about 7000 km/s and a scatter between its individual members of 3000 km/s). This was one of several reasons why 'the theory of the expanding universe can hardly be considered as a completely satisfactory solution of the problem of the redshift'.[112]

Some other non-standard explanations belonged to the tired-light category, where light was assumed to gradually lose energy during its journey through empty space.[113] John Stewart, a Princeton physicist, suggested in 1931 that the frequency decreased exponentially with distance, and six years later Samuel Sambursky from the Hebrew University, Jerusalem, postulated Planck's constant to vary in time as $\exp(-Ht)$, which leads to the same result. Neither Stewart nor Sambursky provided physical reasons for their suggestions except that they evidently wanted to avoid the expanding universe.

William D. MacMillan, a professor of astronomy at the University of Chicago, preferred the old-fashioned Newtonian universe and therefore questioned the interpretation of the redshift as a Doppler effect. In 1932 he argued, as an alternative, that photons lost energy in their travels through space. The same kind of explanation was offered by Walther Nernst, the German physical chemist and Nobel laureate of 1920, who was a devout advocate of a stationary universe. In works from the 1930s, Nernst hypothesized that photons lose energy according to $\mathrm{d}E/\mathrm{d}t = -HE$, where $E = \mathrm{h}\nu$ is the photon energy. It follows that $\ln(\nu_0/\nu) = Ht$ or, if the decrease in frequency $\Delta\nu$ is small compared with the frequency ν, that $\Delta\nu/\nu = Ht$. Since $t = r/c$, the result becomes $\Delta\nu/\nu = Hr/c$, which is Hubble's law. Thus, the cause of the linearity is not that the universe expands, but simply that it takes more time for the photons to reach the Earth, the farther away the galaxy is located. Nernst did not consider the Hubble parameter H to be a cosmical constant, but a quantum decay constant giving the decay rate of photons. The many non-Doppler explanations of the redshifts were not well received, because of their ad hoc nature, and consequently their impact on the development of cosmology was limited.

Nernst and MacMillan agreed that the universe was not governed by the laws of general relativity, and also that it was necessarily stationary and eternal. Their cosmological views had more affinity with those discussed in the late Victorian era than with the mathematical cosmology associated with the theory of relativity. Nernst, MacMillan, and their few followers were uniformitarians, believers in a stationary universe with a never-ending exchange of energy between matter and ether which made the heat death avoidable. According to MacMillan, writing in 1925, 'the universe does not change in any one direction.... It is like the surface of the ocean, never twice alike and yet always the same.'[114]

Nernst was no less committed to this kind of universe and believed that the decay of stellar matter through radioactive processes was balanced by the formation of new matter. Unable to explain the creation of matter physically, he postulated that it occurred in some kind of energy–ether interaction. In 1937 he pictured the ether as consisting of massless neutrons, these being transformed into real, ponderable neutrons by the absorption of radiation energy into the ether's pool of zero-point energy. In this way, neutrons would be created throughout the universe so that interstellar space would include a rarefied gas of neutrons, mixed with electrons and protons arising from the disintegration of the radioactive neutrons. From this cosmic particle gas, heavier elements would be formed.

Not surprisingly, Nernst's ether-based speculations were ignored by most astrophysicists and cosmologists. On the other hand, non-Doppler explanations of the redshift continued to be investigated by a minority of researchers who felt uneasy about the expanding universe. To mention just two examples, Einstein's former collaborator Erwin Freundlich suggested in the 1950s that the redshift could be explained by a hypothesis of photon–photon interaction. His suggestion caused some controversy, but failed to change the attitude of most specialists that the redshifts were caused by the expansion of the universe.[115] The same can be said about a static and eternal cosmological model proposed by Gerald Hawkins in the 1960s, intended to be an alternative to both the Big Bang and the steady-state theories. Hawkins argued that the observed redshifts of distant galaxies were a gravitational effect, and when the cosmic microwave background was discovered in 1965 he interpreted it as the temperature of intergalactic dust grains.

3.4.2 *Quantum and cosmos*

Quantum cosmology before the Second World War? Well, not exactly, but the general idea of bridging or integrating quantum theory and relativistic cosmology was not foreign at this early date. One of the earliest works of this kind was published shortly before the advent of quantum mechanics in the summer of 1925. Lanczos investigated an Einstein-like world model which was periodic with respect to space and time, but in such a way that there was no endless recurrence. In connection with the propagation of waves in his 'spherical ring model', Lanczos was led to introduce quantum theory. For the world period, he arrived at the expression $T = 4\pi^2 mR^2/h^2$, which, with a world radius R of one million light years, gives $T \cong 10^{41}$ years. According to Lanczos, the quantum nature of microphysical phenomena reflected the state of the cosmos, rather than being a feature specific to the atomic world. 'The solution of the quantum secrets are hidden in the spatial and temporal closedness of the world,' he wrote.[116] In the same year, James Rice, a Liverpool physicist, suggested that the fine structure constant $\alpha \equiv 4\pi^2 e^2/h^2 c$ was related to the geometry of the world, and derived a formula which connected α with the radius of the Einstein universe.[117]

The speculations of Lanczos and Rice can be seen as precursors of the research programme that Eddington developed between 1929 and his death in 1944, culminating in the posthumously published *Fundamental Theory*. His interest in joining gravity to quantum theory can be found as early as 1920, in the second edition of his *Report on the Relativity Theory of Gravitation*, where he pointed out that a fundamental length of value 4×10^{-35} m can be constructed from the constants c, h and G, namely, the quantity $\sqrt{hG/c^3}$, later known as the Planck length. Eddington was convinced that the microphysical and cosmical constants of nature were deeply interconnected and that the cosmological constant was needed to make the connection. Moreover, he believed that quantum mechanics was intrinsically cosmological because the wave function of a particle existed throughout the universe. According to Eddington, the mass of an electron was an interchange energy with all the other charges in the universe, and hence of cosmological nature.

The ordinary Schrödinger wave equation of 1926 is non-relativistic, the relativistic generalization is known as the Klein–Gordon equation (named after Oskar Klein and Walter Gordon). This equation fails, however, to describe spinning electrons, and in 1928 Paul Dirac found an alternative quantum wave equation which explains the electron's spin and also satisfies the requirements of the special theory of relativity. Interpreting the Dirac wave equation cosmologically, Eddington found in the 1930s numerical relations that combined cosmological quantities and atomic constants. To give an impression of Eddington's approach, two examples will suffice:

$$\frac{mc}{\alpha\hbar} = \frac{\sqrt{N}}{R} \quad \text{and} \quad \frac{c}{\hbar}\sqrt{\frac{mM}{\Lambda}} = \sqrt{\frac{N}{30}}.$$

Here, N denotes the 'cosmical number'—the number of protons in the universe—and m and M are the mass of the electron and the proton, respectively. The standard symbol $\hbar$ stands for $h/2\pi$. Eddington used his mathematically complex theory to calculate Hubble's constant from the laboratory values of other constants. As mentioned, in 1944 he found $H_0=572$ km/s/Mpc in this way.[118]

Most leading physicists rejected Eddington's theory, which also was coolly received, if received at all, by astronomers. Yet it was not without impact, for it inspired several physicists to pursue work in a similar tradition, more often than not in a numerological style. The German Hans Ertel and the Austrian–American Arthur Erich Haas were particularly active in the field. Contrary to Eddington, Ertel was interested in Lemaître's Big Bang model and investigated the 'Friedmann–Lemaître cosmos' from a quantum-cosmological point of view.[119] Haas published in 1934 *Kosmologische Probleme der Physik*, partly inspired by Eddington, and two years later, after having settled in the United States, he offered 'a purely theoretical derivation of the mass of the universe'. From the assumption that the total energy of the universe was zero, he concluded that the mass of the (Einstein) universe was given by Rc^2/G.[120] Among Haas's contributions to early physical cosmology must also be counted a symposium he arranged at the University of Notre Dame in 1938. The theme of the symposium was 'The Physics of the Universe and the Nature of Primordial Particles' (strikingly modern as it sounds), and among the participants were leading physicists and cosmologists such as Arthur H. Compton, Gregory Breit, Harkins, Shapley, and Lemaître.

Contrary to other leading quantum physicists, Erwin Schrödinger received Eddington's quantum-cosmological programme with enthusiasm. He was convinced, if only for a time, that 'for a long time to come, the most important research in physical theory will follow closely the lines of thought inaugurated by Sir Arthur Eddington'.[121] Schrödinger even added to the basket of Eddingtonian formulae his own piece of micro-macro numerology, suggesting that $R/\delta=\sqrt{N}$, where δ is the mean distance between nucleons in an atomic nucleus. In works of 1939–40, he investigated the proper vibrations of quantum waves in closed and expanding world models in the hope of deriving in this way the mass spectrum of elementary particles as a consequence of the structure of space–time. Examining the solutions of the Klein–Gordon wave equation in expanding space, he found what he called an 'alarming phenomenon', namely that 'There will be then a mutual adulteration of positive and negative frequency terms in the course of time, giving rise to [pair production].'[122] What he found was that, in a universe with accelerated expansion, a pair of particles, such as an electron and a positron, can be formed out of the vacuum, without violating energy conservation. Schrödinger did not pursue this line of work, and soon lost his interest in Eddington's approach. His discovery that a varying gravitational field can create particles was important, but it was totally ignored and had to be rediscovered by a later generation of cosmologists.

3.4.3 *Varying gravity*

If Newton's (or Einstein's) constant of gravitation is allowed to change with time, and hence is not a true constant, the rules and results of the cosmological game change too. This audacious idea was first introduced in the 1930s by two of the pioneers of quantum mechanics, Paul Dirac and Pascual Jordan (in a different way, it had been entertained a little earlier by Milne). Dirac, a newcomer to cosmology, readily accepted Lemaître's picture of the Big Bang, which he took to mean that 'the universe had a beginning about 2×10^9 years ago, when all the spiral nebulae were shot out from a small region of space, or perhaps from a point'.[123] The age of the universe, expressed in the 'natural' time unit e^2/mc^3 (where e is the elementary charge and m the electron mass), is a very large number, about 2×10^{39}. As Weyl and Eddington had pointed out earlier, the ratio of the electrical to the gravitational force between an electron and a proton is of the same order of magnitude.[124] Dirac believed that the relation

$$\frac{T_0}{e^2/mc^3} \cong \frac{e^2}{GmM} \cong 10^{39}$$

could not be a coincidence, but must signify some deep connection in nature between the realms of cosmology and atomic theory. This connection he claimed to be the 'large number hypothesis', namely, that two very large numbers constructed from natural constants (or otherwise turning up in fundamental physics) must be related in a simple way. Since the left part of the equation is a measure of cosmic time, and since Dirac assumed that neither c nor the atomic constants e, m, and M changed with time, it followed *ex hypothesi* that the gravitational constant decreased as $G(t) \sim t^{-1}$. According to Dirac's hypothesis, the relative change of G was given by

$$\frac{1}{G}\frac{\mathrm{d}G}{\mathrm{d}t} = -3H,$$

or about 10^{-11} per year, a change that at the time was much too small to be subject to direct experimental test. In a paper of 1938, Dirac developed a cosmological theory based on Hubble's relation and the varying-G hypothesis.[125] By applying the large number hypothesis also to the inverse mean density of matter in the universe (which is a large number), he found for the expansion rate the relation

$$R(t) \sim t^{1/3},$$

which implied that the age of the universe was related to Hubble's parameter according to

$$t_0 = \frac{1}{3}H_0^{-1}.$$

Moreover, Dirac argued that a universe complying with the large number hypothesis had to be spatially flat, infinite, and with a zero cosmological constant. His argument for space being flat was, characteristically, that the cases of positive and negative curvature were ruled out because they implied very large numbers independent of the cosmic epoch.

Clearly, Dirac's cosmological theory was problematic both empirically and methodologically. Not only did it conflict with Einstein's general theory of relativity (which does not allow G to vary), but it also led to a hopelessly small value of the age of the universe, about 700 million years, which was definitely smaller than the age of the Earth. Yet Dirac convinced himself that these difficulties could be overcome. To his mind, the large number hypothesis was a fundamental principle of such power and beauty that it had to be right. As he explained in a lecture of 1939, it followed from the principle that the laws of nature were evolutionary, not fixed once and for all: 'At the beginning of time the laws of Nature were probably very different from what they are now. Thus we should consider the laws of Nature as continually changing with the epoch, instead of as holding uniformly throughout all space–time.' Dirac even went as far as to suggest that the laws of nature at a given cosmic time were not the same at all places in the universe: 'We should expect them also to depend on position in space, in order to preserve the beautiful idea of the theory of relativity that there is fundamental similarity between space and time.'[126] Although Dirac's theory was enthusiastically taken up by Jordan, and also attracted the interest from a few other physicists (including the young Subrahmanyan Chandrasekhar, the eminent astrophysicist), it was ignored by the majority of physicists and cosmologists. Even Dirac himself lost interest in his theory and only returned to it in the 1970s.

Jordan had an interest in cosmology even before he encountered Dirac's idea of gravitation varying in time. He was one of the early converts to Lemaître's picture of a Big Bang, which he described in 1936 as follows:

> About 10 billion years ago the world diameter, today grown to ten million [*sic*] light years, must have been vanishingly small.... The initially small universe arose from an original explosion. Not only atoms, stars and milky way systems but also space and time were born at that time. Since then the universe has been growing, growing with the furious velocity which we detect in the flight of the spiral nebulae.[127]

In a series of works between 1938 and the early 1950s, Jordan developed his own version of Big Bang cosmology based on Dirac's large number hypothesis. The difference was that he also adopted the idea—originally suggested by Dirac in 1937, but abandoned soon thereafter—of spontaneous matter creation. If ρ denotes the mean density of luminous matter in the universe, the dimensionless ratio $\rho(cT_0)^3/M$ comes out as approximately 10^{78}, the square of the cosmic time in atomic units. According to the large number hypothesis, as understood by Jordan, this meant that the number of particles in the universe must increase as $N \sim t^2$. However, whereas Dirac had originally thought of matter creation in terms of individual protons and electrons spread throughout the depths of the universe, Jordan proposed that stars and galaxies would be formed spontaneously as whole bodies, at first with a density corresponding to that of an atomic nucleus. Although Jordan was strongly influenced by Dirac's thinking, he ended up with a quite different world model: whereas Dirac's universe was infinite and spatially flat, Jordan's was finite and positively curved. Both physicists kept to a zero cosmological constant.

Naturally, the spontaneous creation of supercompact stars and galaxies was a problematic feature, as it seemed to imply a gross violation of energy conservation. However, by developing a suggestion originally made by Haas, Jordan argued that there was no contradiction. The Haas–Jordan argument was that the mass increase was energetically compensated for by the increase in negative potential energy that follows from the expansion of the universe. In this way, the total mass–energy of the universe remained unchanged, namely, equal to zero.[128]

Jordan elaborated his unorthodox cosmic scenario in a paper of 1944 (which, because of the war, was not much noticed). He now concluded that there was no Big Bang in Lemaître's cataclysmic sense, for initially there was no primordial atom to explode. Matter, he suggested, was created along with the expansion. In a word, it was a Big Bang without a bang. According to Jordan, the history of the universe could be traced back to a time when its radius was only about 10^{-15} m and when it consisted of only one pair of newly created neutrons. As space expanded and the neutrons separated, the change in gravitational energy would be balanced by the creation of new matter. Ten seconds after the initial bang, the universe would have grown to the size of the Sun, though with a mass less than that of the Moon. At this early epoch, the universe would consist of about 10^{12} stars of an average mass of one million tons, and a supernova formed at that time (and they would form abundantly) would initially have a radius of only 1 mm. Remarkably, Jordan found his scenario to be in agreement with empirical facts and claimed that it was inductively based on these.

Except for Jordan and his co-workers, little more was heard of varying-G cosmologies in the years after the Second World War. Still, although the idea did not attract much attention, it remained alive and reappeared in the 1970s, when Dirac developed it in various versions in attempts to make it conform to general relativity. None of his versions of varying-G cosmology won much support, but a few researchers of the younger generation found the idea sufficiently interesting to develop it independently. Vittorio Canuto and co-workers constructed a cosmological model based on Dirac's large number hypothesis, which, they argued, could account for both the microwave background radiation and the primordial helium production.[129] On the experimental side, studies in the early 1980s using the Viking landers on Mars showed that if G varied at all, the variation was less than one part in 10^{11} per year, in disagreement with Dirac's prediction. This result caused most cosmologists to dismiss theories with a decreasing gravitational constant. However, the more general idea of natural constants varying in time has recently experienced a spectacular revival (compare Section 5.3).

3.4.4 *Milne's universe*

Edward Arthur Milne, from 1929 to his death in 1950 a professor of mathematics at the University of Oxford, was an eminent astrophysicist who, in the 1920s, made important contributions to the theory of stellar atmospheres and the radiation equilibria of stars. In the spring of 1932 he turned towards cosmology, a field he developed in his own, original way in a number of important articles and monographs. His first systematic exposition of cosmology appeared in a book-length article in 1933, and in 1935 he published his major opus, *Relativity, Gravitation and World-Structure*, followed in 1948 by *Kinematic Relativity*.[130]

Milne's cosmology was unusual in several respects, not least because it was independent of the theory of general relativity, which Milne did not accept (he did accept and use the special theory). He denied that space could have a structure, curved or not, and also that space itself could expand. As merely a system of reference, space could have no physical qualities. His cosmological system was built on two fundamental principles, one being the constancy of the velocity of light, which he considered to be true by convention; the other was what Milne called the cosmological principle, that is, that all observers will see the same things, irrespective of their position and the direction in which they look. As we have seen, in a loose sense the cosmological principle can be traced back a long time, first appearing in Nicholas of Cusa in the fifteenth century; without elevating it into a general principle, Einstein made use of it in his 1917 cosmological model, and it was adopted by most later workers in the field.

In his book of 1935, Milne argued that all the basic laws of cosmic physics could be deduced from a few principles of a kinematic kind. His original theory was restricted to distance and time relations, but he later extended it to cover also dynamics and electromagnetism. Contrary to Eddington, he did not try to integrate cosmology and atomic physics, and quantum theory played no role at all. In Milne's idealized model, galaxies were represented by randomly moving particles, much like the molecules making up a gas. By simple kinematic considerations he could show that the system would evolve in a Hubble-like way, with the fastest-moving particles creating a dense spherical front near a distance ct from the point of origin. At any time, the system was bounded by an impenetrable barrier

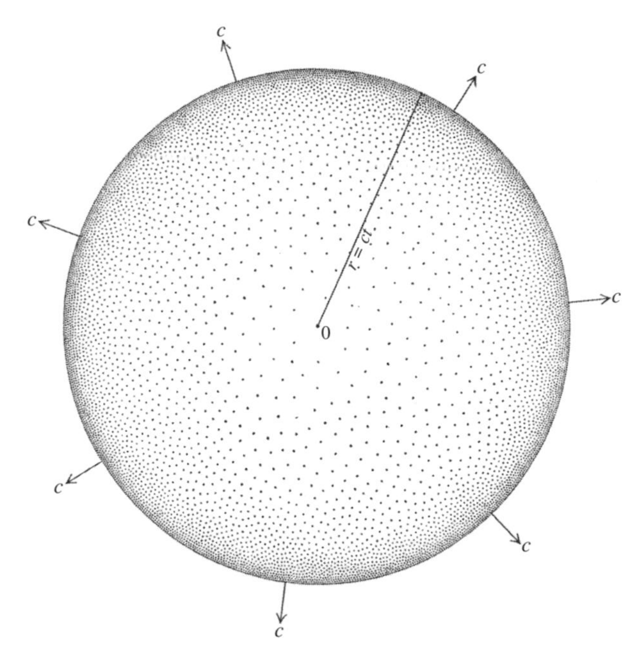

Fig. 3.9 In A. E. Milne's world model, an observer O would see the nebulae as bounded at distance ct at a epoch t by a barrier of infinite density, which he took to represent the creation of the universe. Each dot represents a nebula in outward motion from the observer. The boundary is receding from the observer at the speed of light. Illustration from Milne 1935.

of infinite particle density. In spite of the infinite number of galaxies, by using the Doppler shift formula of the special theory of relativity Milne could show that the total brightness would be finite. Olbers' paradox was not part of his world model.

In Milne's expanding universe, the distance between any two galaxies moving with relative velocity v would increase with time as $r = vt$. If the Hubble constant is identified as the inverse of t, Hubble's law $v = Hr$ comes out. Milne thus explained Hubble's law purely kinematically, without any reference to gravitation. In 1935 he deduced that $4G\rho t^2 \cong 1$, which means (because ρ varies inversely as t^3) that the constant of gravitation increases with the epoch. The result had the advantage that shortly after $t = 0$, when the particle-galaxies were closely packed, there would be no gravitation to brake the rapid expansion; with increasing epoch G would grow, but now the galaxies would be so far apart that gravitation could be neglected. Milne emphasized that the linear relation between G and t did not imply that local gravitation, as in the solar system, increased in strength. In fact, he did not consider the time dependence of G to be subject to experimental test.

Milne operated with two measures of time, called kinematic time (t) and dynamic time (τ), which were connected by the relation

$$\tau = t_0 \log\left(\frac{t}{t_0}\right) + t_0,$$

where the constant t_0 was identified with the present epoch. He argued that the time of Newtonian physics was τ-time, whereas optical phenomena would run according to t-time. On the t-scale the universe is expanding from a point source, but on the τ-scale the world is static. According to Milne, it was not meaningful to ask if the universe was really expanding or not, for the two descriptions were merely two different ways of picturing the same world. It followed from Milne's understanding of these time scales that he was unconcerned with the age paradox, which for him was a pseudo-problem.

However, in 1945 Milne reached the conclusion that kinematic and dynamic time were not, after all, equally valid, but that 'phenomena themselves' were best studied through the more fundamental t-scale. About the origin of the universe, he wrote: 'Just as the epoch $t = 0$ is a singularity in the mechanical t-history of the universe—an epoch at which the frequency of radiation was infinite—so the epoch $t = 0$ is a singularity in the optical history of the universe, namely, an epoch at which the frequency of radiation was infinite, because the wave-length had to be zero.' He further suggested that the presently observed cosmic rays were the fossils of the primitive high-frequency radiation, an idea which had some similarity to Lemaître's scenario, except that Lemaître's fossils were not light quanta but charged particles.[131] About the sudden beginning of the universe, the creation, Milne wrote: 'We can form no idea of an actual event occurring at $t = 0$; we can make propositions in principle only *after* the event $t = 0$. As for why the event happened, we can only say that had no such event happened, we should not be here to discuss it.'[132]

In his great monograph of 1935, Milne included among the many pages densely packed with mathematical equations a section on 'Creation and Deity', in which he stated that the ultimate questions of cosmology needed reference to God. This was more than a casual remark, and in later works he developed his unorthodox ideas of cosmo-theology, which further alienated him from mainstream cosmology.[133]

Milne's kinematic–relativistic cosmology aroused great interest in the 1930s. Many of the responses were critical, but other scientists, mostly in England, considered the theory to be promising and worth developing as an alternative to general-relativistic cosmology. Most of the interest in the theory came from British physicists and astronomers, but initially American astronomers also found it worthwhile to study it. 'Your paper is very widely discussed and we are holding seminars on the subject,' Hubble wrote to Milne in 1933.[134] Hubble found the kinematic model to 'possess unusually significant features', as he wrote in *The Realm of the Universe*. On his side, Milne believed that Hubble's observations provided empirical support for his theory: 'Hubble's observations disclosed a density-distribution of nebulae increasing outwards if recession is adopted, and a homogeneous distribution if recession is denied. This is just what is predicted on the present treatment.'[135]

Milne's ambitious attempt to reconstruct physics and base it on cosmological principles was for a while followed with great interest but also met stiff resistance. Robertson, whose attitude was hostile, studied the theory carefully in a series of papers in 1935–36. At the same time, Arthur G. Walker, who was directly inspired by Milne, compared kinematic relativity and general-relativistic cosmology from a formal and conceptual point of view. His paper on cosmological space–times of constant spatial curvature carried the title 'On Milne's Theory of World Structure'. What is known as the Robertson–Walker metric is the most general form of metric for a space–time satisfying the cosmological principle. As mentioned earlier, this kind of metric was first found by Robertson in 1929, but in its general formulation of 1935 it was indebted to Milne's theory, either positively (Walker) or critically (Robertson).

One reason for the controversies that followed Milne's theory was the philosophy behind it, a mixture of conventionalism, operationalism, and extreme rationalism. The theory might be valuable from mathematical and philosophical points of view, but had it anything to do with physical reality? Milne proudly claimed that world physics was as much, or more, a matter of logic and reason than of observation and experiment. As he wrote:

> The philosopher may take comfort from the fact that, in spite of the much vaunted sway and dominance of pure observation and experiment in ordinary physics, world-physics propounds questions of an objective, non-metaphysical character which cannot be answered by observation but must be answered, if at all, by pure reason; natural philosophy is something bigger than the totality of conceivable observations.[136]

Statements like this, and similar views expressed by Eddington, Dirac, and Jeans, infuriated scientists subscribing to a more empirical and inductivist view of science. Herbert Dingle, in particular, accused Milne and other rationalist cosmo-physicists of perverting the true spirit of science and replacing it with 'a pseudo-science of inveterate cosmythology'. This attack led in 1937 to a heated debate in *Nature*, which engaged many of Britain's most prominent scientists.[137] The controversy spilled over into *The Observatory*, the monthly review published by the Royal Astronomical Society, where McVittie, in a paper of 1940, expressed his and others' objections in emotional terms: 'It is eventually borne in on the puzzled reader that Milne and Walker are not trying to understand Nature but rather are telling Nature what she ought to be. If Nature is recalcitrant and refuses to fall in with their pattern so much the worse for her.'[138] McVittie was at first attracted by Milne's theory, which he tried to relate to observations, but soon reached the conclusion that the approach of kinematical relativity was unsatisfactory and close to pseudo-science, and hence had to be rejected.

Notes

1. On the history of non-Euclidean geometry, see Gray 1979.
2. Quoted in Jammer 1993, p. 147.
3. According to Miller 1972, the story of Gauss's experiment is a myth.
4. Riemann 1873, p. 36.
5. Zöllner 1872, p. 308.
6. Quoted in Jaki 1969, p. 165. A very similar sentence appeared on p. 665 in *Newcomb-Engelmanns Populäre Astronomie* (fourth edition, edited by P. Kempf. Leipzig: Engelmann, 1911), an extensive and updated account of astronomy that was widely read in German-speaking Europe.
7. Schwarzschild's paper exists in an English translation as Schwarzschild 1998, where this quotation is on p. 2542. The original source is *Vierteljahrschrift der Astronomische Gesellschaft* **25** (1900), 337–247. See also Schemmel 2005.
8. H. Poincaré, *Science and Method*. Here from Poincaré 1982, p. 414.
9. Quoted in Torretti 1978, p. 276. Calinon's papers appeared in *Revue Philosophique de la France et de l'Étranger*. His paper of 1889 is translated into English in Čapek 1976.
10. The paper is translated in Einstein *et al.* 1952. Quotation from p. 108.
11. On Soldner's contributions to astronomy and cosmology, see Jaki 1978, which includes an English translation of Soldner's paper. The German physicist Philipp Lenard, who was strongly anti-Einstein and anti-relativity, argued in 1921 that Soldner's result anticipated Einstein's and showed the theory of relativity to be unnecessary.
12. Einstein 1982, p. 47.
13. Einstein 1916 (reprinted in *The Collected Papers of Albert Einstein*, vol. 6).
14. Classical black holes or invisible stars were considered in the late eighteenth century by John Michell and Laplace (see Section 4.4). For a history, see Eisenstaedt 1993.
15. Einstein 1998, document 162. Einstein first calculated the shift of the perihelion of Mercury in 1913, on the basis of the *Entwurf* theory, but obtained a wrong result and chose not to publish it. For the history of the motion of the perihelion of mercury, see Rosevear 1982.
16. Einstein *et al.* 1952, p. 163.
17. On Freundlich and his collaboration with Einstein, see Hentschel 1997.
18. A valuable account of the 1919 eclipse expedition and other early attempts to test the general theory of relativity is given in Crelinsten 2006.
19. In fact, the agreement between theory and observation was fortuitous, as both of the values were incorrect. Only in the 1980s was it shown that the values are four to five times as large as believed in the 1920s. Nonetheless, the agreement between theory and observation remains. This case, an instructive one with respect to the interplay between theory and observation, is discussed in Hetherington 1988, pp. 65–72.
20. In the third edition of a popular book on relativity theory, published in 1918, Einstein referred to Seeliger's work and his proposal of modifying Newton's law. However, Einstein found this unsatisfactory because it had no theoretical foundation. See Einstein 1947, pp. 125–127.
21. Einstein 1998, document 308. At about the same time that Einstein took up the cosmological implications of his theory, or maybe a little earlier, Schwarzschild entertained the possibility of a closed universe as a solution to the field equations of general relativity (Schemmel 2005).
22. Einstein *et al.* 1952, p. 183; Einstein 1917.
23. Einstein *et al.* 1952, p. 188.
24. For a detailed account of Einstein's commitment to Mach's principle and how it shaped his cosmology, see Hoefer 1994.
25. Einstein 1998, document 327.
26. Ibid., document 604.
27. Einstein *et al.* 1952, p. 193.
28. Einstein supposedly called it 'my greatest blunder' in a conversation with George Gamow. For a full historical and analytical account of the controversial cosmological constant, see Earman 2001.
29. Einstein 1998, document 311. In reply, de Sitter declared himself in agreement with Einstein, but only 'if you do not want to impose your conception on reality'. Document 312.
30. A full account of de Sitter's cosmology and its relation to Einstein's can be found in Kerzberg 1989. De Sitter's third paper is reproduced in Bernstein and Feinberg 1986, pp. 27–48.
31. Letter of 24 March 1917. Einstein 1998, document 317. Einstein's belief that cosmology must agree with Mach's principle was more wrong than he could have imagined. In 1951 the mathematician Abraham Taub

found that the Einstein equations, even without the cosmological term, could describe a curved space in the absence of matter.

32. De Sitter 1917, p. 18. However, de Sitter did not find the cosmological constant more mysterious than other constants of nature, and he continued to use it even after the discovery of the expanding universe.
33. Ibid., p. 26.
34. Ibid., p. 28 (postscript added October 1917).
35. Quoted in Smith 1982, p. 174.
36. North 1990, pp. 95–106; Kerzberg 1989.
37. On Weyl's contributions to cosmology, see Kerzberg 1989 and Goenner 2001.
38. Hubble 1926, partly reproduced in Lang and Gingerich 1979, pp. 716–724 (quotation on p. 724). Characteristically, Hubble did not refer to Einstein's work but to a book by the Austrian physicist Erich Haas (*Introduction to Theoretical Physics*, published in 1925) which quoted Einstein's relations.
39. Silberstein 1930, p. 188. Based on globular clusters alone, his result was about 30 Mpc.
40. Borel 1960, p. 227.
41. Whitehead 1922; Eddington 1924.
42. North 1990, pp. 190–196. On later theories of gravitation of the Whitehead type, see Schild 1962.
43. According to Barrow and Tipler 1986, pp. 191–194, Whitehead's cosmology foreshadowed the ideas of many-universe models that appeared in the 1980s.
44. Eddington 1923a, p. 161.
45. Ibid., p. 167. Eddington's number 3×10^{42} was basically the same as the ratio between the electrical and gravitational forces of two electrons, a number that Weyl had called attention to in 1919. See also Eddington 1920, pp. 178–179, where it is suggested that this number equals the ratio between the radius of the universe and the electron's radius, yielding a radius of the universe of about 2×10^{11}parsecs.
46. Lemaître 1925, p. 188.
47. Robertson 1928, p. 844.
48. On Friedmann's life and career, see Tropp, Frenkel, and Chernin 1993.
49. Friedmann 1922, English translation in Lang and Gingerich 1979, pp. 838–843; and Bernstein and Feinberg 1986, pp. 49–58 (which also includes translations of Einstein's responses and Friedmann's 1924 paper).
50. Friedmann 2000, pp. 110–111.
51. Ibid., p. 109.
52. Friedmann referred in 1922 to Eddington 1920, which included a brief account of Slipher's measurements of the receding motions of spiral nebulae. Friedmann did not mention the redshifts in his book of 1923.
53. Robertson 1929, reproduced in Bernstein and Feinberg 1986, pp. 68–76.
54. Bernstein and Feinberg 1986, p. 67. In a draft of his reply to Friedmann 1922, Einstein remarked that 'a physical significance can hardly be ascribed to it', but decided not to publish the remark. See Kerzberg 1989, p. 335. Einstein's draft manuscript is reproduced in facsimile in Renn 2005, p. 185.
55. For Lemaître's life and career, see Lambert 2000 and Stoffel 1996. His contributions to cosmology are also dealt with in Godart and Heller 1985 and Kragh 1996a.
56. As recalled many years later in Lemaître 1958.
57. Lemaître 1927, p. 50. Together with other classics of early relativistic cosmology, the paper is reproduced in Luminet 1997. On Eddington's instigation, the paper was translated into English in *Monthly Notices of the Royal Astronomical Society* (vol. 111, 1931, pp. 483–490), unfortunately rather inaccurately and with part of the original text left out. The 1931 translation is reproduced in Bernstein and Feinberg 1986 and Lang and Gingerich 1979.
58. In his draft version, reproduced in Stoffel 1996, pp. 41–55, Lemaître first wrote *variable*, which he then crossed out and replaced with *croissant* ('increasing').
59. Eddington 1928, p. 166.
60. Hubble 1929, reprinted in Lang and Gingerich 1979, pp. 726–728 and Bernstein and Feinberg 1986, pp. 77–83. See also Smith 1982, pp. 180–183.
61. Hubble 1929, p. 172. On the different interpretations of the de Sitter effect in the 1920s, see North 1990, pp. 92–104. Hubble did not cite any theoretical cosmologists, but he was in contact with Richard Tolman (with whom he later collaborated) and may have known about the predictions of a linear redshift-distance relation. On the other hand, in early 1929 neither he nor Tolman knew about Lemaître's prediction.
62. See the discussion in Kragh and Smith 2003.
63. Quoted in Smith 1982, p. 192.
64. Humason 1929, p. 168.
65. Hubble and Humason 1931, p. 80.

66. Tolman 1929, p. 246.
67. Quoted in Lambert 2000, p. 107.
68. Eddington 1930, p. 675.
69. Letter of 17 April 1930, quoted in Smith 1982, p. 187. On the reception of Lemaître's theory, see also Kragh 1996a, pp. 31–35 and Lambert 2000, pp. 106–108.
70. Einstein 1931.
71. Heckmann 1932; Robertson 1933; Tolman 1934.
72. Eddington 1933, p. vii.
73. Bronstein 1933; Lemaître 1934.
74. Holmes 1913, pp. 120–121. For the entropic paradox, see Section 2.4.
75. Knopf 1914, p. 983.
76. Weyl 1924, p. 349.
77. Dingle 1924, pp. 399–401.
78. Jeans 1928, pp. 698–699, repeated almost verbatim in Jeans, *The Universe Around Us* (New York: Macmillan, 1929), pp. 316–317.
79. See Kragh 1996a, pp. 42–44 and 81–101, where references to the primary sources are given.
80. Lenz 1926. The idea of a zero-point energy, being the energy of a system in its lowest state, goes back to Planck in 1912, but was only vindicated by studies of molecular spectra in 1924 and subsequently by quantum mechanics.
81. Tolman 1931, p. 1641.
82. Lewis 1922, p. 309. On the astrochemical and cosmochemical tradition, see Kragh 2001a. As some scientists explored the astronomy-chemistry borderland, others called attention to the astronomy-geology borderland. See, e.g., Eddington 1923b.
83. Eddington 1923b, p. 19.
84. Eddington 1931. This section builds on Kragh 1996a, pp. 44–55 and Kragh 2003.
85. Lemaître 1931a. In a manuscript version of the note to *Nature*, Lemaître ended with the sentence 'I think that everyone who believes in a supreme being supporting every being and every acting, believe also that God is essentially hidden and may be glad to see how present physics provides a veil hiding the creation.' Lemaître crossed out the sentence before he sent the note to *Nature*. Lemaître Archive, Louvain-de-la-Neuve.
86. Lemaître 1931b, p. 406. Reprinted in Luminet 1997, pp. 215–238.
87. Lemaître 1949; Kragh 1996a, pp. 53–54; Earman 2001. Apart from Lemaître, the main protagonists of the cosmological constant were Eddington and McVittie.
88. Lemaître 1933, p. 84. Published in an obscure journal, the paper was not well known among cosmologists. An English translation appears in *General Relativity and Gravitation* **29** (1997), 641–680.
89. Barnes 1933, p. 96.
90. Einstein 1931, p. 236.
91. Einstein and de Sitter 1932. Reproduced in Lang and Gingerich 1979, pp. 849–850.
92. Undated draft in Lemaître Archive, Louvain-de-la-Neuve, Belgium. Lemaître and Einstein met at Mount Wilson in early 1933. According to *The Literary Digest*, Einstein said, 'This is the most beautiful and satisfactory explanation of creation to which I have ever listened.' See Kragh 1996a, p. 55. Einstein returned to cosmology only in 1945, when he made it clear that he was in favour of a model with a definite beginning in time (Einstein 1945).
93. Plaskett 1933, p. 252.
94. Tolman 1934, p. 486.
95. Eddington 1933, p. 125.
96. Hubble and Tolman 1935, p. 335. Hubble's cosmological work is considered in Hetherington 1982. For surveys of observational cosmology since the 1930s, see North 1990, pp. 234–254 and Sandage 1998. Many of Hubble's papers, as well as works on Hubble, can be found on the Internet; see www.phys-astro.sonoma.edu/BruceMedalists/Hubble/HubbleRefs.html.
97. Quoted in Berendzen, Hart, and Seeley 1984, p. 208.
98. Hubble 1942.
99. Hubble 1936, pp. 201–202.
100. McVittie 1939; Heckmann 1942.
101. Hubble 1937, p. 62.
102. For more details, see North 1990, pp. 224–226 and 386–389, and Kragh 1996a, pp. 73–79.
103. Tolman 1949, p. 377.
104. Einstein 1945, p. 132.

105. Weizsäcker 1938, as translated in Lang and Gingerich 1979, pp. 309–319. On the early nuclear-archaeological research programme in cosmology, see also Kragh 2001b.
106. Heckmann, although finding the proposal attractive, objected that the mechanism suggested by Weizsäcker was unable to explain the recessional velocities of the faintest galaxies, which he took to be one-fifth of the speed of light. Heckmann 1942, p. 100.
107. Quoted in Mason 1992, p. 102.
108. Goldschmidt 1937 (although dated 1937, the volume appeared only in 1938). On Goldschmidt, see Kragh 2001a.
109. North 1990, pp. 229–234. For a contemporary discussion, see Bruggencate 1937.
110. Gheury de Bray 1939 suggested that $c = 299.774 - 173T$ km/s, where T is in millions of years.
111. Eddington 1946, p. 8.
112. Zwicky 1935.
113. Karl Popper, the famous philosopher of science, was among those who advocated (in 1940) a tired-light explanation for the redshifts. See Kragh 1996a, p. 246.
114. Quoted in Kragh 1995, where more details of the Nernst–MacMillan alternative can be found.
115. See Hentschel 1997, pp. 141–146. A survey of alternative interpretations of redshifts is provided in Assis and Neves 1995.
116. Lanczos 1925, p. 80.
117. Rice 1925.
118. Eddington 1944. On Eddington's research programme, see Kilmister 1994. In 1935 Eddington concluded that $H_0 \cong 850$ km/s/Mpc, a conspicuously large value that was not taken seriously by his astronomer colleagues.
119. Ertel's works are analyzed in Schröder and Treder 1996.
120. Haas 1936.
121. Schrödinger 1937, p. 744. For a careful examination of Schrödinger's views, see Rüger 1988.
122. Schrödinger 1939, p. 901.
123. Dirac 1937, p. 323. An account of Dirac's cosmology, from 1937 to about 1980, is given in Kragh 1990, pp. 223–246.
124. The magical number 10^{39}, signifying the ratio between the electrical and gravitational forces, is usually said to go back to a paper Hermann Weyl wrote in 1919. However, it was noticed earlier. Thus, the British physicist Owen Richardson, a Nobel laureate of 1928, called attention to 'the smallness of gravitational attraction compared with the forces betweens the electrons composing the attracting matter' and gave the ratio as 4×10^{40}. O. Richardson, *The Electron Theory of Matter* (Cambridge: Cambridge University Press, 1916), p. 609.
125. Dirac 1938.
126. Dirac 1939, p. 139.
127. Jordan 1944, p. 183 (English translation of book published in German in 1936). See also Jordan 1952. The 'ten million light years' for the diameter of the world is a misprint for 'ten billion light years'. An evaluation of Jordan's cosmological programme is given in Kragh 2004, pp. 175–185.
128. Basically the same idea has been utilized in much later cosmological theories; compare Kragh 2004, pp. 182–182.
129. For a contemporary review, see Wesson 1980.
130. Much has been written on Milne and his cosmology. See, for example, North 1990, pp. 149–185, Urani and Gale 1994, and Kragh and Rebsdorf 2002, which includes further references.
131. See Kragh 2004, p. 224.
132. Milne 1952, p. 58.
133. The fullest exposition was Milne 1952, published posthumously. Milne's views on the cosmos and God were as unorthodox theologically as they were scientifically. See Kragh 2004, pp. 212–229.
134. Quoted in Kragh 2004, p. 211. Likewise, when Hubble gave the Halley Lecture of 1934 on the subject of nebular redshifts, he did not refer to relativistic models of the universe. He did, however, refer to 'Professor Milne's fascinating kinematical theory of the expanding universe'.
135. Milne 1938, p. 344.
136. Milne 1935, p. 266.
137. The controversy is analysed in Kragh 1982 and Urani and Gale 1994. See also George Gale's account on http://plato.stanford.edu.entries/cosmology-30s/
138. McVittie 1940, p. 280. On McVittie as an empiricist cosmologist, see Sánchez-Ron 2005.

4

THE HOT BIG BANG

4.1 Cosmology—a branch of nuclear physics?

The Big Bang model of the universe has a curious history, as it was proposed three times, largely independently, over a period of more than thirty years. Lemaître's primeval-atom hypothesis of 1931 played no role for George Gamow in his development of a nuclear-physical theory of the early universe in the late 1940s. Likewise, when Robert Dicke, James Peebles, and others developed their version of the hot-Big-Bang theory in 1965 and the following years, they did not build on either Lemaître's or Gamow's earlier work. From a priority point of view, there can be little doubt that Lemaître was the true originator of Big-Bang cosmology; but also, there can be little doubt that it was Gamow and his collaborators who first developed the theory in a quantitative and physical way. Given that the hot-Big-Bang scenario existed in a highly developed form in 1953, and that it predicted the existence of the cold microwave radiation found in 1965, it is most remarkable that it was simply forgotten and had to be reinvented in the mid 1960s.

It should be noted that Big Bang cosmology, in whatever of the three versions, was not a theory *of* the Big Bang, but a theory of what happened after the hypothetical explosive act in which the universe supposedly came into existence. This continued to be the situation with the versions of Big Bang theory developed later in the century.

4.1.1 *Gamow's exploding universe*

From about 1940 to the early 1950s, a few physicists in the United States developed a new framework for early-universe cosmology, a theory or research programme founded on nuclear archaeology and the expansion of the universe as explained by general relativity. During this work, nuclear and particle physics was for the first time firmly introduced as an indispensable ingredient into cosmology, and the result was the first modern version of what came to be known as the Big Bang model. The key figure in this development was undoubtedly George Gamow, a Russian-born theoretical physicist who in 1933 emigrated to the United States. Gamow was a pioneer in the young and exciting field of nuclear physics, which he realized was of primary importance also to the study of the energy production in stars and the ways in which elements are formed in stellar processes. It was this nuclear–astrophysical route which led him to cosmology.[1] This was an unusual route, for most physicists and astronomers at the time came to cosmology either through the theory of general relativity or through some of the rival theories of space and time—and a few, such as Hubble, through observational astronomy.

Gamow was well acquainted with Weizsäcker's idea of a prestellar, highly compressed state of the universe, which reappeared in a paper Gamow published in 1939 together with fellow nuclear physicist Edward Teller. The two physicists concluded from the Friedmann equations for the expanding universe that the galaxies had been much closer together in the past; to understand their formation, they found it necessary to assume an infinite, ever-expanding space. Although the Gamow–Teller paper did not presuppose any explosive event in the past, it was to serve as a kind of blueprint for Gamow's further contributions to cosmology. During a conference in Washington, DC in 1942, Gamow and the other participants discussed how the heavier elements could be built up by nuclear reactions, and they concluded that this could not be accounted for by means of equilibrium processes, but required an irreversible, cataclysmic event, something corresponding to the origin of the universe. According to the conference report, 'It seems . . . more plausible that the elements originated in a process of explosive character, which took place at the "beginning of time" and resulted in the present expansion of the universe.'[2] By autumn 1945 at the latest, Gamow had reached the conclusion that the problem of the origin of the elements could only be solved by combining the relativistic expansion formulae with the rates of nuclear reactions, as he reported in a letter to Bohr of 24 October.

In a paper published in the *Physical Review* in late 1946, Gamow proposed that the elements were formed in a brief period of time in a high-density state of the early universe consisting of a gaseous soup of primeval neutrons. In a letter to Einstein, he wrote: 'It is important to remember that in order to explain the present relative abundance of the chemical elements one must agree that in "the Days of Creation" the mean density and temp. of the Universe was 10^7 gm/cm^3 and 10^{10} K.'[3] In this first attempt, Gamow imagined that the fast expansion would cause neutrons to coagulate into neutronic complexes, which by subsequent emission of beta particles would turn into the known chemical elements. In this way, he believed that the qualitative features of Goldschmidt's abundance curve might be explained, but he soon decided that the essential building-up process was instead neutron capture by protons and nuclei formed by proton–neutron reactions (the protons would be formed by decaying neutrons).

Until that time Gamow had worked alone, but subsequently he was assisted by Ralph Alpher, who, under Gamow's supervision, prepared his doctoral dissertation on the formation of elements in the primeval universe. In early 1948, Gamow and Alpher had ready an improved version of the Big Bang model which offered a new picture of the early universe and indicated the route to be followed in further research.[4]

According to the theory of Gamow and Alpher, 'nuclear cooking' had to take place within the first half hour of expansion. Its basic mechanism was neutron capture, which required a very high neutron density. The questions from where the primordial neutrons came, and what caused them to decay about two billion years ago, were left as unanswered in the Gamow–Alpher theory as they were in Lemaître's earlier theory. Right after the Big Bang, neutrons would decay into protons, and some of the protons would combine with electrons and form neutrons. With continued expansion and decrease in temperature and density, the latter process would become rare and soon stop, whereas neutrons would continue to decay with a constant rate. The protons generated from radioactive decay would then combine with neutrons to form deuterons, and higher nuclei would be built up by further neutron capture processes followed by beta decay. The nuclear building-up process was assumed to begin about twenty seconds after $t = 0$. By working out this scenario semi-quantitatively, but without making detailed calculations of the thermonuclear processes, Gamow and Alpher found a reasonably close fit to Goldschmidt's abundance curve.

In a later and more detailed paper of 1948, based on his dissertation, Alpher made full use of the data on neutron capture cross-sections (reaction probabilities) that had recently been made public by Argonne National Laboratory. He also introduced for the primordial soup the term 'ylem', an ancient name for the original substance of the world, which (unknown to Alpher) had been used much earlier by theologians, alchemists, and chemists.[5] Alpher assumed an open, ever-expanding model of the universe, but this assumption was of no crucial importance. Although he mentioned the universal expansion, it was not fully worked into the nuclear-physical calculations, where, essentially, the only adjustable parameter was the density of matter.

In his article of 1948, Alpher estimated the temperature and density of matter at the starting time of element formation to be about 10^9 K and 0.001 g/cm^3. He noted that at this temperature, the radiation density as given by the Stefan–Boltzmann law would be about 10 g/cm^3 and thus dominate over the matter contribution, but did not elaborate on the consequences. He did, however, refer to a result obtained earlier by Tolman, that whereas

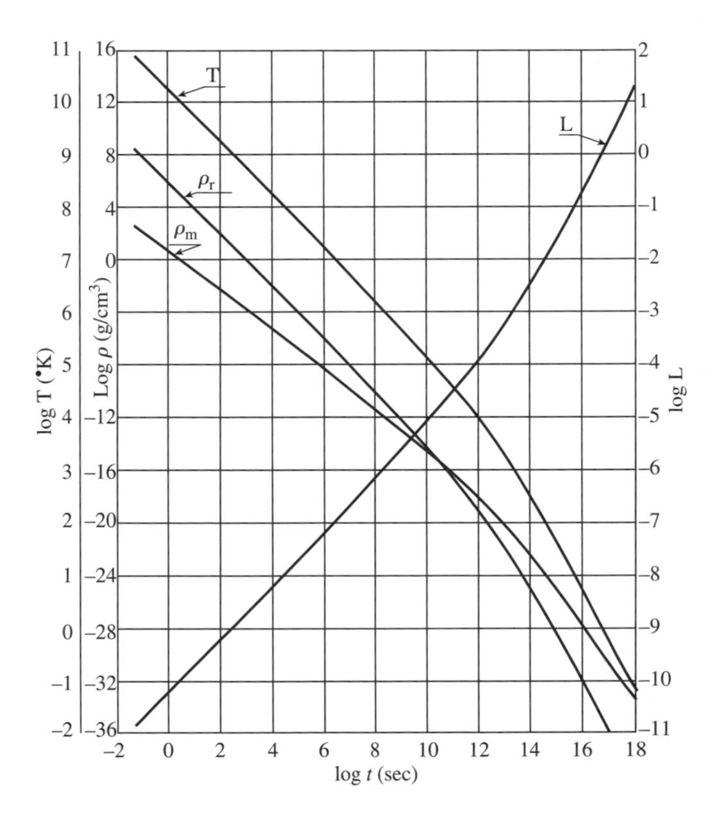

Fig. 4.1 One of Alpher and Herman's 'divine creation curves' of 1949. This figure shows the expansion of the universe (the curve marked *L*) and the variation with time of the matter and radiation densities (ρ_m and ρ_r), as well as the temperature *T*. Reprinted with permission from R. Alpher and R. Herman, 'Remarks on the evolution of the expanding universe', *Physical Review* **75** (1949), 1089–1099. Copyright by the American Physical Society.

the density of matter decreases with time as $t^{-3/2}$, in a model filled with black-body radiation the variation proceeds as t^{-2}. According to Alpher and Gamow, in the early radiation-dominated universe the temperature (in absolute degrees) would decrease as

$$T = \frac{1.52 \times 10^{10}}{\sqrt{t}},$$

where t is measured in seconds. A much fuller discussion was included in papers from 1948–51 written in collaboration with Robert Herman, a physicist who had studied cosmology under H. P. Robertson and in 1948 joined forces with Gamow and Alpher in their attempt to understand the early universe.

4.1.2 *Nuclear Big Bang theory*

1948 was a great year in the annals of cosmology. Not only did it witness the introduction of the steady-state model, it was also in this year that the hot-Big-Bang model emerged. At about the same time as Alpher was doing so, Gamow considered a radiation-dominated, hot universe and noted that since the radiation density (ρ_r) decreased faster than the matter density (ρ_m), there must have been an era when the two densities were equal. He believed this crossover or decoupling time was important because it gave the right conditions for galaxy formation, a problem which greatly interested Gamow.

In the autumn of 1948, Alpher and Hermann took up the same problem in a more careful treatment. Whereas earlier studies had dealt with either a radiation-filled or a matter-filled early universe, these authors now realized that they had to take account of both quantities in the expanding universe. In this way, they were led to the result that the product $\rho_r \rho_m^{-4/3}$ remained constant during the expansion. For the present mean density of matter, they adopted Hubble's value of 10^{-30} g/cm^3 and estimated the values at the time of element formation to be $\rho_m \cong 10^{-6}$ g/cm^3 and $\rho_r \cong 1$ g/cm^3, the latter value 'for purposes of simplicity'. As a rough value for the present radiation density, they thus got 10^{-32} g/cm^3, corresponding to a temperature of about 5 K. In the words of Alpher and Herman, 'This mean temperature for the universe is to be interpreted as the background temperature which would result from the universal expansion alone.'[6] However, they also noted that the temperature of space would increase because of starlight, thereby creating the impression that the two sources might not be observationally separable. It needs to be emphasized that the prediction of the cosmic heat radiation was due to Alpher and Herman, and that Gamow had no share in it. In fact, at first Gamow did not believe the prediction to be true, or, if it was, that the radiation could be observed. Only after several years did he take it seriously, and even then he understood it in a different way from Alpher and Herman.

The prediction of a cosmic microwave background radiation—for that is what is was—failed to attract the interest of physicists and astronomers, even though Gamow, Alpher, and Herman mentioned it in their publications seven times between 1948 and 1956. The reasons for the neglect are hard to tell, but it cannot have helped that the Washington physicists suggested values for the radiation temperature in a wide range between 3 K and 50 K (both values were due to Gamow). As early as 1940–41 the Canadian astrophysicist Andrew McKellar had suggested that a hitherto unidentified interstellar absorption line was due to a quantum transition corresponding to an excitation temperature of about 2.7 K. It was later realized that the source of excitation was the cosmic heat bath, but that insight dates from

1966, at a time when the microwave background had been detected by other means. In the late 1940s no one thought of a cosmological explanation for the excitation temperature, which thus remained for many years just one more spectroscopic fact of astrophysics. Between 1948 and 1954, Alpher and Herman made a number of attempts to get observational astronomers interested in the radiation and, if possible, to detect it, but their efforts bore no fruit.[7] There are reasons to believe that it would have been technologically possible to observe the faint microwave background even in the early 1950s, and if it had been detected, the history of cosmology might have taken a different course. But history is about what happened, not about hypothetical pasts.

The early calculations of the Gamow group did not involve detailed thermonuclear reactions. Such calculations attracted the interest of a few experts in nuclear physics, among them Enrico Fermi, who, together with Anthony Turkevich, took up the problem in 1949, but without ever publishing their results. Fermi and Turkevich considered a large number of possible nuclear reactions and arrived, after lengthy calculations, at the result that half an hour after the initial neutronic state about 24% of the mass would have transformed into helium; the rest was mainly hydrogen, with small amounts of neutrons and helium-3 nuclei. Their work seemed encouraging at first, not least because of the calculated hydrogen-to-helium ratio (about 6.7 in terms of atoms), but it failed to account for the heavier elements. The problem was the lack of atomic nuclei with mass numbers of 5 and 8, which were necessary to build up elements heavier than helium by means of neutron capture. This 'mass gap problem' deeply concerned Fermi, Gamow, and other nuclear physicists, but whatever suggestions they made, they failed to bridge the gap in a realistic manner. By 1953 at the latest it was clear that the problem was a serious one indeed, unsolved if not necessarily unsolvable. Whatever the true status of the mass gap problem, it was widely seen as a grave difficulty for the Big Bang theory, which in Gamow's version was inextricably connected with element formation. If the main purpose of the theory was to explain the distribution of elements, and if it could only account satisfactorily for helium, what good was it?

As far as helium was concerned, the Fermi–Turkevich result was in broad agreement with what the Japanese physicist Chusiro Hayashi found in 1950. Hayashi suggested an even higher initial temperature than did Alpher and Herman, and also that electrons and positrons would contribute importantly to the nuclear processes in the very early universe. He concluded that the original matter of the universe, the ylem, could be neither purely neutronic nor very much dominated by neutrons at the time element formation began, but would be composed of a mixture of neutrons and protons in the ratio 4 : 1. According to Hayashi's calculations, helium would be built up in the early inferno of protons and neutrons in such a way that the present hydrogen–helium ratio would be close to 6 : 1. Admitting the crudeness of his calculations, Hayashi found the predicted value to agree satisfactorily with the rather uncertain observational data.

Inspired by Hayashi's work, Alpher and Herman, together with James Follin, developed their theory in 1953 into a much improved version which resulted in a present hydrogen–helium ratio between 10 : 1 and 7 : 1, or a helium content between 29% and 36% by weight. However, although the prediction was impressive, it did not count much as a confirmation, because of the uncertainty in the observed amount of helium in the universe. It was to take until the early 1960s before astrophysicists could claim with some confidence

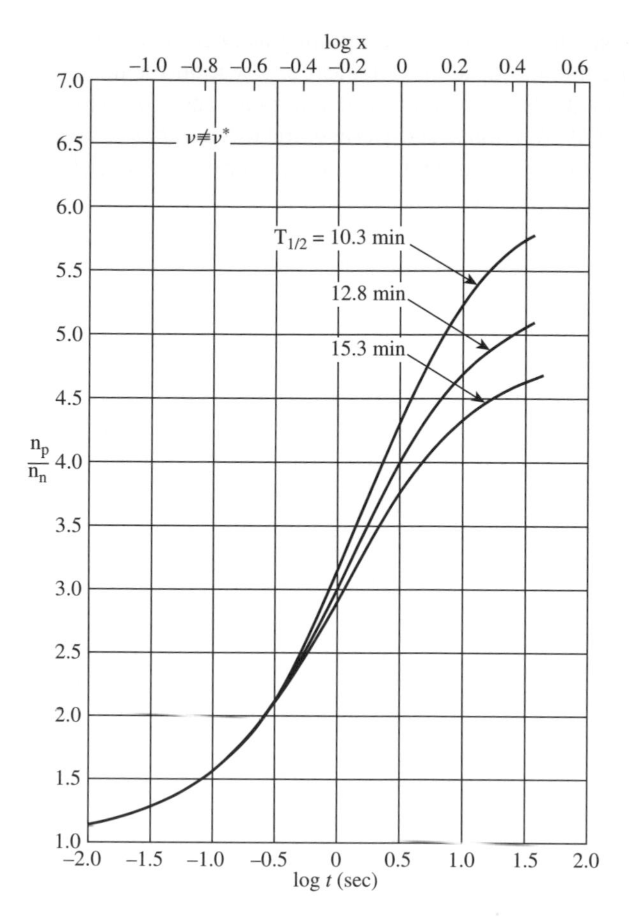

Fig. 4.2. A calculation of how the ratio of protons and neutrons would develop in the early universe. The ratio depends on the half-life of the neutron, which at the time had been measured to be about 12.8 minutes (the modern value is 14 minutes 47 seconds). The upper horizontal axis refers to the temperature ($x = mc^2/kT$). Reprinted with permission from Alpher, Follin, and Herman 1953. Copyright by the American Physical Society.

that about 30% of the matter in the universe consisted of helium, most of the rest being hydrogen.

The Gamow theory of the hot Big Bang reached its climax in the 1953 paper by Alpher, Herman, and Follin.[8] These three authors made innovative use of the most recent advances in nuclear and particle physics, for example by taking into account elementary particles such as neutrinos, muons, and pions. They started their calculations at a time only 10^{-4} seconds after the Big Bang, corresponding to a temperature of nearly 10^{13} K, and argued that for even earlier times the field equations of general relativity would not be valid. The earliest universe was no longer the ylem made up of neutrons, as imagined a few years earlier, but now consisted of photons, neutrinos, electrons, positrons, and muons, with only traces of neutrons and protons. After a few minutes' expansion, most electrons and positrons would have annihilated, leaving mainly photons and neutrinos. Alpher, Herman, and Follin made detailed calculations until 600 seconds after the Big Bang, when the half hour of nuclear genesis began. After 100 million years had elapsed, they found the temperature to have dropped to 170 K and the mass density to 10^{-26} g/cm^3. The paper was a theoretical tour de force, but most physicists and astronomers found it unconvincing. The mass gap problem remained unsolved, and the theory's only prediction, of the helium content of the universe, could not be compared with reliable observations.

4.1.3 *A failed research programme?*

The Big Bang theory of Gamow and his small group of collaborators did not arouse much attention, especially not outside the United States. After 1953 it largely ceased to be the object of further research. Given that the theory was later to be revived in the highly successful standard Big Bang theory, it is remarkable that for more than a decade only a single scientific paper was devoted to it. In Japan, Hayashi and his collaborator Minoru Nishida suggested in 1956 that if the primordial universe was as dense as 10^7 g/cm^3 three helium nuclei might combine and make possible the formation of heavier nuclei in spite of the mass gaps related to the missing elements of mass 5 and 8. The mechanism was unable to reproduce the distribution of elements, though, and was therefore of academic interest only.

Although Gamow's theory failed to attract active interest from physicists and astronomers, around 1950 it was well known and often referred to. One noteworthy example is provided by a book published that year, *Geochemistry*, the first comprehensive textbook on geochemistry written in a Western language. The two Finnish authors, Kaleva Rankama and Ture Sahama, included substantial chapters on astrophysics and cosmology related to the formation and distribution of chemical elements. They referred approvingly to the new Big Bang theory, which they found to promise an understanding of how the elements had been originally formed. 'It is generally assumed that the elements of a large part of the whole of the Universe were created during a momentary, catastrophic happening,' they wrote, exaggerating the popularity of Gamow's theory.[9]

The Gamow–Alpher–Herman theory or research programme was very much of a nuclear-physical rather than an astronomical nature. It dealt with the building up of nuclei in the early universe, whereas it did not have much to say about the later development of the universe. Correspondingly, Gamow and his collaborators published their results in the physics journals, primarily the *Physical Review*, not in the astronomy journals, such as the *Astrophysical Journal* and the *Astronomical Journal*.[10] Most astronomers were foreign or instinctively hostile to Gamow's kind of theory, which they scarcely recognized as belonging to their science, if to science at all. Whereas attempts to account for the formation of elements were accepted parts of astrophysics, theories which supposedly dealt with the creation of matter were considered with suspicion. As the French astrophysicist Evry Schatzmann put it, 'One should not introduce the idea of creation in the elaboration of the theories of formation of the elements. The problem is to study under which conditions the actually observed abundance of the elements have been produced, and not to invent a state of the universe completely different from the one of its actual state.'[11] If Schatzmann had the Big Bang theory in mind, his objections were in part misplaced. To be sure, the theory was certainly based on 'a state of the universe completely different from the one of its actual state', which was the very essence of the Big Bang idea, whether in Gamow's or

Lemaître's version. On the other hand, none of the versions dealt with the cosmic *creation* of matter or energy. Strictly speaking, they were evolution theories, not creation theories, but this important distinction was not always recognized, neither then nor later.

Gamow's style of cosmology—which was shared by Alpher and Herman—included a robust, no-nonsense approach to the study of the universe which was thoroughly permeated with instrumentalist ideas of science adopted from his work in nuclear physics.[12] When it came to the theoretical foundation of cosmology, Gamow and his collaborators were orthodox believers in Einstein's theory of general relativity, which they took to be unproblematical and applied in the form of the Friedmann equations. They intentionally disregarded alternative theories such as Jordan's and Milne's. With regard to the 'beginning', they simply started their calculations briefly after the the magical moment $t = 0$, in a pre-existing mini-universe, and they did not concern themselves with the difficult question of what 'happened' at $t = 0$.[13] Theirs was a creation cosmology of a kind, but not in the *creatio ex nihilo* sense, only in the sense of explaining matter and radiation as a creation from an earlier state.

In accordance with his non-philosophical, almost engineer's, attitude to cosmology, Gamow considered the very early universe merely to be an extremely hot and compact crucible, an exotic laboratory for nuclear-physical calculations. His approach was conservative in the sense that he saw no need to introduce new principles of physics. The known laws would do. Gamow did not deny that cosmology posed difficult problems of a conceptual and methodological nature, but he found it unprofitable to dwell on these as long as progress could be obtained by the tested methods of physics. This was the approach of what he called 'factual cosmology', which was to be contrasted with 'postulary cosmology' in the rationalist style of Milne and the steady-state theory. It is indeed possible, if of course only roughly, to place most of the cosmologists of the period from about 1935 to 1965 in a kind of methodological spectrum that ranges from extreme pragmatism to extreme rationalism.[14]

A somewhat similar classification was suggested by McVittie, who in 1961 distinguished between empiricist and rationalist schools of cosmology. Yet McVittie conceived of cosmology in quite a different light from the way Gamow did. According to McVittie, cosmology was essentially an interplay between the general theory of relativity and astronomical observations. He was an empirical cosmologist, but hardly a pragmatic in Gamow's sense, and he did not feel at home with Gamow's emphasis on the physical properties of the early universe. As he correctly pointed out, the general theory of relativity predicts no Big Bang, no nuclear explosion, it merely predicts (when supplemented with astronomical observations) that the expansion of the universe began from a state in which all matter was concentrated at a single point. Without mentioning Gamow's name, he referred scornfully to 'imaginative writers' who had woven fanciful notions such as the Big Bang round the predictions of general relativity.[15]

4.2 The steady-state challenge

At a time when Big Bang cosmology was still a somewhat immature research programme, it was challenged by an entirely different theory of the universe, the steady-state model. According to this theory, the large-scale features of the universe had always been, and would always be, the same, which implied that there was neither a cosmic beginning nor a cosmic end. The controversy between the new theory and the class of relativistic evolution theories with a finite age continued for more than fifteen years and deeply influenced the cosmological scene in the 1950s. The steady-state alternative turned out to be wrong, but no one could foresee that in the mid 1950s. It is important to realize that the steady-state theory, wrong as it is, was a serious rival and that it did much to advance cosmological knowledge in the period. Moreover, from a methodological point of view, it was highly attractive and in many ways superior to the Big Bang model. (But truth and methodological virtues are not necessarily related.)

The controversy is of particular interest because it involved a heavy dose of debate concerning fundamental questions and the philosophical foundations of cosmology as such. The stakes were high, and they forced cosmologists to take up issues that are rarely discussed in phases of normal science. In this section, the steady-state theory is followed from its birth in 1948 until about 1960, at a time when it was under pressure but still very much alive. Five years later it was practically dead, as will be recounted in the next section.

4.2.1 *An everlasting universe*

The idea of the universe being in a steady state, with energy dissipation being balanced by some kind of creative processes, was advocated by several scientists in the 1920s and 1930s, such as those mentioned in Section 3.4. MacMillan not only assumed that the distribution of matter throughout the universe was uniform on a large scale, he also denied 'that the universe as a whole has ever been or ever will be essentially different from what it is today'.[16] That is, he effectively formulated what in the later steady-state theory was called the perfect cosmological principle. In 1940, Reginald Kapp, a British professor of electrical engineering, further developed this kind of world picture in a book entitled *Science versus Materialism*. According to Kapp, the hypothesis of a universe with a finite past was alien to the spirit of science because it postulated a mythical past with processes entirely different from those observed today. To keep an everlasting and uniform universe, he not only assumed that matter was continually created but also that it continually disappeared. Kapp's theory was basically philosophical and he never developed it into a mathematically formulated cosmological model.

In spite of qualitative similarities, the steady-state theory of 1948 owed nothing to the older tradition of cosmological thinking associated with names such as MacMillan, Nernst, and Kapp. It appeared in two papers in *Monthly Notices of the Royal Astronomical Society*, written by three young Cambridge physicists. Fred Hoyle was trained in theoretical physics but had by 1940 moved into astronomy, in which science he published on a variety of subjects, at the same time as he was engaged in military radar research. In relation to this research, he came to know two young physicists of Austrian descent, Hermann Bondi and Thomas Gold, and after the war, when they returned to Cambridge, the three physicists continued to meet and discuss problems of physics and astronomy. One of these problems was relativistic cosmology, a field with which none of them had previously been occupied but which they agreed was in an unsatisfactory state. Bondi had an interest in the general theory of relativity and published in 1948 a review article on cosmology in which he stressed the principles and qualitative features of the various theories rather than their mathematical details. The article can in some respects can be seen as a methodological prologue to the steady-state theory which appeared later the same year.[17]

During their freewheeling discussions of 1947, Bondi, Gold, and Hoyle were led to the conclusion that the standard evolution cosmology based on Einstein's field equations had to be replaced by a better theory. They had no problem with the expansion of the universe, which they took for granted, but found it methodologically objectionable that the relativistic models could accommodate almost any observation, and hence had little real predictive power. Moreover, they considered the age paradox a grave problem, which it was illegitimate to get rid of by introducing special assumptions such as the cosmological constant in Lemaître's model. For scientific as well as philosophical reasons, the three Cambridge physicists wanted an unchanging yet dynamic universe in agreement with Hubble's observations. To make the two desiderata meet, they were forced to postulate that matter was continually created throughout the universe, an idea which originally came from Gold.

The discussions of Bondi, Gold, and Hoyle did not result in a joint paper but in two papers in which the steady-state theory was presented rather differently, one by Hoyle and another by Bondi and Gold.[18] Still, it was not a case of two different theories, but two different versions of the same theory. Both versions adopted as a starting point what Bondi and Gold called the 'perfect cosmological principle' and Hoyle called 'the wide cosmological principle', namely, that the universe is not only spatially but also temporally homogeneous. That is, the large-scale appearance of the universe is the same at any location *and* at any time. The perfect cosmological principle naturally implied an eternal universe and hence eliminated the age paradox. A stationary universe may seem to conflict with the recession of the galaxies, but Bondi, Gold, and Hoyle countered the problem by introducing continual creation of matter at such a rate that the mean density of matter in the universe would remain constant. This feature of steady-state cosmology was most controversial and often seen as the main characteristic of the theory, which was consequently sometimes referred to as 'continuous creation cosmology'.

It follows from simple considerations that the rate of matter creation needed to maintain a stationary universe is $3\rho H$, where ρ is the mean density of matter and H is Hubble's constant. With the values of the constants known at the time, the creation rate becomes 10^{-43} g/s/cm^3 or about three new hydrogen atoms per cubic metre per million years, much too small to be directly detectable. Neither Hoyle nor Bondi and Gold could predict in what form the new matter would appear, but for reasons of simplicity it was assumed to be hydrogen atoms or perhaps protons and electrons.

It is important to realize that matter creation in the steady-state theory was *ex nihilo*, not creation out of energy. This implies violation of one of the most fundamental laws of physics, the principle of energy conservation, a price that Bondi and Gold were willing to pay because they believed the perfect cosmological principle to be even more fundamental. Although they admitted that energy was approximately conserved, they saw no reason to conclude that it was absolutely conserved on a cosmic scale: 'In the conflict with another principle [the perfect cosmological principle] which is much more far-reaching and capable of making more statements about the nature of the universe and the applicability of physical laws, there is no reason for upholding the principle of continuity to an indefinite accuracy, far beyond experimental evidence.' Because of the continual formation of matter, new galaxies would be formed at such a rate that the density of galaxies remained constant in spite of the expansion of the universe. According to the steady-state theory, in any large volume of space there would be old and young galaxies, the ages being distributed in accordance with a certain statistical law which implied an average age of $T/3$, where T is the Hubble time.

From the assumptions of the steady-state theory followed not only the creation rate of matter but also the average density of matter, the metric of space, and the expansion rate. Hoyle constructed his version of the steady-state theory in close resemblance to the field theories of general relativity, which he did by adding a term (called the C-field) representing the spontaneous creation of matter. With this contraption, he was able to deduce an expression for the constant average density of matter in the universe. The expression happened to be exactly the same as the critical density in the Einstein–de Sitter model, $\rho = 3H^2/8\pi G$ or, numerically, about 5×10^{-28} g/cm^3. The predicted mass density was thus considerably larger than the observed density, but Hoyle argued that the discrepancy was not a problem, since only a small part of the matter might be visible and exist in the form of stars and galaxies. As to the metric and the expansion of space, it followed that the steady-state universe was flat and expanding exponentially, just as in the old de Sitter model (except that this model contained no matter). In the expression for the expansion, $R(t) = R_0 \exp(Ht) = R_0 \exp(t/T)$, the inverse Hubble constant was no longer a measure of the age of the universe but just a characteristic timescale.

Altogether, the new cosmological theory which appeared in 1948 was remarkably precise and yielded a number of definite consequences. Bondi and Gold expressed this methodological quality by contrasting it with the situation in relativistic evolution cosmology:

> In general relativity a very wide range of models is available and the comparisons [between theory and observation] merely attempt to find out which of these models fits the facts best. The number of free parameters is so much larger than the number of observational points that a fit certainly exists, and not even all the parameters can be fixed.

The scenario of the heat death obviously disagrees with the perfect cosmological principle, upon which the steady-state theory rested. Hoyle argued that there was no problem with the second law of thermodynamics, for although entropy would increase locally, the creation of matter would prevent any approach of a global heat death. In his book *Cosmology* of 1952, Bondi similarly argued that the creation process, together with the expansion of the universe, countered the increase of entropy towards a maximum value. As the universe expanded, radiation energy would be lost, but it would be replenished by the formation of new stars which would generate fresh radiation. 'High-entropy energy (in the form of radiation) is constantly being lost through the operation of the Doppler shift in the expanding universe, while low-entropy energy is being supplied in the form of matter.'[19]

4.2.2 *The reception and development of the steady-state programme*

A large part of the astronomy community responded to the steady-state theory by ignoring it. Outside Great Britain the theory received little attention, and mainstream cosmologists such as Gamow, Robertson, Tolman, Lemaître, and Heckmann rejected it without examining it seriously. They found matter creation to be preposterous and claimed that the theory, apart from being artificial, disagreed with recent observations of the reddening of galactic spectra. Joel Stebbins and Alfred Whitford, two American astronomers, reported in 1948 that they had found a reddening in the spectra of distant galaxies in excess of that expected from the ordinary, velocity-dependent redshift. Whereas this could be explained as an age effect in the evolutionary view, it seemed to contradict steady-state cosmology, where the average age of galaxies does not depend on time. Gamow, Heckmann, and other relativist cosmologists used the Stebbins–Whitford effect to discredit the steady-state theory, but in a critical review of the data Bondi, Gold, and Dennis Sciama showed in 1954 that the effect was spurious. After a couple of years' confusion, it was agreed that there was no age-determined colour effect to contradict the steady-state theory.

It was mostly in England that the steady-state theory attracted followers, and it was also here that it met with the strongest opposition. Shortly after its appearance it was criticized by Milne, McVittie, and Dingle, who argued that the hypothesis of continual creation of matter violated fundamental principles of science. On the other hand, the new cosmological theory received some support from the Astronomer Royal, Harold Spencer Jones, who was impressed by the theory's high degree of testability. As he concluded, correctly, in a review article, 'it has at least one great advantage in that it can be tested by observation without any additional hypothesis.'[20]

It was response to the opposition, rather than potentialities in the steady-state programme itself, that caused most of the development of the theory, which consequently occurred in a somewhat incoherent way. The defence of steady-state theory took place along two lines of argument. One strategy was to modify the theory in order to counter various philosophical and observational objections to it, in particular by incorporating the objectionable creation of matter within the framework of existing physical theory. Because of the rigid structure of the Bondi–Gold version, which allowed virtually no change in the original scheme, these modifications were all developments within Hoyle's field-theoretical version. The second strategy was to furnish indirect support for the steady-state theory by attempting to weaken its main rival, relativistic evolution cosmology of the Big Bang type, either by emphasizing the inadequacies of the latter or by showing that its successes could be matched by arguments based on the steady-state theory.

William Hunter McCrea, a London theoretical physicist and former collaborator of Eddington and Milne, was favourably inclined to the steady-state theory from its very beginning. In papers between 1951 and 1953, he endeavoured to develop Hoyle's theory so as to make it accord better with general relativity theory and to satisfy some kind of energy conservation.[21] Whereas in Hoyle's theory the pressure was zero, McCrea found that the theory could be reformulated by introducing a uniform negative pressure, $p = -\rho c^2$. In Hoyle's model the expansion was caused by an outward pressure produced by the created matter, whereas McCrea explained the expansion as a result of the negative pressure, which was itself unobservable because of its uniformity. McCrea's reinterpretation remained within the steady-state programme and led to the very same results as Hoyle's theory. But instead of making use of the creation process itself as the primary postulate, the introduction of a zero-point stress in space shifted the focus of the theory away from the mysterious creation of matter, which was no longer seen as a genuine *creatio ex nihilo* process, but rather as a kind of transformation. In the new interpretation, the creation process was a consequence of space being endowed with negative pressure.

McCrea's alternative version of the steady-state theory inspired several other British physicists to take up similar work and attempt to give some kind of explanation of the troublesome concept of creation out of nothing. Thus, elements of McCrea's theory were incorporated into further developments of the *C*-field theory made in 1960–68 by Hoyle and his young collaborator Jayant Narlikar. According to the extended *C*-field theory of Hoyle and Narlikar, matter creation took place without violating the principle of energy conservation, as the energy of the created particles would be compensated by the negative energy of the *C*-field. Although the Hoyle–Narlikar theory led to interesting mathematics and promised a unification of particle physics and gravitation, most scientists found it barren as a physical and cosmological theory.

In relation to McCrea's theory, Bondi and Raymond Lyttleton suggested in 1959 a remarkable cosmological theory based on the idea that McCrea's stress was a manifestation of the electromagnetic field owing to a slight inequality between the numerical charges of the proton and the electron. They found that with a charge excess of the hydrogen atom of $2 \times 10^{-18}e$ (where e is the elementary charge), the Hubble expansion law could be explained as the result of electrostatic repulsion. If matter was created out of nothing, so was electrical charge, which meant that they had to modify Maxwell's field equations. The Bondi–Lyttleton theory not only led to an explanation of Hubble's law and a derivation of the creation rate of matter in agreement with the result obtained in 1948, but it also provided McCrea's negative pressure with a physical cause, namely, the hypothetical charge excess. Bondi, Lyttleton and also Hoyle (who contributed to the idea of the electrical universe in 1960) had high hopes for a time in this ingenious theory. However, it rested on the postulate of a minimum charge excess, which turned out to disagree with experiments.[22] Consequently, the electrical version of the steady-state theory was abandoned after less than a year's life. Nothing more was heard of it. The fate of the theory is worth noting, as it may have been the first time a theory of the universe was shot down by ordinary laboratory experiments.

It was agreed that a good cosmological theory should be able to explain the formation and distribution of galaxies, a problem that was studied from the viewpoint of both of the contending theories, the relativistic evolution theory and the steady-state theory. Lemaître's theory of galaxy formation, developed in the 1930s, depended on the assumption of a non-zero cosmological constant, and Gamow developed another theory associated with his Big Bang cosmology. However, in 1956–57 the British physicist William Bonnor criticized Gamow's theory and concluded that the ordinary Big Bang theory was unlikely to lead to an understanding of galaxy formation. Within the framework of steady-state theory, a very different theory of galaxy formation was proposed by the young Dennis Sciama, who was impressed by the theory of Bondi, Gold, and Hoyle (and more so by the Bondi–Gold version) and belonged to its staunch advocates for more than a decade. Sciama's model of 1955 was based on the accretion of cosmic material by already existing galaxies, and another theory, also based on the steady-state theory, was proposed by Hoyle and Gold in 1958. By the early 1960s the general impression was that steady-state cosmology offered better explanations of the formation of galaxies than the theories based on relativistic cosmology, but also that the entire problem was too complex to warrant any definite conclusion with regard to the two rival conceptions of the world.[23]

In stark contrast to Gamow's theory, which was completely based on nuclear physics, the steady-state theory had no direct connection to nuclear particles and their interactions. But of course it was realized that the problem of element formation through nuclear reactions belonged as much to steady-state theory as to Gamow's theory. Since a cosmic origin was ruled out *ex hypothesi*, the formation of *all* elements had to take place in existing sources such as stars and novae. It was not really a cosmological problem, but one to be solved astrophysically. As mentioned, it was a stumbling block to the development of Big Bang theory that heavier elements could not be produced cosmologically, because of the mass gaps at 5 and 8. In 1952, the Austrian-American physicist Edwin Salpeter found a mechanism by means of which three helium nuclei could unite and form carbon ($3\alpha \rightarrow {}^{12}C$) at a sufficiently high rate under the physical conditions governing the interior of some stars. The cosmological significance was that the mechanism would not work under the conditions assumed to govern the early Big Bang universe.

Hoyle developed Salpeter's triple-alpha process in a paper of 1953 and started collaborative work on an ambitious theory of element formation with the American William Fowler and the two British astrophysicists Margaret Burbidge and Geoffrey Burbidge. The result of their work, the so-called B^2HF theory, was published in 1957.[24] By making use of a variety of complex nuclear processes, it gave a satisfactory explanation of the abundances of almost all the elements. Although the theory was not explicitly associated with the steady-state theory, it agreed with it, and it was definitely not a Big Bang theory, since it made no use of a hypothetical hot, compact past of the universe. For this reason, the success of the B^2HF theory was implicitly a success of steady-state cosmology and it reduced the motivation to develop the primordial theory of Gamow and his co-workers. On the other hand, the abundances of the very light elements, such as helium and heavy hydrogen, were still best explained within a Big Bang context, and so nucleosynthesis did not unambiguously distinguish between the two cosmological theories.

4.2.3 *Redshifts and other observations*

The age paradox was a major reason why Big Bang models with a zero cosmological constant were met with some scepticism. Conversely, since the paradox does not turn up in the steady-state theory, it was seen, at least in some quarters, as support for this kind of cosmology or at least a reason to take it seriously. The accepted value of the Hubble time in about 1950 was 1.8 billion years, corresponding to a recession constant of 540 km/s/Mpc, with an estimated uncertainty of 10% or so. But how reliable was Hubble's value? It rested on the calibration of the cepheid distance scale, going back to Shapley's work in the late 1910s, and this calibration was generally believed to be authoritative. That this might not be the case was first suggested by the French astronomer Henri Mineur in 1944, and seven years later Albert Behr, a German astronomer, concluded that all intergalactic distances had to be increased by a factor of about 2.2. Behr consequently estimated that the Hubble time was about 3.8 billion years, an age which 'agrees with the determinations of the age of the world from the abundance of radioactive substances and their decay products'.[25]

Behr's result did not attract much attention, and it was only when the German–American astronomer Walter Baade, working at the recently completed 200-inch Hale telescope, took up the question that things changed. Baade had left Germany for a position at Mount Wilson Observatory in 1931, and in the early 1940s he found that all the stars in the Andromeda Nebula belonged to one or other of two different classes, or populations. The problem was that Hubble had used cepheids from one population to determine the distance to Andromeda, while Shapley had used cepheids from the other population in his period–luminosity calibration. Baade therefore found it necessary to recalibrate the period–luminosity curve, and during the 1952 meeting of the International Astronomical Union in Rome he announced that the cepheids used as distance indicators had a considerably

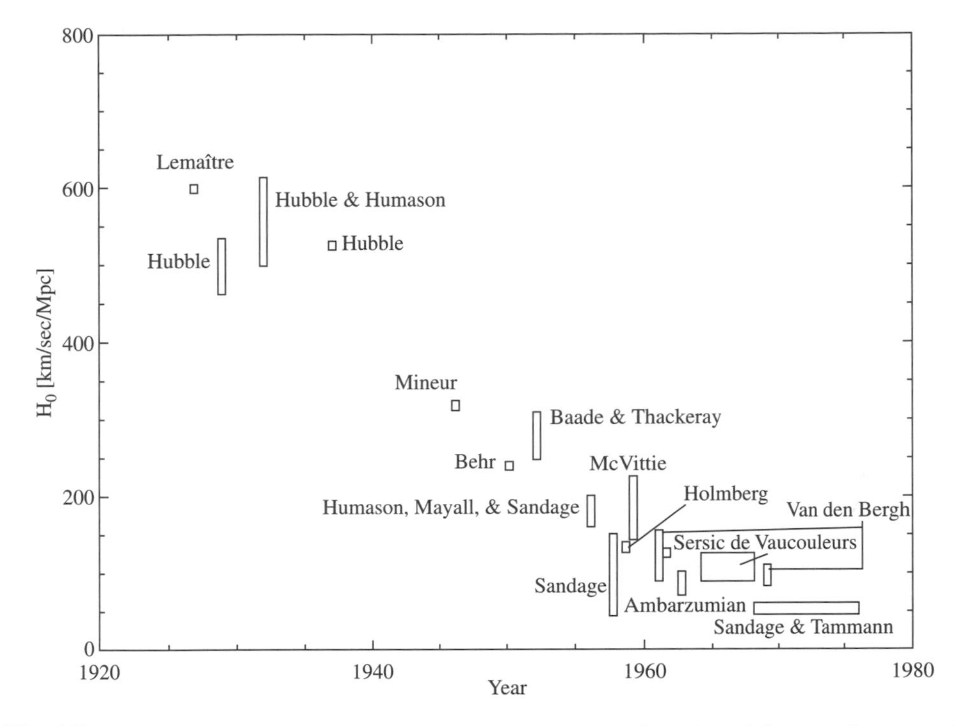

Fig. 4.3 Values of the Hubble constant from 1927 to the 1970s. The dimensions of the rectangles are rough estimates of the uncertainties. This figure originally appeared in Trimble 1996. Copyright 1996. Astronomical Society of the Pacific; reproduced with permission of the Editors.

higher luminosity than hitherto assumed. This had drastic consequences, for it implied that the cosmic distance scale, and the timescale as well, had to be doubled—as Baade concluded, 'Hubble's characteristic time scale for the Universe must now be increased from about 1.8×10^9 years to about 3.6×10^9 years.'[26]

Support for Baade's new Hubble time followed quickly, and new measurements increased it further. These were mainly due to Allan Sandage, who after Hubble's death took over the observational programme at the Hale telescope, which he gave a clearer cosmological orientation. In 1956, Sandage, Humason, and Nicholas Mayall determined the best value of the constant to be 5.4 billion years, and two years later Sandage concluded that the age of the universe, based on the flat Einstein–de Sitter model, was between 6.5 and 13 billion years.

The revised timescale was good news for the Big Bang theory, although it did not make the age paradox disappear. For one thing, the best determinations of the age of the Earth continued to increase until the value stabilized at 4.55 billion years in 1956, a result chiefly due to the Chicago geochemist Clair Patterson. Adopting an Einstein–de Sitter universe and 5.4 billion years for the Hubble time, this means a universe younger than the Earth. For another thing, most stars were agreed to be considerably older than the Earth, some of them much older. Theories of stellar evolution indicated ages of the oldest stars of more than

15 billion years, far older than even the most optimistic age of the Big Bang universe. Still, many astronomers preferred to put the blame for the discrepancy on the admittedly uncertain stellar models rather than consider it a real problem for the evolutionary universe. Whether for good or bad reasons, by the late 1950s the age paradox stopped being an important part of the cosmological controversy, and it was rarely mentioned as an argument in favour of the steady-state theory.[27]

In order to discriminate between cosmological models, and especially between relativistic expansion models and the steady-state model, analysis of the relationship between the redshifts and apparent magnitudes of galaxies (or clusters of galaxies) moved into the focus of observational cosmology. The method had roots in the work of Hubble and Tolman in the 1930s but was greatly improved in the 1950s, when the deceleration parameter was introduced as an important cosmological quantity. This dimensionless parameter is a measure of the rate of slowing down of the expansion and is defined as

$$q_0 = -\left(\frac{R''}{RH^2}\right)_0,$$

where the subscript refers to the present time. In a paper of 1958, the German astronomer Walter Mattig derived a redshift–magnitude relation which was valid for any value of the redshift and deceleration parameters. In an approximate form, it can be written as

$$m = M + 5\log(cz) + 1.0861(1-q_0)z + \text{const.},$$

where z is the redshift. The same relation, but in a different notation, can be found as early as 1942, in Heckmann's textbook *Theorien der Kosmologie*. It was further known that in Friedmann models with zero cosmological constant, q_0 is related to the space curvature, as given by the curvature constant k, by

$$\frac{kc^2}{R^2} = H_0^2(2q_0 - 1).$$

For such models, the deceleration parameter is given by $q_0 = \Omega/2$, meaning that the relation may be written in terms of the critical density Ω rather than q_0. For models with a cosmological constant, the relation is

$$q_0 = \frac{\Omega}{2} - \frac{\Lambda c^2}{3H_0}.$$

The point is that if q_0 can be found from redshift–magnitude measurements, it yields a determination of the space curvature and hence indicates which world models are ruled out and which not.[28] The deceleration parameter of the steady-state theory was as small as -1, which distinguished it from most Friedmann models. This insight was discussed in a joint paper by Hoyle and Sandage from 1956, which included the following summary:[29]

Exploding models ($\Lambda = 0$)		Steady state model
$q_0 > 1/2$;	$k = +1$	
$q_0 = 1/2$;	$k = 0$	$q_0 = -1$; $k = 0$
$0 < q_0 < 1/2$;	$k = -1$	

The idea worked nicely in principle, but what about observational practice? Passing over all the uncertainties and problems related to the redshift–magnitude test, in 1956 Sandage,

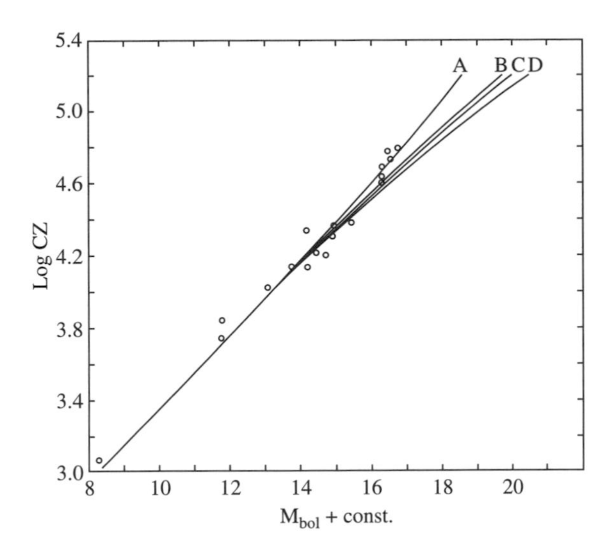

Fig. 4.4 The relationship between redshifts and apparent magnitudes for four cases, corresponding to different values of the deceleration parameter q_0. The cases A, B, and C represent Big Bang models, and case D represents the steady-state model with $q_0 = -1$. Observational data for 18 clusters of galaxies are included. This figure originally appeared in Hoyle and Sandage 1956. Copyright 1996. Astronomical Society of the Pacific; reproduced with permission of the Editors.

Humason, and Mayall reported a best value for the deceleration parameter of 2.5 ± 1. They realized that this was a provisional result but nonetheless felt confident that it ruled out the $q_0 = -1$ predicted by the steady-state model. That measurements of distant galaxies were problematical was confirmed the following year, when William Baum reported that q_0 was likely to be somewhere between 0.5 and 1.5. This was still incompatible with the steady-state theory, but Hoyle and his followers interpreted the numbers as indicating the inconclusiveness of the test, not that they refuted their favoured cosmological model. In this situation one could only hope for better, more reliable data. However, in spite of much work, it proved disappointingly difficult to get an unambiguous value for the deceleration parameter. Sandage, McVittie, and most other astronomers believed by the mid 1960s that the accumulated redshift–magnitude observations spoke against the steady-state alternative, but the alleged refutation was not clear enough to convince those in favour of the theory.

The most serious challenge to steady-state cosmology came from the new science of radio astronomy which had evolved in Great Britain since James Hey had detected the first discrete 'radio star' in 1946, fourteen years after Karl Jansky in the United States had discovered radio noise signals from the Milky Way.[30] Originally radio astronomy was not considered relevant to cosmology, but after it was recognized around 1954 that most radio sources were extragalactic, some astronomers realized that the new science might contribute significantly to the solution of cosmological problems. The leading radio astronomer of the time, Martin Ryle of Cambridge University, consciously used the 2C Cambridge survey to draw cosmological conclusions. Ryle was not a 'radio cosmologist', though, and as an observational astronomer he had little respect for the work of cosmologists. As he wrote in notes for a course he gave in about 1953,

> Cosmologists have always lived in a happy state of being able to postulate theories which had no chance of being disproved—all that was necessary was that they should work in the observable universe out to regions where the velocity is about ½*c*. . . Now we do seem to have some possibility of exploring these most distant regions. Even if we never actually succeed in measurements with sufficient accuracy to disprove any cosmological theory, the threat may discourage too great a sense of irresponsibility.[31]

The method that Ryle and his co-workers used was to count the number N of radio sources with a flux density larger than a certain value S. If it is assumed that the sources are distributed uniformly in a static flat space, the two quantities will be related according to

$$\log N(\geq S) = -1.5 \log S + \text{const.}$$

Cosmological models with different geometries and expansion rates will lead to different predictions, meaning that a number count results in a log N–log S distribution which can be compared with a straight line with slope -1.5. From such a comparison, it might be possible to infer whether the prediction agreed with observations. However, because of evolution effects, the programme could not be used for relativistic models, but only for the steady-state model, which predicted that the radio sources would lie beneath a line of slope -1.5 in the log $N-$log S plot.

Having examined the nearly 2000 radio sources in the 2C survey, Ryle and his group concluded in 1955 that the main part of the sources corresponded to a line of slope -3. This did not confirm any of the relativistic evolution models, but it strongly disagreed with the steady-state prediction, in conformity with Ryle's hope. His conclusion was unambiguous but also, it soon turned out, premature. Gold and other steady-state advocates suggested systematic errors, and they were unexpectedly supported by Bernard Mills and his group of radio astronomers in Sydney, who got results from the southern hemisphere quite different from those obtained in Cambridge. According to the Sydney group, the major part of the log N–log S curve had a slope of -1.8, and in 1958 the figure had come down to -1.65. Although the Cambridge radio astronomers had to admit that they had misinterpreted their data, Ryle pressed on with his belief in the cosmological significance of that data, which created a rift between the two groups.[32] In a paper in the not widely read *Australian Journal of Physics* from 1957, Mills and O. Bruce Slee compared the two sets of data and concluded that 'discrepancies, in the main, reflect errors in the Cambridge catalogue, and accordingly deductions of cosmological interest derived from its analysis are without foundation. . . . There is no clear evidence for any effect of cosmological importance in the source counts.'[33]

The objections made by the Australians did not mean that the steady-state theory was vindicated, but they did mean that the case for an evolutionary universe had not been proven. The confusing situation was discussed at the 1958 International Astronomical Union meeting in Paris and also at the Solvay congress held in Brussels the same year, when the topic was 'The Structure and Evolution of the Universe'. In Brussels, Bernard Lovell, the leading radio astronomer and director of the Jodrell Bank Observatory, gave an address in which he concluded that, so far, radio astronomical observations had failed to distinguish between the competing models of the universe. Most experts agreed and looked forward to more and better data.

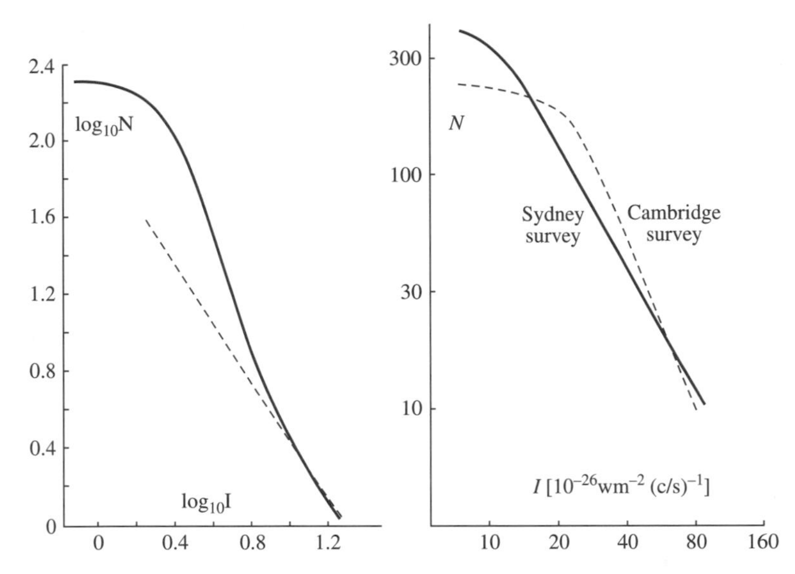

Fig. 4.5 A. C. Bernard Lovell's presentation of radio cosmology at the eleventh Solvay congress on physics held in Brussels, June 1958. The curve on the left compares the Cambridge 2C survey with the straight broken line of slope −1.5 as obtained from the steady-state prediction of a uniform spatial density of sources. The figure on the right compares the log N−log S plots for the Cambridge and Sydney surveys. From R. Stoops, ed., *La structure et l'évolution de l'univers* (Brussels: Solvay Institute, 1958), p. 189 and p. 193. Reproduced with permission of Instituts Internationaux de Physique et de Chimie.

4.2.4 *Wider issues*

A major reason why the steady-state theory became a subject of debate also in the public arena was the publication in 1950 of Hoyle's *The Nature of the Universe*, a small popular book based on a series of BBC talks given the previous year.[34] Not only did Hoyle attack the hypothesis of the exploding universe (for which he coined the name 'Big Bang'), but he also drew consequences from the two rival cosmological systems to the realms of ethics, politics, and religion. Hoyle did not hide his disdain for organized religion, and suggested that there existed an unholy alliance between Christianity and the Big Bang theory; conversely, there was no room for Christian belief in a universe governed by the steady-state theory. Understandably, his provocation caused reactions and much concern in the British religious community and among Christian scientists. Hoyle and a few other cosmologists, including William Bonnor, intimated that theories of the Big Bang type were implicitly religious as they relied on a first event—a miracle—that only made sense if caused by a supernatural being. Steven Weinberg, the leading particle theorist and contributor to Big Bang cosmology, asserted in 1967 that 'The steady state theory is philosophically the most attractive theory because it *least* resembles the account given in Genesis. It is a pity that the steady state theory is contradicted by experiment.'[35]

The accusation of an alliance between Big Bang cosmology and Christian theology was generally unfair, but it was not without substance, as was illustrated by an address that Pope Pius XII gave to the Pontifical Academy of Sciences in Rome in the autumn of 1951.

Reviewing recent developments in astrophysics and cosmology, the Pope concluded that modern cosmologists had arrived at the same truth that theologians had known for more than a millennium. The scientific picture of the development of the universe was in complete harmony with Christian belief. Modern science, the Pope claimed, 'has confirmed the contingency of the universe and also the well-founded deduction as to the epoch when the world came forth from the hands of the Creator. Hence, creation took place. We say: therefore, there is a creator. Therefore, God exists!'[36] Of course, this direct association between scientific theory (Big Bang cosmology) and religious belief (Christian creation theology) was questioned by many theologians and even more scientists.[37] Lemaître, for one, was unhappy with the address of his pontiff. He denied that there was any direct connection between a particular cosmological model and what Christian religion was really about. Yet in the 1950s many scientists, theologians, and commentators felt it natural to engage in discussions concerning the relationship between cosmology and religion. Although this is an interesting chapter of cultural history, the many discussions did not influence the scientific development of the rival cosmologies to any extent.

Much of the discussion concerning cosmology in the fifteen or so years following 1948 was of a philosophical nature, in particular concerning methodological issues. This was at a time when it was still a matter of some debate whether cosmology should be counted as a science in the first place. And, if it was, which of the two main conceptions was the most scientific? According to what criteria? In the early phase of the controversy, Dingle resumed the crusade against rationalistic cosmo-physics which he had fought in the 1930s against Milne and Eddington. He tended to see cosmology as such as scientifically dubious and was particularly annoyed with the steady-state theory and its foundation in the perfect cosmological principle. Such an a priori principle (as he claimed it was) had nothing to do with science, he charged, and would lead to nothing but the 'unscientific romanticizing' that characterized the Bondi–Gold–Hoyle conception of the universe. Other philosophers, including Milton Munitz and Mario Bunge, focused on the spontaneous creation of matter out of nothing, a problematic feature of the steady-state theory which they found difficult to reconcile with accepted standards of science.

The lack of professional maturity in cosmology was a major issue in a public discussion of 1954 in which Bondi and Gerald Whitrow discussed 'Is Physical Cosmology a Science?' in the pages of the *British Journal for the Philosophy of Science*. Whitrow, a theoretical astronomer with philosophical interests, argued that cosmology was not truly scientific and that it was unlikely ever to become so. Because of its peculiar object, the universe, it was destined to remain on the borderline between science and philosophy. Bondi, on the other hand, suggested that cosmology had already become a respectable science, a field based on observations and mathematical models which had largely replaced the philosophical arguments that were so important in the past. He stressed that the criteria for choosing between theories were the same as in the other physical sciences, namely the rejection of a theory whose predictions disagreed with observations. According to Bondi, a cosmological theory must be falsifiable, and the higher degree of falsifiability the better. By this standard, he found the steady-state theory to be eminently scientific, much more so than the Big Bang alternative.

Bondi's view of cosmology, and to some extent also those of other steady-state theorists, was directly inspired by Karl Popper's philosophy, which presented science as a hypothetical–deductive system and highlighted the falsification (rather than verification) of hypotheses as the hallmark of science. In a 1959 review of Popper's main work *The Logic of Scientific Discovery*, Bondi linked Popper's falsificationist philosophy directly to the controversy in cosmology: 'For here the correct argument has always been that the steady state model was the one that could be disproved most easily by observations. Therefore, it should take precedence over other less disprovable ones until it has been disproved.'[38] On several later occasions, Bondi described himself as a follower of Popper and praised his philosophy of science.

Bondi's claim that philosophical, foundational, and aesthetic arguments no longer played a role in cosmology was wishful thinking, as is evident from the literature of the 1950s, where such arguments occurred abundantly. Bonnor, who was in favour of relativistic evolution theory (but not the Big Bang version), objected that steady-state cosmology was merely a phenomenological, not an explanatory theory. To Bonnor, this was a grave deficiency, but to Bondi it was a necessity of any cosmological theory, because the object of such a theory, the universe, is unique and cannot be subjected to law-like explanation like other objects of science. As Bonnor and Bondi disagreed over the role of explanations in cosmology, so they disagreed over the meaning and function of simplicity. Bonnor felt that the perfect cosmological principle was complex and uneconomical, designed for a particular model of the universe, whereas Bondi defended the principle. Characteristically, he did so in terms of Popper's philosophy:

> It is the purpose of a scientific hypothesis to stick out its neck, that is to *be* vulnerable. It is because the perfect cosmological principle is so extremely vulnerable that I regard it as a useful principle. . . .[Bonnor's] views of what constitutes science must differ markedly from mine. I certainly regard vulnerability to observation as the chief purpose of any theory.[39]

Popper's ideas regarding testing and demarcation may have been favoured in particular by advocates of the steady-state theory, but their appeal was not limited to this group. Ever since the 1950s, Popperian philosophy has played an important role among astronomers and cosmologists, sometimes explicitly and sometimes implicitly. Several modern cosmologists have felt attracted by Popper's emphasis on risky and falsifiable hypotheses, which to some extent has guided their scientific practice.[40]

There is little doubt that philosophical and even subjective factors played a role in cosmology in the early 1950s, although it is hard to tell how important they were. Certainly, not all scientists engaged in cosmological research shared Bondi's view that it was now a matter of comparing theory with observations. According to Martin Johnson, a British physicist inspired by Milne's line of thinking, the choice between cosmological models was 'an aesthetic or imaginative choice'. The Estonian–Irish astronomer Ernst Öpik disliked the steady-state theory, which he believed 'can at present be addressed only [1954] from the standpoint of esthetic value'. To mention but one more example, the Swedish physicist and cosmologist Oskar Klein stated in 1953 that cosmology was a field where 'personal taste will greatly influence the choice of basic hypotheses'. As to the perfect cosmological principle, it was just one of those basic hypotheses, 'a matter of taste'.[41]

Cosmology in the first few decades after the Second World War was not confined to the controversy between Big Bang and steady-state theories, nor to attempts at understanding the large-scale features of the real universe. A couple of examples will suffice to illustrate the variety of cosmological thinking in the period. In a paper of 1949, the Austrian–American mathematician and logician Kurt Gödel studied an anisotropic solution to the cosmological field equations which corresponded to a universe rotating around some centre with a period of about 10^{11} years.[42] The surprising idea of a rotating universe had first been suggested by Gamow in 1946, who pointed out that one might in this way avoid the initial singularity, but Gödel's proposal seems to have been independent of Gamow's. The model of 1949 was remarkable in several respects. Gödel did not make use of a common cosmic time, and he assumed a negative cosmological constant corresponding to a positive pressure ($\Lambda \sim -\kappa\rho$). The most surprising feature of the model was that it allowed round trips in time, including causality-violating space travels into the past, with all the fascinating paradoxes that follow. Gödel realized that his 'toy model' yielded no redshift and for this reason could not be a candidate for the real universe, but he suggested that 'it is not impossible' that our world was an expanding version of the rotating model. His science-fiction-like solution has been further investigated by later cosmologists, and of course it has also attracted a good deal of philosophical attention.

The cosmological theories of the 1950s and 1960s had obvious implications for astronomical and astrophysical problems. It is less well known that there were also connections to geology, a science which, after all, is very different from cosmology. Jordan believed that his varying-G cosmology had important geophysical consequences, and in his *Schwerkraft und Weltall* of 1952 he even used it to explain Alfred Wegener's still controversial hypothesis of drifting continents. Jordan later developed his views into an elaborate geophysical theory of an expanding Earth, which he also used to explain volcanoes, lunar craters, the ice ages, the formation of planets, etc.[43]

The idea of an expanding Earth was independently proposed by the Australian geologist Samuel Warren Carey, who argued that it offered a better explanation of the surface of our globe than the plate-tectonical theory that emerged in the 1960s. In order to justify the hypothesis physically, he drew on cosmologies of the Dirac–Jordan type, which, in his view, allowed a continual production of new matter in the interior of the Earth. Cosmology was an integrated part of Carey's wide-ranging attempt to reform the Earth sciences, although in an unorthodox form that combined features of the Dirac–Jordan hypothesis with the steady-state theory. 'The implications of Hubble's law', he wrote retrospectively, 'forced me to the conclusion that everything in the universe has suffered the same accelerating increase in mass. Therefore to understand the Expansion of the Earth, we must seek to understand the expansion of the universe.'[44] Carey's unorthodox cosmo-geophysical theory attracted considerable interest in the 1960s but never succeeded in seriously challenging the successful theory of plate tectonics and sea-floor spreading. With the advent of standard Big Bang cosmology, his theory became even more unorthodox.

If the Gödel universe deserves the science fiction label, so do 'wormholes', hypothetical tunnels or bridges that short-cut one region of space–time to another. The wormhole phenomenon is described by certain solutions to Einstein's field equations. These were first studied by the Austrian physicist Ludwig Flamm as early as 1916, and in a paper of 1935 Einstein and his collaborator Nathan Rosen examined the solutions in a different context, namely, with the aim of constructing an atomistic theory of matter on the basis of the gravitational and electromagnetic equations. During this work they discovered the 'Einstein–Rosen bridge', which they identified with an elementary particle. The name 'wormhole' was introduced by John Wheeler in 1955, at a time when bridges in space–time were still thought to be a mathematical curiosity. In the late 1980s Kip Thorne and others began to examine new kinds of wormholes which were 'traversable', meaning that hypothetical objects could move along them, going from one universe to another and also back and forth in time.[45] Naturally, science fiction writers loved the idea.

4.2.5 *Cosmology and political ideology*

Until about 1960, research into cosmological models was largely limited to Western Europe and North America. In the Soviet Union, the pioneering work of Friedmann was not followed up to any extent, and after the mid 1930s the political situation made it increasingly difficult to work with cosmology in the Western style.[46] Communist ideologues and scientists considered the very application of physical theories to the universe as a whole to be suspect and against the spirit of dialectical materialism, the state philosophy based upon the thoughts of Marx, Engels, and Lenin. In conformity with the view of socialists and materialists in the late nineteenth century (compare Section 2.4), it was claimed that the universe must necessarily be infinite in space as well as time. Models such as Milne's and Lemaître's Big Bang universes were particularly dangerous—a 'cancerous tumor that corrodes modern astronomical theory and is the main ideological enemy of materialist science', as one Soviet astronomer warned in 1953.[47] Of course, it didn't help that Lemaître was a Catholic priest and that the Pope had recently supported the Big Bang theory as evidence for the truth of the Bible.

Soviet physicists and astronomers in the Stalinist era did not generally see the theory of the expanding universe to be ideologically problematical, but models of a finite age were either ignored or categorically rejected. They were *teoria non grata*. According to official Soviet astronomy, the models proposed by Lemaître, Milne, and Gamow were nothing but bourgeois mythology, 'astronomical idealism, which helps clericalism', as expressed in a declaration of December 1948.[48] Other astronomers tended to reject cosmology as such, in whatever model it was dressed. The eminent astrophysicist Victor Ambarzumian argued that cosmology proper was unscientific, a myth which relied on unjustified extrapolations of observations and theories based on the empirically accessible part of the universe. If the Big Bang theory was considered to be reactionary and bourgeois, the rival steady-state theory did not fare better. The theory of Hoyle and his allies was deemed politically incorrect because it operated with continual creation of matter.

Although there was no official ban on cosmology—it wasn't needed—the effect of the ideological pressure was that for a couple of decades cosmology was practically non-existent in the Soviet Union. In 1947, the chief ideologue Andrei Zhdanov explicitly warned against the reactionary 'fairy tale' of the finite-age universe. Soviet astronomers by and large conformed to the dogma of the authoritarian communist regime by giving up the study of the universe as a whole. The situation only changed in the early 1960s, when Soviet cosmology under the leadership of Yakov Zel'dovich emerged as a vigorous research programme within international cosmology.

The sad story of cosmology's clash with communist ideology did not end with the Stalinist era in the Soviet Union. In the People's Republic of China, left-wing Maoist ideologues developed their own version of dialectical materialism which implied that modern cosmology became a forbidden science in Mao Zedong's empire. Fang Lizhi, a Chinese physicist, got caught up in the frenzy of the Cultural Revolution, starting in 1966 and ending about a decade later. After having spent a year in jail he began to pursue astrophysics and in 1972 he published a theoretical paper on Big Bang cosmology, the first of its kind in the People's Republic. Enraged radical Marxists rallied against Fang's heresy and its obvious betrayal of the true spirit of dialectical materialism and proletarian science. Between 1973 and 1976, some thirty papers were published against the Big Bang theory. An article of 1976 in *Acta Physica Sinica* summarized what was wrong with this kind of theory:

> Materialism asserts that the universe is infinite, while idealism advocates finitude. At every stage in the history of physics, these two philosophical lines have engaged in fierce struggle . . . with every new advance in science the idealists distort and take advantage of the latest results to 'prove' with varying sleights of hand that the universe is finite, serving the reactionary rule of the moribund exploiting classes . . . We must ferret out and combat every kind of reactionary philosophical viewpoint in the domain of scientific research, using Marxism to establish out position in the natural sciences.

The notion of a finite universe born ten billion years ago was 'linked up with all sorts of idealist philosophy, including theology'. The party line was to deny cosmology scientific legitimacy, much as materialists and positivists had argued in the nineteenth century. Questions about the universe at large could not be answered scientifically, only on the basis of the 'profound philosophical synthesis' of Marxism–Leninism:

> The dialectical-materialist conception of the universe tells us that the natural world is infinite, and it exists indefinitely. The world is infinite. Both space and time are boundless and infinite. The universe in both its macroscopic and microscopic aspects is infinite. Matter is infinitely divisible.[49]

Fortunately, the campaign against Fang and Big Bang cosmology came at a time when the Cultural Revolution was in decline, and consequently it did no lasting harm. In the autumn of 1975, when Fang and his colleagues were allowed to defend themselves, they stated that 'whether the big bang is a correct theory or not, recent developments such as radiotelescopy had made cosmology an experimental science, to be approached through the usual scientific methods rather than through philosophical discourse'.[50] Fang's troubles with the political authorities were not over, though. He developed into China's most prominent political dissident, the country's parallel to Sakharov in the Soviet Union. In 1987 he was expelled from the Chinese Communist Party for the second time, and in the turmoil following the Tiananmen Square massacre of June 1989 he took refuge in the US Embassy in Beijing to avoid being arrested as a traitor and class enemy.

4.3 Relativistic standard cosmology

The 1960s was a decisive decade in the history of modern cosmology. With improved data for the distribution of radio sources and the new quasars, and particularly with the discovery of the cosmic microwave background, steady-state cosmology was largely abandoned and left the cosmological scene to the now victorious hot-Big-Bang theory. By the end of the decade, this theory, consisting of a large class of models sharing the assumption of a hot, dense beginning of the universe, had become a standard theory accepted by a large majority of cosmologists. In fact, it was only from this time that cosmology emerged as a scientific discipline and 'cosmologist' appeared as a name for a professional practitioner of a science, on a par with terms such as 'nuclear physicist' and 'organic chemist'. Although rival cosmologies did not disappear, they were marginalized. Not only was the Big Bang now taken to be a fact, rather than merely a hypothesis, it was also taken for granted that the structure and development of the universe were governed by Einstein's cosmological field equations of 1917.

In spite of the new-born confidence in relativistic Big Bang cosmology, it was realized that there were still many unsolved problems. One of them was the nature and status of the initial singularity; another was the amount and distribution of matter in the universe. To the astronomers' surprise, it turned out that there was more dark matter in the universe than visible matter. This was a puzzle, but it was good news to those cosmologists who were in favour of a universe which was nearly critical in density. By the end of the 1970s it seemed that a critical, flat universe without a cosmological constant might be a good candidate for the real world.

4.3.1 *Radio waves and microwaves*

In the late 1950s, the situation in radio cosmology was undecided. Martin Ryle's original claim that the 2C data contradicted the steady-state theory was weakened by the discordant data from the Sydney radio astronomers, who concluded that the slope of the log N–log S curve was -1.65, a value not grossly conflicting with the steady-state prediction of -1.5. With results from the new 3C survey and complementary results from the southern hemisphere, a growing consensus emerged and it appeared that radio astronomy was, after all, capable of distinguishing between cosmological models. The point of no return was reached in the beginning of 1961, when Ryle presented improved data that were clearly incompatible with the steady-state theory, even when the largest permissible variation in source luminosity was included. Unlike the 2C survey, the new Cambridge results remained stable and were not seriously questioned by other radio astronomers. By 1963 Ryle had determined the slope of the log N–log S diagram to be -1.8 ± 0.1, which agreed excellently with what the Sydney group found the following year, namely -1.85 ± 0.1. From this time at the latest, there was consensus among specialists in radio astronomy that the slope could not possibly be -1.5, the favoured value from the point of view of the steady-state theory.

The radio-astronomical consensus did not kill the steady-state theory, but it weakened it considerably and left the theory in a bad shape, as a rather unattractive alternative to the relativistic evolution models. As to these models, they were only indirectly supported by the number counts of radio sources, namely by ruling out the steady-state theory. Radio astronomy did not help in determining which of the evolution models was the best candidate for the real universe, except that some of the relativistic models, such as the flat-space Einstein–de Sitter model, were compatible with the radio data. By and large, the cosmological use of the radio source counts was restricted to the negative level, to rule out the steady-state theory.

The long waves from radio sources were not the only kind of electromagnetic radiation that made life difficult for advocates of the steady-state theory. Although the 1948 Alpher–Herman prediction of a cosmic microwave background was ignored, it was not completely forgotten.[51] In 1963 two Russian astrophysicists, Andrei Doroshkevich and Igor Novikov, discussed some observations of reflected microwave signals made by Edward Ohm, a physicist employed by Bell Laboratories. Ohm had found an excess temperature of 3.3 K in his antenna, a result Doroshkevich and Novikov related to the microwave background that was to be expected according to the hot-Big-Bang theory of Gamow and his collaborators. However, the two Russians misunderstood parts of Ohm's report and consequently concluded that if Ohm had unknowingly detected a cosmic microwave background, it must have a temperature close to absolute zero. They discussed the matter with their senior colleague Yakov Zel'dovich, but nothing further happened and no Soviet experiment was set up to detect the radiation.

The discovery of the cosmic background radiation was serendipitous, in the sense that the original experiments were not aimed at cosmological questions and were performed by two scientists who did not even know about the possibility of a fossil radiation from the Big Bang. Arno Penzias and Robert Wilson, two other researchers at Bell Laboratories, had in 1963 started radio-astronomical work with a horn antenna originally designed for communication purposes. During this work, they noticed to their surprise an excess temperature in their antenna of some 4 K, which they could neither understand nor get rid of. The noise, or surplus radiation, was the same in all directions and therefore indicated a cosmic origin, but Penzias and Wilson had no idea of what this meant. It was only in March 1965 that they came to hear of work going on in Princeton University that offered an explanation of the excess temperature. Then, and only then, did they realize that they had made an important cosmological discovery. Thirteen years later they were awarded the Nobel Prize in physics for their discovery, 'after which cosmology is a science, open to verification by experiment and observation', as it was phrased in the presentation speech in Stockholm.

The Princeton physicist Robert Dicke had an interest in cosmology and general relativity, but was unaware of (or had forgotten about) the prediction of Alpher and Herman. In about 1963, while thinking of the consequences of an oscillating universe with many big bangs and big squeezes, he concluded that there might exist today a relic black-body radiation that originally had been very hot but had cooled off as the universe expanded. His idea was that during the collapse phase starlight would be shifted to the blue, becoming more energetic, and part of the blueshifted light would photodissociate the elements formed in the stars, leaving a fresh supply of hydrogen after the bounce.

In 1964 Dicke suggested to James Peebles, a 29-year-old Canadian-born physicist and former student of his, to look at the problem and calculate the properties of the assumed radiation. Peebles's first answer was a temperature of about 10 K, a result he arrived at without knowing of the earlier work of Alpher, Herman, and Gamow. At that time, in the early months of 1965, Dicke and Peebles started a collaboration with their Princeton colleagues Peter Roll and David Wilkinson, who had constructed a radiometer to measure thermal radiation at a wavelength of 3 cm (Penzias and Wilson had used 7.4 cm). When they learned about the Penzias–Wilson excess temperature, they realized that the cosmic background radiation had already been found. The Bell and Princeton physicists published their works as companion papers in the July 1965 issue of the *Astrophysical Journal*. Penzias and Wilson reported their finding of an excess temperature of 3.5 ± 1.0 K at $\lambda = 7.3$ cm, without mentioning its implication for cosmology. This was left for the four Princeton physicists, who argued that the observed radiation was indeed part of the black-body radiation remaining from the primordial decoupling of matter and radiation.

The Big Bang interpretation of the Penzias–Wilson experiment created a sensation and was quickly accepted by the majority of astronomers and physicists. Still, to count as a full confirmation, and conversely as a refutation of the steady-state theory, the background

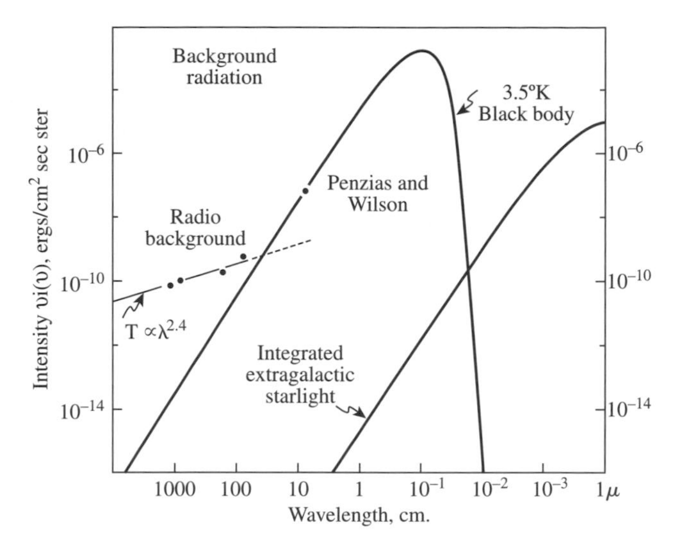

Fig. 4.6 Spectrum of the cosmic microwave background radiation, as presented by Peebles in a talk of March 1965. This was shortly after he had come to know of the Penzias–Wilson experiment, which had not yet been published. Reproduced from Peebles 1993, © 1993 Princeton University Press, p. 148. Reprinted by permission of Princeton University Press.

radiation would have to be detected at more than a single wavelength. Only if the spectrum was in fact black-body-distributed and isotropic would it confirm the predictions of the Big Bang theory and vindicate the early work of Alpher and Herman. The first confirmation came in early 1966, when Roll and Wilkinson reported a measurement of a 3.0 K radiation at a wavelength of 3.2 cm, and later the same year it was pointed out that the energy of the background radiation corresponded to the excitation energy of CN molecules that McKellar had measured back in 1940. During the following years, the spectrum was pieced together by many individual measurements. The result was a confirmation of the black-body distribution and a more precise value of the background temperature, which by the mid 1970s had been narrowed down to 2.7 K.

As the black-body shape was established by accumulated data, so did sensitive measurements rule out any discrete sources or other fine-scaled anisotropy of the background radiation. It was, however, realized that the background radiation should exhibit a kinematic anisotropy due to the motion of the Earth (as part of the Milky Way) relative to the average rest frame of the universe. By the end of the 1970s this anisotropy, corresponding to a motion of the Earth relative to the radiation at a velocity of 390 km/s, had been detected in experiments using balloons and aeroplanes.

How did the steady-state theorists respond to the bad news from radio source counts and cosmic microwaves? In general, none of the leading advocates of steady-state cosmology admitted that the new discoveries amounted to a refutation of their favoured theory. As to the data from radio astronomy, they either questioned the reliability of the measurements or produced alternative explanations within the steady-state framework. For example, Sciama developed in 1963 a model based on the assumption that about half of the radio sources were located inside the Milky Way, and in this way, and by making use of a number of additional hypotheses, he managed to reproduce the observed slope of −1.8 without admitting an evolving universe. Almost all astronomers found his theory to be artificial and ad hoc, a desperate and unnecessary attempt to deny the obvious—that the universe was evolving. All the same, in an article of 1965 Sciama boldly maintained that 'the steady-state model remains in the field, bloody but unbowed'.[52]

The response to the discovery of the microwave background in 1965 followed a similar path, first by questioning whether the waves were really isotropic and black-body-distributed and next by devising alternative explanations consistent with a steady-state universe. From the mid 1960s onwards Hoyle, partly in collaboration with Narlikar and Chandra Wickramasinghe, developed an alternative theory based on the idea of thermalization, the assumption that a major part of the stellar radiation was somehow converted into long-wavelength components of a black-body-like form. If so, the microwave background would not be cosmic, but starlight in disguise. Hoyle and his colleagues argued that the thermalizing mechanism consisted of tiny interstellar grains, but the hypothesis attracted little interest. The same was the case with a model Sciama proposed in 1966, which was based on the assumed existence of a new kind of radio source that collectively would produce the observed background radiation. Whatever alternative the steady-state theorists came up with, they failed to convince the large majority of astronomers and physicists that the theory was still alive and worth considering.

The determination of the temperature of the cosmic microwave background had important implications for the calculations of nuclear processes in the early universe leading to the formation of helium and other light elements. In 1966 Peebles calculated the helium abundance on the assumption of a radiation temperature of 3 K and arrived at 26–28% helium, depending on the value of the present density of matter. The result agreed

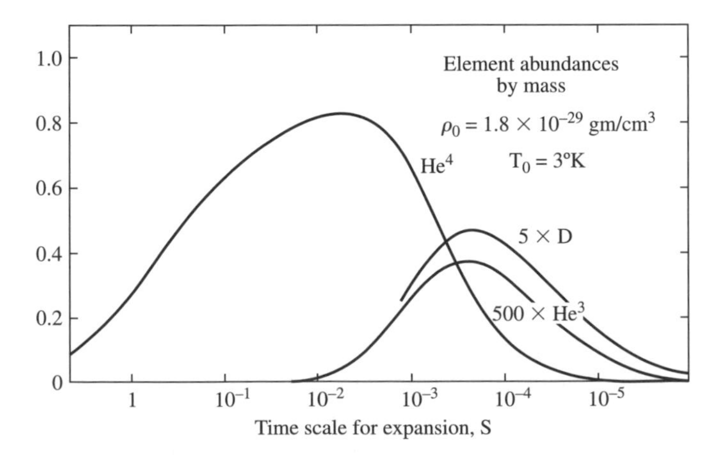

Fig. 4.7 One of Peebles's calculations of 1966, showing the formation of light atomic nuclei in the early universe. From P. J. E. Peebles, 'Primordial helium abundance and the primordial fireball, II', *Astrophysical Journal* **146** (1966), 542–552. Reproduced by permission of the American Astronomical Society.

excellently with observations, which Peebles took to be one more confirmation of the hot-Big-Bang model (or 'fireball' model, as he and Dicke called it).

Even before the discovery of the cosmic microwave background, Hoyle and Roger Tayler had found that a helium abundance of this magnitude required physical conditions corresponding to a hot Big Bang, whereas stellar processes would not do. However, in their 1964 paper, they did not conclude in favour of the Big Bang: Hoyle preferred the alternative of hypothetical, supermassive objects, which would result in similar amounts of helium. Much more detailed calculations of nucleosynthesis were published in 1967 by Hoyle,

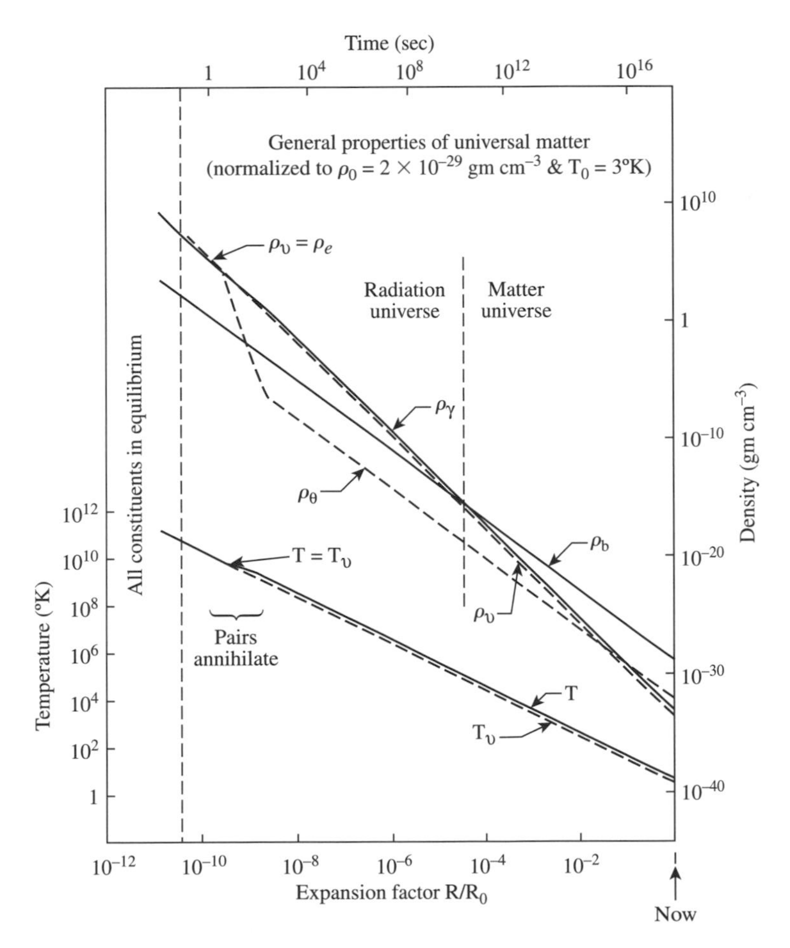

Fig. 4.8 The early universe according to Wagoner, Fowler, and Hoyle. Ironically, although this 'divine creation curve' provided a fine representation of Big Bang cosmology in 1967, Hoyle dismissed the theory. The figure shows the variation of the densities (matter and radiation) and the temperature of the universe (photons and neutrinos) during the early phase of expansion. A similar curve was given by Dicke, Peebles, Roll, and Wilkinson in 1965, and by Alpher and Herman in 1949 (compare Fig. 4.1). From R. Wagoner, W. A. Fowler, and F. Hoyle, 'On the synthesis of elements at very high temperatures', *Astrophysical Journal* **148** (1967), 3–49. Reproduced by permission of the American Astronomical Society.

Fowler, and Robert Wagoner, who concluded that under Big-Bang-like conditions the observed abundances of the light elements could be reproduced satisfactorily, but that most heavier elements had to be the results of stellar processes.[53] Hoyle continued to favour supermassive objects over the Big Bang, but few followed him. As calculations were further refined, a consensus emerged that deuterium and the helium isotopes had their origin in the Big Bang from which the universe evolved. By 1970, the range of helium abundance by mass had been computed to be between 22.5% and 27.5%, depending on the mass density and the exact half-life of the neutron.

4.3.2 *Quasars and other cosmic novelties*

If quasars are to be assigned a definite discovery date, 5 February 1963 is a good choice.[54] On this day the Dutch–American astronomer Maarten Schmidt studied the unusual spectrum of a recently identifed 'radio star' known as 3C 273 (that is, number 273 in the 3C Cambridge survey). The line spectrum puzzled him until he realized that some of the lines were spaced like the familiar Balmer lines in the hydrogen spectrum, only shifted towards the red end. Schmidt concluded that it was indeed a Balmer spectrum redshifted by 16 per cent, which implied that the faint star had a recessional velocity of almost 48 000 km/s. He and his colleagues Jesse Greenstein and Thomas Matthews subsequently turned their attention to the radio source 3C 48, which Sandage had found as early as 1960, and they similarly explained its spectrum as arising from an object of redshift $z = 0.37$.

During the next few years several other star-like objects with high redshifts were identified. These strange objects were referred to as quasi-stellar radio sources or *quasars*, a name which dates back to 1964. In fact, quasars have nothing to do with stars and emit most of their energy outside the radio range, but this was not known at the time. The new heavenly objects immediately caught the attention of the astrophysicists, who sought to understand how they could emit such enormous amounts of energy. If their redshifts were of cosmological origin—due to the expansion of the universe—they were very distant objects with a power much larger than the power of giant galaxies. Could this amazing output of energy be explained by ordinary nuclear processes? Or perhaps by gravitational collapse of massive superstars? (The latter mechanism had been discussed by Hoyle and Fowler shortly before Schmidt's discovery in 1963.)[55]

The cosmological significance of quasars was related to their great distances rather than their great energy output. If the redshifts were really cosmological, these quasars must have existed only very far away and long ago, which contradicts the perfect cosmological principle, the very heart of steady-state cosmology. The question was discussed in the years around 1965, and most astronomers agreed that there was compelling evidence that quasars were indeed at cosmological distances, more than about 300 Mpc away. Moreover, blueshifted quasars were conspicuously absent, suggesting that they could not be nearby objects with redshifts due to some local velocity effect.

If the redshifts were due to the cosmic expansion, some kind of Hubble relation (or relation between redshifts and magnitudes) would be expected. In a paper of 1966 Sciama and his student Martin Rees, the later Astronomer Royal, examined data for 35 quasars and concluded that their distribution in a redshift flux density diagram disagreed with the prediction of steady-state theory. Sciama took the result to imply a refutation of the steady-state model that he until then had staunchly defended: 'If the red-shifts of quasars are cosmological in origin, then the present red-shift–flux density relation for quasars rules out the steady state model of the universe.'[56] Expectedly, Hoyle denied that the data from quasars were incompatible with the steady-state theory, but by 1970 almost all other astronomers sided with Sciama and agreed that quasars were one more nail in the coffin of the steady-state theory.

Quasars were not the only strange animals discovered in the cosmic zoo in the 1960s. In 1967, Jocelyn Bell (later Bell Burnell), a 24-year-old research student at Cambridge University, was measuring scintillating radio sources when she noticed some noise signals that did not quite look like noise. The pulse-like signals came from a definite part of the sky, and she and her thesis adviser, Anthony Hewish, found that they oscillated very accurately, with a period of about one and a half seconds. She had discovered the first *pulsar* and within a few months she found a second one, with a period only slightly less than that of the first object.[57] Bell had made a discovery of radio-emitting, pulsating objects, but what were they? Hewish sought to explain the phenomenon in terms of neutron stars, and in 1968 Thomas Gold argued that pulsars were rapidly rotating neutron stars with a strong magnetic field, a hypothesis which won general acceptance. A similar mechanism had been suggested by Franco Pacini in 1967, shortly before the discovery of pulsars.

Bell's discovery and its interpretation by Gold and others implied that neutron stars, until then purely hypothetical objects, had finally been detected. These superdense celestial objects have a history which goes back to the 1930s, when Baade and Zwicky introduced the concept and name in a paper of 1934, suggesting that a supernova explosion was caused by the conversion of a normal star into a neutron star. Five years later, Robert Oppenheimer and George Volkoff studied the physics of neutron stars by means of quantum mechanics and general relativity theory and estimated the upper mass limit for a neutron star to be about 70 per cent of the Sun's mass.[58] However, since the radius of a neutron star was expected to be only about 10 km, there seemed to be no prospect of detecting the object and consequently neutron stars remained for a long time an interesting hypothesis studied by theoretical astrophysicists. Only with the unexpected discovery by Bell and Hewish did they become observationally real. The discovery in 1968 of a pulsar with a period of only 0.033 in the centre of the Crab Nebula was of special importance because of its relation to the earliest known supernova, reported in Chinese sources from 1054. (The Crab Nebula was definitely identified with the remnants of the 1054 star by Mayall and Jan Oort in a review of 1942.)

Pulsars were not of direct cosmological significance, but have been very important for relativistic astrophysics. For one thing, because of the extreme density of neutron stars, they represent matter in a state near gravitational collapse, which caused researchers to take black holes more seriously (see below). For another thing, binary pulsars—two neutron stars orbiting around each other in eccentric orbits—have been used to confirm the general theory of relativity to very high accuracy, and have resulted in convincing evidence that gravitation propagates as waves with the speed of light. The first binary pulsar, called PSR 1913+16, was discovered in 1974 by the American astrophysicist Joseph Taylor and his research student Russell Hulse, and in 1993 they were awarded the Nobel Prize for 'the discovery of a new type of pulsar, a discovery that has opened up new possibilities for the study of gravitation'. Using the technique of pulsar timing, Taylor and others were able to show that if the constant of gravitation varied with time, the variation was less than 10^{-11} or even 10^{-12} per year, which ruled out cosmological models of the Dirac type.[59]

Astronomy investigating celestial X-rays and gamma rays took off in the early 1960s by means of detectors carried by rockets and satellites. William Kraushaar and George Clark, two physicists from the Massachusetts Institute of Technology, used the Explorer XI satellite to measure the flux of high-energy gamma rays that might have resulted from annihilation of matter and antimatter in space. Ever since Dirac's prediction of antimatter in 1931, a few scientists had speculated about the possible existence of cosmic antimatter made up of antiprotons, positrons, and antineutrons, possibly in an amount equal to that of ordinary matter. Dirac himself had, in his Nobel lecture of 1933, speculated about stars and planets made up of antimatter, and in 1956 the American nuclear physicist Maurice Goldhaber proposed that the universe might consist of a cosmos and an 'anticosmos' separated from the very beginning of their existence. However, Kraushaar and Clark found in 1962 that the cosmic flux of high-energy gamma rays due to proton–antiproton annihilation was so small that it disproved the hypothesis of a symmetric matter–antimatter universe.

Later experiments with satellites and balloons showed that there exist a large variety of objects that emit X-rays and gamma rays and that these rays also exist in the form of diffuse background radiation. Data from the Vela satellites proved the existence of mysterious, short-lived bursts of gamma rays, a phenomenon first reported in 1973 but detected six years earlier. The Vela satellites were launched by the US Air Force to detect signals from nuclear weapons being tested in space, which was banned by a treaty of 1963. Nuclear bombs emit a large amount of gamma rays, and it was by examining gamma ray records from 2 July 1967 that it was realized that the rays did not originate from bombs, but from celestial sources. For the first several years, the data were classified.

It was unknown what these gamma bursts were, and from where they came, until it turned out in 1997 that they were at cosmological distances. One gamma burst detected that year was located in a galaxy with redshift 3.42, corresponding to a distance of about 3600 Mpc, which implied that gamma bursts must be much more powerful than even supernovae and quasars. The duration of the bursts is typically a few seconds, and the energy of the emitted photons may be as high as 10 GeV (a gamma photon from a radioactive substance has an energy of about 1 MeV). In 2005 an Italian-led group of astronomers measured gamma radiation from a burst with $z = 6.3$, making it one of the most distant objects known in the universe. These highly energetic phenomena were poorly understood, but many astronomers thought they might be caused by fusion of two neutron stars or a neutron star and a black hole. When, in the spring of 2003, a gamma burst was observed simultaneously with a supernova, this provided support for a third explanation, that gamma bursts are essentially caused by heavy stars collapsing into a black hole.

Finally, the Indian-born Shiv Sharan Kumar at the University of Virginia proposed in 1963 the existence of star-like objects with too little mass to sustain thermonuclear reactions. These 'brown dwarfs', as they were baptized in 1975, were originally known as black or infrared stars. They remained hypothetical for many years but are now believed to be real constituents of the universe, as abundant as ordinary stars.[60] Consequently, they may account for a sizeable fraction of the dark mass in the universe. In the 1980s there were some reports of brown dwarfs, but none of them were confirmed. The situation changed in the mid 1990s, when several candidates for high-mass brown dwarfs were detected. The first indisputable brown dwarf (called Gliese 229B) was reported in 1995, and in the following year several Jupiter-size substellar bodies were inferred to exist. Gliese 229B was a million times fainter than the Sun, had a surface temperature of about 1000 K, and a mass 30 to 40 times that of Jupiter.

4.3.3 *Developments in general relativity*

From about 1925 to 1955, Einstein's theory of general relativity was decidely unfashionable and was cultivated only by a small group of physicists, mathematicians, and astronomers.[61] The only area of research in which the theory played a major role was cosmology, but at that time many physicists and astronomers found cosmological studies to be on or even beyond the periphery of science. From about 1955 this situation began to change, and work in general relativity and gravitation gained greater respectability. The 'renaissance' was to a large extent caused by advances in experimental physics which made it possible to test predictions of general relativity in the laboratory. Moreover, the rejection of steady-state cosmology was widely seen as a triumph of Einstein's theory, and new astronomical discoveries greatly stimulated the application of relativity to astrophysical problems, as illustrated by quasars and pulsars.

The rise of general relativity as a strong research area of physics and astronomy indirectly implied an enhanced status also of relativistic cosmology. The renaissance was reflected in the venerable Solvay congresses, which from their start in 1911 until 1958 had not included gravitational physics. But from 1958 to 1973, three of the Solvay congresses focused on gravitation-related topics (1958, 'Astrophysics, Gravitation, and the Structure of the Universe'; 1964, 'The Structure and Evolution of Galaxies'; 1973, 'Astrophysics and Gravitation'). These and many other conferences were followed by specialized journals, such as *General Relativity and Gravitation* (1970), and authoritative textbooks began to appear. Noteworthy examples were *Relativistic Astrophysics* (1971) by Zel'dovich and Novikov, *Gravitation and Cosmology* (1972) by Steven Weinberg, and *Gravitation* (1973) by Charles Misner, Kip Thorne, and John Wheeler.

From Einstein's first prediction in 1911 of the bending of light by massive objects, there may seem to be but a small step to conceive of such objects as 'gravitational lenses' in analogy to optical lenses.[62] Indeed, as early as 1912 Einstein had the basic idea of gravitational lensing, but he decided not to publish it. In the following years, the idea appeared a couple of times in the scientific literature. Eddington mentioned the possibility of observing a double image by way of gravitational lensing in his *Space, Time and Gravitation* of 1920, and in 1924 the Russian physicist Orest Chwolson discussed the idea in *Astronomische Nachrichten*. Nothing much further happened until 1936, when Einstein met a Czech amateur scientist, Rudi Mandl, who tried to convince him about the lensing effect and its consequences (which, according to Mandl, included the evolution of life). Einstein, who seems to have forgotten his own earlier investigation, hesitatingly agreed to look at the problem and submitted a short note to *Science* on 'Lens-Like Action of a Star by the Deviation of Light in the Gravitational Field'. Here he derived exactly the same result for the image magnification that he had found 24 years earlier!

Einstein did not believe that the phenomenon would be observable, but his note had the effect that the lensing effect was taken up by other researchers, notably Zwicky, who, in 1937, realized that galaxies would serve much better than stars as gravitational telescopes. Contrary to Einstein, Zwicky was optimistic that gravitational lenses might be detected. It was only in the early 1960s that the idea was developed into more realistic versions, for instance by the young Norwegian astronomer Sjur Refsdal in papers of 1964. Zwicky was right in his optimism but did not live to experience the detection of the lensing effect. The first gravitational lens was discovered in 1979 by Dennis Walsh, Robert Carswell, and Raymond Weymann, who found two quasars separated by 6″ to be a double image of the same background quasar. Since then, gravitational lensing has developed into a hot area of relativistic astrophysics with important implications for cosmology. For example, in 1995 a team of German and Estonian astronomers used gravitational lensing to determine the value of the Hubble constant to be no more than 70 km/s/Mpc.

The concept of black holes has roots back in the late eighteenth century.[63] John Michell, a prominent natural philosopher in the Newtonian tradition, hypothesized that light particles would be retarded in the sense that the speed of light emitted from a very heavy star would be less than that of sunlight. He suggested that a star of the same density as the Sun, but 155 times its diameter, would emit light with a speed five per cent less than the ordinary speed. However, observations failed to confirm this interesting prediction.[64] What if the light-emitting star was even heavier? In 1784, Michell wrote another paper based on the universality of the law of gravitation and the assumption that corpuscles of light carried mass. It was then easy to calculate that light would be unable to escape from a sufficiently massive body: 'If there should really exist in nature any bodies whose density is not less than that of the Sun, and whose diameters are more than 500 times the diameter of the Sun, since their light could not arrive at us . . . we could have no information from sight.'[65] Michell realized that such bodies, although invisible, would not be undetectable, for stars and other luminous bodies in their neighbourhood would be strongly perturbed by their gravitational field and might thus reveal the existence of these dark stars. A similar suggestion was made by Laplace in his influential *Exposition du système du monde* of 1796 and repeated in the second edition of 1799 (but not in the third edition of 1808, nor in the fourth and fifth editions of 1813 and 1824). The idea of dark stars was taken up by Soldner, who in 1800 suggested that Laplace's argument was inadequate because it presupposed the speed of light to be constant. According to Soldner, gravity would slow down the speed of the light particles, with the result that even small stars might be invisible.

With the acceptance of the wave theory of light in the 1820s such considerations disappeared from physics and astronomy, only to reappear in the twentieth century, inspired by Einstein's theory of general relativity. In an address of 1921, the 70-year-old Oliver Lodge (who did not accept Einstein's theory) developed the scenario of black holes, still on the assumption that light had weight. He found that a mass capable of retaining its light must have a density and radius satisfying $\rho R^2 = 1.6 \times 10^{27}$ g/cm. 'If a mass like that of the sun (2.2×10^{33} grammes) could be concentrated into a globe about 3 kilometres in radius, such a globe would have the properties [of a black hole]; but concentration to that extent is beyond the range of rational attention. The earth would have to be still more squeezed, into a globe 1 centimetre in diameter.'

Lodge referred to work published by Schwarzschild in 1916, which is today recognized to be one of the founding papers of black holes. As Lodge interpreted Schwarzschild's metric, it allowed 'the speed of light to be zero in . . . the neighbourhood of a mass so great that $2M/R = c^2/G$ and in that case light cannot altogether escape from the body.'[66] Schwarzschild provided the exact solution to Einstein's field equations for a uniform spherical mass and found that there exists an 'exterior singularity' at a distance from the centre given by $R_S = 2GM/c^2$. The same solution was independently derived by a Dutch physicist, Johannes Droste, whose work also appeared in 1916. The Schwarzschild or Schwarzschild–Droste radius happens to be the same expression as that found by Laplace, but of course the explanation is radically different. Schwarzschild's paper was almost entirely mathematical and he did not consider the meaning of the singularity appearing in his formulae.

The nature of the singularity, whether it was a mathematical artefact or physically real, was much discussed during the next decades. Einstein, for one, believed it was physically irrelevant, and in 1933 Lemaître showed that it was apparent only and could be removed by a change of coordinates. Its relation to the gravitational collapse of massive stars was pointed out by Oppenheimer and Hartland Snyder in a paper of 1939, which, more than Schwarzschild's, qualifies as the pioneering work of black-hole physics. By using Schwarzschild's result, the two American physicists showed that the radius R_S does not correspond to a singularity, but defines a surface, or horizon, from which light cannot escape to infinity. The result of the collapse of a sufficiently massive star was, they wrote, that the star 'tends to close itself off from any communication with a distant observer; only its gravitational field persists'.[67] It was only in about 1960 that the subject attracted wide interest and astrophysicists began to speculate whether black holes might be constituents of the universe. The name 'black hole' was introduced by John Wheeler in a talk of 1967: 'What was once the core of a star is no longer visible. The core like the Cheshire cat [in *Alice in Wonderland*] fades from view. One leaves behind only its grin, the other, only its gravitational attraction . . . Moreover light and particles incident from the outside emerge and go down the black hole only to add to its mass and increase its gravitational attraction.'[68]

In 1971, Stephen Hawking at Cambridge University proved 'the second law of black hole mechanics', that the area of the event horizon can never decrease. The analogy with thermodynamics—the area corresponds to the entropy—was obvious, and according to Jacob Bekenstein, a young Princeton physicist, it was not merely formal but showed that black holes can be ascribed entropy. The suggestion was developed by Hawking, who in a paper of 1974 argued that black holes must have a temperature and therefore emit radiation as if they were hot bodies. If so, they would eventually evaporate, although the decay time might be exceedingly long. Although at first controversial, within a few years Hawking's proposal was generally accepted and the 'Hawking radiation' was added to the theoretical tool box not only of the astrophysicists but also the cosmologists.

The first serious searches for black holes started in the early 1970s, and in 1973 it was suggested that the X-ray source Cygnus X-1 consisted of a visible star orbiting around a black hole. Another and much better candidate was identified in 1994, when British astronomers analysed the binary system V404 Cygni and interpreted it as a star orbiting around a black hole 12 times as heavy as the Sun. The mass of the companion star was estimated to be 70 per cent of the Sun's mass. By that time, few astronomers doubted that black holes were plentiful in the universe. It is also believed that an enormous concentration of dark mass near the centre of the Milky Way, established by the German astronomer Reinhard Genzel and co-workers in the late 1990s, is a supermassive black hole.

So-called white holes, time-reversed versions of black holes, are hypothetical bodies emerging spontaneously from a singularity. They were first considered by Novikov and Yuval Ne'eman in the mid 1960s, when some physicists thought they might be useful in explaining the recently discovered quasars. However, contrary to their black counterparts, white holes are being investigated as hypothetical objects only. They are not believed to be part of the fabric of the universe.

The problem of an initial space–time singularity in the cosmological field equations was a matter of some concern in the 1930s, when it was investigated by Lemaître, Tolman, and a few others. Einstein offered his opinion in *The Meaning of Relativity*, published in 1945:

> One may not . . . assume the validity of the equations for very high density of field and matter, and one may not conclude that 'the beginning of the expansion' must mean a singularity in the mathematical sense. . . . This consideration does, however, not alter the fact that 'the beginning of the world' really constitutes a beginning, from the point of view of the development of the new existing stars and systems of stars, at which those stars and systems of stars did not yet exist as individual entities.[69]

Ten years later the question was reconsidered from a more general, mathematical point of view by the Indian physicist Amalkumar Raychaudhuri, who argued that the cosmic singularity was not an artefact of assumptions of homogeneity and isotropy but a consequence of general relativity. On the other hand, in the Soviet Union, Evgeny Lifschitz and Isaac Khalatnikov concluded in the early 1960s that the general case of an arbitrary distribution of matter did not lead to the appearance of a singularity. To many cosmologists this was a reassuring conclusion, but in early 1965 the British mathematician Roger Penrose spoiled the comfort by proving that a gravitationally collapsing star will inevitably end in a space-time singularity, that is, turn into a black hole. Half a year later, Hawking extended the result to apply also cosmologically, thereby disproving the optimistic conclusion of Lifschitz and Khalatnikov. Further work by Penrose, Hawking, George Ellis, and Robert Geroch resulted in a comprehensive singularity theorem, the essence of which was that a universe governed by the classical theory of general relativity must necessarily have a space–time singularity.

The Penrose–Hawking theorem does not really imply that the universe started in a singularity at $t = 0$, but only that there is a singularity somewhere. The theorem proves the existence of a singularity, but says nothing about its properties or whether it belongs to the past or the future. Moreover, in the very early universe, quantum effects cannot be ignored, which means that the validity of the Penrose–Hawking theorem cannot be assumed without qualifications. All the same, the theorem brought the embarrassing singularity at the birth of the universe back on stage and was sometimes taken to be a proof of it.

Whereas most cosmologists since Lemaître's time wanted to avoid the conceptually troublesome initial singularity, a few chose to see it as a positive and useful element in cosmological thinking. Erwin Freundlich, Einstein's early collaborator, compared in a book of 1951 the fear of the singularity to the *horror vacui* of Aristotelian physics and suggested that it was 'the last reminder of a subconscious yearning for a harmonious universe'. This was also the opinion of the American physicist Charles Misner, who in 1969 proposed that 'We should stretch our minds, find some more acceptable set of words to describe the mathematical situation now identified as "singular," and then proceed to incorporate this singularity into our physical thinking until observational difficulties force revision on us.'[70] Misner's stretching of his mind led him to suggest a redefinition of time, to substitute the ordinary time parameter t by $\log t$, essentially the same time recalibration that Milne had used in the 1930s. With such a concept of time, the original singularity would disappear into the infinite past, and Misner could state that 'The universe is meaningfully infinitely old because infinitely many things have happened since the beginning.' Misner's reinterpretation was conceptually satisfactory but did not affect the development of cosmology any more than Milne's attempt had done.

4.3.4 *Matter in the universe*

Dark cosmic matter became a big topic in the 1970s, but in a general sense the concept was familiar also to astronomers of the pre-Einstein era.[71] The first example of dark matter in the history of cosmology is probably Philolaus' invisible counter-Earth from the fifth century BC, which we encountered in Section 1.1. The dark stars that Michell and Laplace considered in the late eighteenth century provide another example of invisible but gravitationally detectable matter. The idea lived on, as illustrated by Agnes Clerke's *Problems of Astrophysics* of 1903, which included a chapter on 'Dark Stars', in which she concluded: 'Unseen bodies may, for aught we can tell, predominate in mass over the sum-total of those that shine; they possibly supply the chief part of the motive power of the universe.'[72] Clerke described the hypothetical dark stars as 'suns in a state of senile decay'.

With the first estimates of the mass density of the Milky Way in the early part of the twentieth century, the concept of dark matter started its slow march from speculation to observationally supported hypothesis. Jeans suggested in 1922 that there must be about three times as many dark stars as bright stars in our galaxy. As to the observed part of the universe, Hubble's 1926 estimate of the average density of the universe, based on counts of galaxies, was about 10^{-31} g/cm^3, a value which by the 1930s had increased by a factor of ten or more. In *The Realm of the Nebulae* of 1936, Hubble stated 10^{-30} as a lower limit and 10^{-28} as an upper limit. These figures are to be compared with the critical density of the Einstein–de Sitter universe, originally based on a value of the Hubble constant that was too high, which in the 1932 paper by Einstein and de Sitter was given as 4×10^{-28} g/cm^3. With the revision of the Hubble constant in the 1950s, it became clear that the universe contained too little visible matter to make it critical ($\Omega = 1$), that is, to conform to either the Einstein–de Sitter model or the steady-state model. In an address delivered to the 1958 Solvay congress, the Dutch astronomer Jan Oort concluded that $\rho \cong 4 \times 10^{-31}$ g/cm^3 ($\Omega \cong 0.01$), on the basis only matter contained in galaxies. He realized that the real density might be higher. From the perspective of the steady-state theory there was no reason to worry, since this theory assumed large amounts of neutral hydrogen in intergalactic space as a candidate for dark matter.

It was not, however, arguments related to theoretical cosmology that caused dark matter to make its entry into cosmology. Instead, in connection with investigations of clusters of galaxies, Zwicky suggested in 1933 that the gravitation from visible matter was not enough to keep the clusters together. He studied the velocities of galaxies in the Coma cluster and found a dispersion that necessitated the presence of large amounts of dark matter. 'The average density of the Coma system', he wrote, 'must be at least 400 times greater than what is derived from observation of the luminating matter. Should this be confirmed, the surprising result thus follows that dark matter is present in a very much larger density than luminating matter.'[73] As to the cosmological consequences, he pointed out that although measurements of 'luminating' matter led to $\Omega \ll 1$, it was 'not unreasonable' to assume that the dark matter might increase the matter density to $\Omega \cong 1$.

Zwicky's argument was not much noticed, and it took about forty more years until observational evidence clearly showed that there must be a large amount of invisible, gravitating matter in the discs of rotating galaxies; if not, the galaxies would fly apart because of a too weak gravitational attraction. This conclusion, based on theoretical studies of the dynamics of galaxies, was reached in 1974 by Peebles, Jeremiah Ostriker, and Amos Yahil in the United States, who showed that the masses of spiral nebulae were roughly proportional to their radii, which indicated the presence of dark matter. They made the general claim that about 90–95% of the mass of the universe is in a dark form. Jaan Einasto and a group of Soviet astronomers arrived independently and almost simultaneously at a similar conclusion. The work of the two groups had the result that dark matter was now considered to be a real and most important constituent of the universe. The new status was further emphasized when Vera Rubin and other astronomers found in 1978 compelling evidence in the rotation curves of spiral galaxies that the mass of a galaxy was about five times the mass of its constituent visible stars. Moreover, it followed from their work that the mass, unlike the luminosity, was not concentrated near the galactic centre.[74] By the late 1970s dark matter had been discovered, not just hypothesized, and in amounts much larger than visible matter. The mass density of the universe, including the new estimates of dark matter, led to $\Omega \cong 0.2$–0.6. But what was this dark matter?

As possible candidates for dark celestial bodies made up of ordinary (baryonic) matter, astronomers considered what they called MACHOs, an acronym for 'massive compact halo objects'. These objects were non-luminous but might reveal their existence by their

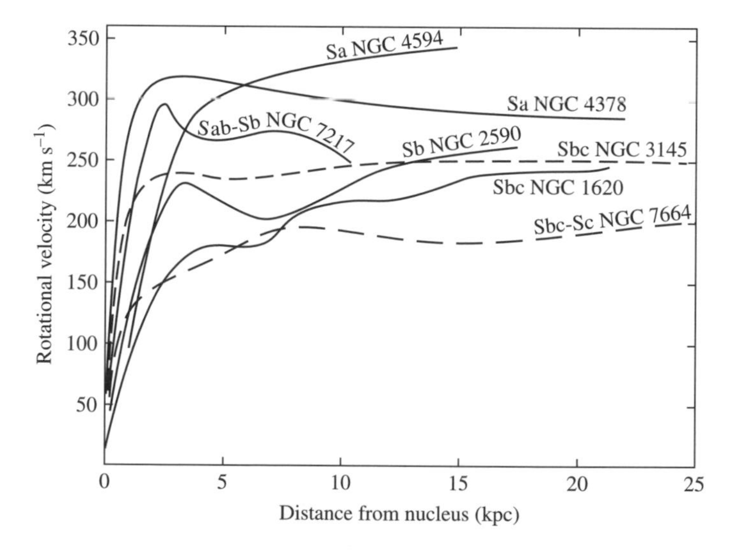

Fig. 4.9 Rotation speeds for some galaxies, as determined by Vera Rubin and collaborators in 1978. Predictions based on the visible mass resulted in curves which at large distances would decrease rather than staying approximately constant. The discrepancy between observations and predictions was interpreted as evidence for large amounts of dark matter in the haloes of the galaxies. From V. C. Rubin *et al.*, 'Extended rotation curves of high-luminosity spiral galaxies', *Astrophysical Journal* **225** (1978), L107–L111. Reproduced by permission of the American Astronomical Society.

gravitational actions. They might be massive black holes, Jupiter-like planets, or brown dwarfs. A collaboration called the Macho Project started in 1993 to look for the MACHOs by means of the gravitational lensing effect and soon found many indications of their existence. This and other projects have demonstrated the existence of dark matter of the MACHO type, but its nature is not yet clear.

4.3.5 *Cosmological models*

Shortly after the discovery of the cosmic background radiation, there emerged a consensus among most cosmologists on the main problems of cosmology and how to solve them. The hot-Big-Bang relativistic theory acquired a nearly paradigmatic status, and alternative theories were marginalized. At the same time cosmology experienced a strong quantitative growth, as is shown by bibliographical data. Whereas the annual number of scientific articles on cosmology had on average been about 30 in the period 1950–1962, between 1962 and 1972 the number increased from 50 to 250. Still, compared with other fields of physics and astronomy, cosmology remained a small and loosely organized science, split between physics and astronomy.[75]

At the same time that cosmology became cognitively institutionalized, it achieved a social institutionalization that made the subject a full-time professional occupation rather than a part-time hobby for astronomers, mathematicians, and physicists. There was a growing integration of the field into university departments, and courses and textbooks on cosmology became common. For the first time, students were taught standard cosmology and brought up in a research tradition with a shared heritage.[76] With the professionalization and status as a respected field of science there followed the exclusion of philosophers and amateur scientists; in some sense there was a narrower view of what cosmology was about. The national differences that had characterized earlier cosmology also disappeared. Originally, physical Big Bang theory had been an American theory, steady-state theory had belonged to the British, and the Russians had for political and ideological reasons hesitated about doing cosmology at all; but now the field became truly international. It was no longer possible to tell an author's nationality from the cosmological theory he or she advocated.

In spite of the new-born confidence in the hot-Big-bang theory, most researchers stressed that the theory included many problems and uncertain features. It was considered a research programme or framework for organizing observations, a starting point for further research rather than the final answer to how the universe had evolved and was structured. It was generally agreed that the universe should be described by Big Bang solutions to Einstein's field equations without the cosmological constant. After the demise of the steady-state theory, confidence in the general theory of relativity grew to an almost dogmatic level. Zel'dovich and Novikov were among those who stressed that no new physics was needed to understand the universe:

> There are no observational data suggesting a limitation on the applicability of GTR [general theory of relativity] to the scales of the Universe. Therefore, the assumption that a change in GTR is needed in applications to cosmology is unfounded. Thus, the aggregate of theoretical, experimental, and observational facts stand in favor of the applicability of the physical laws and GTR to a description of the evolution of the Universe from *almost* the very beginning of the expansion. They apply from times when the matter density is much greater than the density of nuclear matter, $\rho > 10^{14}$ g/cm^3, up to the present time.[77]

Sandage's observational programme was to measure the Hubble constant and the deceleration parameter with sufficient precision to pin down the best candidate for a model of the real universe. In about 1970 he concluded tentatively that $q_0 = 1.2 \pm 0.4$ and that the Hubble time for $\Lambda = 0$ was between 7 and 19.5 Gyr, corresponding to a closed universe with an age in the range from 4.2 to 11.7 Gyr (Gyr = gigayear = one billion years). But these values were contradicted by other measurements which suggested an open and perpetually expanding universe. Indeed, in 1979 Sandage argued from a wide range of data that the preferred values were $H_0 = 50$ km/s/Mpc and $q_0 = 0.02$, corresponding to an open universe of age 19 Gyr.[78] During the 1970s and onwards, Sandage, who at the time was collaborating with the Swiss astronomer Gustav Tammann, became engaged in a protracted controversy with Gérard de Vaucouleurs and others concerning the value of the Hubble constant. Whereas Sandage and Tammann found values about 50 km/s/Mpc, de Vaucouleurs found 100 km/s/Mpc.

Sandage's programme was fine in principle, but did not really deliver what it promised. In the absence of unambiguous observational guidance, many cosmologists preferred the Einstein–de Sitter model, not because it was confirmed by observation but because it was simple and not clearly ruled out. There was always the joker in the game, the cosmological constant, and although it was generally unwelcome, models of the Lemaître type did receive some attention. Around 1970 they were reconsidered by Vahe Petrosian and others, who found such models useful in explaining certain quasar data. From considerations of the Hubble diagram and other evidence, James Gunn and Beatrice Tinsley suggested in 1975 that the most plausible model was a closed, ever-accelerating universe with $\Lambda > 0$ and a density so high that no deuterium was produced during the Big Bang.[79] Mindful that their conclusion was controversial, they presented it as a puzzle in need of better data.

The model favoured by Gunn and Tinsley did not survive for long, and they could not have known that the accelerating universe with a positive cosmological constant would return at full power some twenty years later. The uncertainty continued for many years, and still in the late 1980s the question was completely open, leading the American astronomer Virginia Trimble to comment, 'Those of us who are not directly involved in the fray can only suppose that the universe is open ($\Omega < 1$) on Wednesday, Friday, and Sunday and closed ($\Omega > 1$) on Thursday, Saturday, and Monday. (Tuesday is choir practice.)'[80]

Although by 1970 cosmology was on its way to becoming a normal science, governed by the relativistic Big Bang paradigm, there was plenty of room for unconventional ideas, some of them moderate and others decidedly unorthodox.[81] The idea of 'many universes' in a scientific context goes back to 1934 when Tolman, developing work by Lemaître from 1933, investigated non-homogenous, spherically symmetric solutions to the cosmological field equations and found the possibility that the entire universe may contain independent homogeneous regions, sub-universes of different density. Some of these, he speculated, might expand while others might contract, and there might be a multitude of big bangs spread over time and space. A vaguely similar picture was suggested by Hoyle and Narlikar in 1966 within the framework of a much modified version of the steady-state theory. In the Hoyle–Narlikar theory, the universe was likened to a pulsating bubble, consisting of many separate and continuously formed bubble-universes which would either expand or contract, at different rates. This idea, conflicting as it does with the perfect cosmological principle, departed radically from the original steady-state theory. Another of the former steady-state theorists, Thomas Gold, proposed in 1973 a somewhat similar idea of multiple universes, but without relating it to concepts of the steady-state theory.[82] These theories were largely ignored, but within a few years multiple universes would become much discussed within the framework of inflation cosmology (see Section 5.1).

The Hoyle–Narlikar theory of 1966 may be seen as a precursor of the last stand of the steady-state theory, the QSSC (quasi-steady-state cosmology) theory developed by Hoyle, Narlikar, and Geoffrey Burbidge in the 1990s. QSSC operated with creation of matter in 'little big bangs' in epochs of high density, and pictured the expansion of the universe as superimposed on oscillations on a very large timescale, say 10^{11} years. About the only thing it had in common with the original theory of 1948 was continuous creation of matter in an eternal universe. Hoyle and his colleagues developed their theory in great detail, which included an explanation of the cosmic background radiation and the abundances of light elements, but it failed to make impression on the large majority of cosmologists, who worked within the framework of Big Bang theory.[83]

Another kind of alternative cosmological theory which for a brief period attracted some attention was the plasma cosmology proposed by the Swedish physicist (and Nobel laureate of 1970) Hannes Alfvén in the 1960s. According to this theory, the universe at large consisted of equal amounts of matter and antimatter separated by electromagnetic fields, and annihilation processes were responsible for the huge explosion which in the past made our part of the universe expand. This explosion made a 'big bang', but it was not the birth of the universe, for according to Alfvén and his followers the universe had always existed. Very few cosmologists considered Alfvén's plasma cosmology a serious alternative, for other reasons, namely because there was no observational evidence for large amounts of antiparticles in the universe.

During the 1970s, Paul Dirac attempted to breathe new life into his old idea of a cosmology based on the large number hypothesis and a gravitational constant varying with time (compare Section 3.3). He developed the idea into several versions, but none of them were able to explain the cosmic background radiation, which at the time was a must for any cosmological theory. According to Dirac, the large number hypothesis ruled out the possibility that the microwave background had its origin in a primeval decoupling of matter and radiation. Although Dirac's theory was a failure, the general idea of a time-varying constant of gravitation was taken up by a few researchers, including Paul Wesson and Vittorio Canuto, who developed it into versions more in line with mainstream cosmology.[84] The possibility of 'constants' of nature varying in time continued to fascinate and is still being investigated by some cosmologists, if in ways widely different from Dirac's original ideas.

Whereas the theories of Hoyle, Alfvén, and Dirac belonged to the periphery of cosmology, the alternative proposed by Dicke and his research student Carl Brans in 1961 was considered an interesting and serious challenge to the standard theory. The Brans–Dicke theory was not merely an alternative cosmological theory of the Big Bang type, but an alternative to Einstein's general theory of relativity. It operated with a scalar field, acting as a gravitational constant, which would in general change with time, although not in a specific way, since it depended on the value of a certain external parameter. The cosmological consequences of the new theory did not differ much from those of relativistic evolution theory, except that in general it led to a different expansion rate of the universe and therefore also to a different amount of helium being produced in the early universe.

The theory of Brans and Dicke had implications for geophysics and astrophysics which were much discussed in the 1960s, one of these being that the Sun's oblateness would differ from that predicted by general relativity. Later observations showed that the shape of the Sun did not agree with Dicke's prediction, which was a major reason why interest in the Brans–Dicke theory greatly diminished.

Notes

1. On Gamow and his role in the development of cosmology, see Frenkel 1994, Kragh 1996b, Harper *et al.* 1997, and Harper 2001. References to the primary literature can be found in these sources. Alpher and Herman 2001 offers a semi-historical and partly autobiographical account of the development.
2. Gamow and Fleming 1942, p. 580.
3. Letter of September 1946, quoted in Kragh 1996a, p. 110. Einstein seems to have appreciated Gamow's approach. In a letter to Gamow of 4 August 1948, he wrote: 'I am convinced that the abundance of elements as function of the atomic weight is a highly important starting point for cosmogonic speculations. The idea that the whole expansion process started with a neutron gas seems to be quite natural too.' Ibid., p. 117.
4. Alpher, Gamow, and Bethe 1948, reproduced in Lang and Gingerich 1979, pp. 864–865. The paper, sometimes known as the 'αβγ paper', had Bethe as a co-author, which was a joke on Gamow's part. In fact, Bethe did not contribute and did not know about the paper until he happened to see the manuscript (and he then decided to let the joke pass).
5. The Greek *yle* or *hyle* means 'stuff'. As mentioned in Section 2.3, the term was used by Bernard Sylvester in the mid twelfth century and also by Roger Bacon in the thirteenth century. It was resurrected by the chemist William Prout in 1815, and again by William Crookes in 1886, in the form *protyle* (proto-yle), meaning 'earlier than ylem'. See Crookes 1886, p. 568. While Prout believed that the protyle was hydrogen, later scientists (including J. J. Thomson in 1897) considered the electron to be the true protyle.
6. Alpher and Herman 1948, p. 774, reproduced in Bernstein and Feinberg 1986, p. 868. The sentence also appeared in a more detailed article in *Physical Review* in 1949.
7. See Alpher and Herman 2001, pp. 118–120 and also Weinberg 1977, p. 130.
8. Alpher, Follin, and Herman 1953, reproduced in Bernstein and Feinberg 1986, pp. 159–200.
9. Rankama and Sahama 1950, p. 14. Nuclear physics and cosmology had for some time been of interest to geochemists; see Kragh 2001a.
10. In 1956 Gamow published a paper on the Big Bang theory in the recently founded yearbook *Vistas in Astronomy*, but this was a survey of the theory, not a research article.
11. Quoted in Kragh 1996a, p. 141.
12. On Gamow's style and other styles of cosmological research, see Kragh 2005.
13. In some of his works from about 1950 Gamow imagined a hypothetical collapse that had preceded the present expansion, as allowed by the Friedmann equations, but he was careful to point out that the idea was purely speculative. For the near-destruction of the collapsing universe, he coined the term 'big squeeze' (Gamow 1952).
14. I have done so in Kragh 2005, p. 184.
15. McVittie 1961. See also Sánchez-Ron 2005.
16. MacMillan 1925, p. 99.
17. Bondi 1948. On the birth of the steady-state theory, see Kragh 1996a, pp. 162–179 and Terzian and Bilson 1982, pp. 17–65. See also Hoyle's autobiography, Hoyle 1994.
18. Bondi and Gold 1948; Hoyle 1948. See also the presentation in Bondi 1952.
19. Bondi 1952, p. 144.
20. Spencer Jones 1954, p. 31.
21. On McCrea's work and other attempts to reconcile steady-state theory and general relativity, see Kragh 1999.
22. For the interplay between the electrical-universe theory and the charge experiments, see Kragh 1997.
23. See the survey in Gribbin 1976.
24. Burbidge, Burbidge, Hoyle, and Fowler 1957. See also the non-technical survey in Hoyle 1965. Of course, many other physicists contributed to the development. Important work was done by Ernst Öpik in Estonia and Alastair Cameron in Canada, among others.
25. Mineur's Hubble time would have been 3 billion years, but he did not state this value or relate his recalibration to topics of cosmology. On the changes in the determinations of the distance scale and the Hubble constant, see Fernie 1969 and Trimble 1996.

26. On Baade's celebrated 'doubling the size of the universe', see Osterbrock 2001, pp. 162–176. See also Gray 1953.
27. It could even be argued that the new cosmic timescale caused serious problems for the steady-state theory; the American astronomer Ivan King (see Kragh 1996a, pp. 275–276) did so.
28. The programme is described in Sandage 1970 and, retrospectively, Sandage 1998.
29. Hoyle and Sandage 1956.
30. On the history of radio astronomy in Great Britain, see Edge and Mulkay 1976.
31. Quoted in Sullivan 1990, p. 321.
32. For the Australian contributions to radio astronomy and cosmology, see Robertson 1992 and Edge and Mulkay 1976. Of the almost 2000 sources listed in the 2C catalogue, only about 10 per cent have stood the test of time.
33. Mills and Slee 1957, p. 194. Reprinted in Sullivan 1982.
34. On Hoyle's book and the reactions it evoked, see Mitton 2005, pp. 108–141. Gamow published in 1952 his own version of popular cosmology, *The Creation of the Universe*, which was entirely based on the nuclear-physical Big Bang model (Gamow 1952).
35. Unpublished lecture given at the Massachusetts Institute of Technology. Quoted in Tipler, Clarke, and Ellis 1980, p. 110.
36. The address is reproduced in McLaughlin 1957, pp. 137–147 and can also be found on the Internet address www.papalencyclicals.net/PIUS12EXIST.HTM. On the relationship between cosmology and religion in the 1950s, see also Kragh 1996a, pp. 251–259.
37. Erich Mascall, an Oxford philosopher of religion, pointed out that the steady-state theory was easily reconcilable with the Christian view: 'An uninstructed person, when told that Hoyle is a fervent advocate of the view of continuous creation, might not unreasonably suppose that Hoyle was a convinced theist who saw the hand of God in every being and event in the world's history.' Mascall 1956, p. 158.
38. Quoted in Kragh 1996a, p. 246.
39. Bondi *et al.* 1960, p. 45.
40. For examples from post-1990 astronomy and cosmology, see Sovacool 2005.
41. For references to the opinions of Johnson, Öpik, and Klein, see Kragh 1996a, pp. 222–223.
42. Gödel 1949. The idea of a universe in rotation appeared as early as 1716, in the Leibniz–Clarke correspondence, where Clarke suggested that a rotating universe would generate a centrifugal force, an observable effect (Alexander 1956, p. 101).
43. Jordan 1971.
44. Carey 1988, p. 328. Carey, who started developing his theory of the expanding Earth in the late 1950s, did not accept what he called 'the big bang myth'. There is a good deal of similarity between the cosmological controversy of the 1950s and the later controversy concerning the expanding-Earth theory.
45. Thorne 1994.
46. The physicist Matvei Bronstein and the astronomer Boris Gerasimovich were among the many who were executed in the late 1930s for being spies and class enemies. The charges against them included support of Einsteinian cosmologies. On cosmology in Stalin's Russia, see Kragh 1996a, pp. 259–268 and Graham 1972, pp. 139–194.
47. Quoted in Kragh 1996a, p. 260.
48. Ibid., p. 262.
49. Lizhi 1991, pp. 309–313. The author of the article was Liu Bowen. See also Williams 1999, p. 75 and Hu 2005, pp. 167–169.
50. Quoted in Hu 2005, p. 168.
51. Among the many historical accounts of the discovery of the cosmic background radiation, Weinberg 1977, Bernstein 1984, and Brush 1993 are to be recommended. For a participants' history, see Wilkinson and Peebles 1990. Peebles 1993, a widely used textbook, includes on pp. 131–151 a detailed account of the discovery process.
52. Quoted in Kragh 1996a, p. 331, which includes details about the responses of the steady-state theorists to the discoveries of the 1960s.
53. Wagoner 1990. A review of the primeval-helium problem, as of 1970, is presented in Peebles 1971, pp. 240–276.
54. For histories of the discovery of quasars, see Edge and Mulkay 1976, pp. 202–212, Harwit 1981, pp. 137–140, and Schmidt 1990.
55. The presently favoured model of quasars is that they are supermassive black holes in the centres of active galaxies. Since the 1960s, about 6000 quasars have been catalogued, some of them being so far away as to correspond to a redshift of about 5.

56. Sciama and Rees 1966.
57. The discovery of pulsars has been somewhat controversial. For informative (if not necessarily objective) accounts, see Wade 1975 and Hewish 1986. The pulsar story, as seen from the Jodrell Bank Observatory, is told in Lovell 1973, pp. 121–166. Hewish (but not Bell) was awarded the 1974 Nobel Prize for his part in the discovery process.
58. The papers of Baade and Zwicky (1934), Oppenheimer and Volkoff (1939), Pacini (1967), and Gold (1968) are reproduced in Lang and Gingerich 1979, which also includes the paper in *Nature* of 1968 where Hewish and his group reported the discovery of pulsars.
59. Taylor 1992.
60. Basri 2000.
61. On the low-water mark of general relativity and its reconstitution, see Eisenstaedt 1989 and Kaiser 1998.
62. Renn and Sauer 2003.
63. On black holes and their history, see Israel 1987 and Thorne 1994.
64. Michell communicated his idea to his friend, the chemist and natural philosopher Joseph Priestley. See Priestley 1772, pp. 786–792, reprinted 1978. In a paper of 1911, Einstein concluded that the speed of light depends on the local gravitational field, but soon abandoned the idea (Einstein *et al.* 1952, p. 105).
65. Quoted in Israel 1987, p. 203. Michell's paper, published in the *Philosophical Transactions of the Royal Society* (London), is reprinted in Detweiler 1982.
66. Lodge 1921, pp. 551 and 555.
67. Oppenheimer and Snyder 1939, p. 456.
68. Israel 1987, p. 250. Wheeler's talk was published in 1968.
69. Einstein 1945 (6th edn, 1956, p. 126). A careful examination of the history of singularity theorems is given in Earman 1999. See also Earman and Eisenstaedt 1999 and Tipler, Clarke, and Ellis 1980.
70. Misner 1969; Finlay-Freundlich 1951, p. 49.
71. Brief accounts of the early history of dark matter appear in Trimble 1990 and Van den Bergh 2001. Dark matter has sometimes been referred to as 'missing mass' or 'missing matter' (especially when relating to undetected matter needed for theoretical reasons). Many scientists find 'missing mass' to be a misleading name, and today it has been largely abandoned.
72. Clerke 1903, p. 400.
73. Zwicky 1933, p. 125.
74. For a non-technical account, see Rubin 1983.
75. For bibliographical data and literature on cosmology, see Ryan and Shepley 1976.
76. A sociological perspective on workers in relativistic cosmology considered as a scientific community is presented in Copp 1982 and Copp 1983.
77. Zel'dovich and Novikov 1983 (translation of Russian original of 1975), p. xxi.
78. Sandage 1970; Tammann, Sandage, and Yahil 1979.
79. Gunn and Tinsley 1975.
80. Trimble 1988, p. 389.
81. See the review in Ellis 1984.
82. For references to early many-universe ideas, see Kragh 1996a, pp. 366–368. The Lemaître–Tolman inhomogeneous model is often known as the Tolman–Bondi model because Bondi investigated it in 1947.
83. Hoyle, Burbidge, and Narlikar 2000. For objections to the theory, see Peebles *et al.* 1991.
84. See, for example, Canuto and Lodenquai 1977. This kind of Dirac cosmology claimed to be able to account for the background radiation and not to contradict any observational fact.

5

NEW HORIZONS

What by 1970 had become the standard Big Bang theory was not so standard that it could not change considerably over the next few decades, and by the end of the century it had transmuted into an almost unrecognizable form. Big-Bang-based understanding of the early universe celebrated great triumphs with the ever-tighter symbiosis between cosmology and particle physics, which in the early 1980s reached a new phase with the invention of the inflationary scenario for the very early universe. On the theoretical front, inflation was the most important advance in late-twentieth-century cosmology, and it quickly gave rise to a variety of models, some of which differed drastically from the 'old' Big Bang theory by not accepting *the* Big Bang. Inflation theory also stimulated, and perhaps legitimized, lines of cosmological thinking that were almost purely hypothetical and often highly speculative, such as in the broad arena of quantum cosmology. Remarkably, with the anthropic principle, teleological and semi-philosophical considerations returned to cosmology.

Later historians may single out new observations, rather than new and bold theories, when they have to decide what constituted the most important changes in modern cosmology. By the turn of the millennium, evidence had accumulated to the effect that we live in a universe which expands at an ever faster rate, a picture very different from the standard picture of the 1970s. The acceleration is driven by a force which is still not well understood, but may be represented by the cosmological constant, long thought to be a mistake. The development during the last three decades has been stormy and also confusing. One cannot describe it in a proper historical perspective, but also one cannot ignore it in a historical account of cosmology.

5.1 Early-universe cosmology

The idea of using nuclear and elementary-particle physics to inform and constrain cosmological theories goes back to the 1930s, and played an important role in the pioneering work of Gamow and his group, especially in the Alpher–Herman–Follin theory of 1953. With the explosive development of high-energy physics in the 1960s and 1970s, the close connection between cosmology, astrophysics, and particle physics strengthened. Enthusiastic physicists (more than astronomers) tended to see the connection as giving birth to a new and—finally!—scientific cosmology. According to two of the leading physicists in the new research area, the Americans David Schramm and Gary Steigman,

> cosmology has become a true science in the sense that ideas not only are developed but also are being tested in the laboratory . . . This is a far cry from earlier eras in which cosmological theories proliferated and there was little way to confirm or refute any of them other than on their aesthetic appeal.[1]

Moreover, the confidence in the hot-Big-Bang model led researchers to use this model to get insight into particle physics at the extreme energies of the very early universe, energies far beyond what any human-made accelerator could produce. The early universe became 'the poor man's accelerator', as it was often called, and high-energy physics and cosmology entered an increasingly symbiotic relationship. Whereas nuclear astrophysics was a well-established field in the 1950s, particle astrophysics with applications to cosmology emerged as a vigorous new science in the 1970s. One indication of the symbiosis was the foundation in 1992 of the journal *Astroparticle Physics*.

One of the first and most important results in particle physics arising from Big Bang cosmology related to the number of neutrino species. In the mid 1970s, two types of neutrinos had been detected, one related to the electron and the other to the heavier muon. As far as laboratory experiments could tell, there might be many more neutrinos, but how many they could not tell. In 1977 Steigman, Schramm, and James Gunn used cosmological data and theory to come up with an answer superior to that of the experimental physicists. It was known that the expansion rate of the early universe depended on the total energy density, which in turn depended on the number of particle species at a time when the universe consisted mainly of photons, electrons, and neutrinos. Because the production of helium-4 is very sensitive to the expansion rate, it was possible to constrain the number of neutrino species by comparing calculations with the measured amount of helium. Steigman, Schramm, and Gunn found in this way that there could be no more than six species of neutrinos, and subsequent calculations sharpened the bound to four, and then to three. This prediction, based solely on cosmological arguments, was confirmed in 1993 when results from CERN, the European centre of high-energy physics, indicated that there were indeed three species of neutrinos. A few years later, Schramm proudly pointed out that 'this was the first time that a particle collider had been able to test a cosmological argument, and it also showed that the marriage between particle physics and cosmology had indeed been consummated'.[2]

In addition to this success, by means of similar cosmological arguments it proved possible to put limits on the neutrino masses that were more stringent than those provided by laboratory experiments. In the case of the muon neutrino, the cosmological upper limit (50 eV) was about 10 000 times less than the laboratory limit. Physicists had assumed the existence of a third neutrino, known as the tau neutrino, since about 1975, and calculations in 1991 showed that primordial nucleosynthesis required the mass of the particle to be less than 0.5 MeV. This constraint agreed with, but was finer than, the one obtained experimentally, which afforded yet another test of the early Big Bang scenario and boosted confidence in the 'inner space–outer space connection'.[3]

The focus in the 1960s on primordial production of helium-4 resulted in agreement between observations and predictions which strongly supported the Big Bang model. However, it was realized that the yield of helium-4 in the early universe is not very sensitive to the baryon density, and for this reason the interest of the nuclear cosmologists turned towards the cosmological origin of deuterium (and also helium-3), which constrains the baryon density much more tightly. The determination of the amount of deuterium (D) came to serve as a 'cosmic baryometer'. Prior to the 1970s, the abundance of deuterium in the universe was not well known. Only with the launching in 1972 of the Copernicus satellite did it become possible to measure the amount of interstellar deuterium, which John

Rogerson and Donald York reported to be D/H $\cong 1.4 \times 10^{-5}$. Since deuterium was produced in the Big Bang, it was possible to use the experimental result to infer the value of the present density of the universe. The two astronomers derived $\rho = 1.5 \times 10^{-31}$ g/cm^3, which indicated an open, ever-expanding universe.[4] After much work in this area, convincing evidence for a primeval abundance of deuterium relative to hydrogen of D/H $\cong 3.2 \times 10^{-5}$ was obtained. By the mid 1990s this was taken to imply a baryon density of 3.6×10^{-31} g/cm^3, or $\Omega_B = 0.02h^{-2}$, which provided firm evidence for the existence of large amounts of non-baryonic dark matter.[5]

High-energy physics also threw new light on one of cosmology's old and important questions, why antimatter is practically absent from the universe. In the very early universe, antinucleons were assumed to be as abundant as nucleons, and almost all of these would annihilate into high-energy photons until the universe had expanded to such a size that annihilation became rare. The result would be a nearly complete annihilation of matter, leaving a photon-to-nucleon ratio of about 10^{18}, as was shown by Zel'dovich in 1965. The problem was that the observed ratio was much smaller, about 10^9. How had the 'annihilation catastrophe' been prevented? The discrepancy could be explained by assuming a slight excess of matter over antimatter in the early universe, but not all physicists found this a satisfactory explanation. They objected that this was merely pushing back the asymmetry to the initial condition of the universe, and hence not a real explanation.

A much better explanation became possible with the emergence in the mid 1970s of the first grand unified theories (GUTs), a class of theories which gave a unified account of the electromagnetic, weak, and strong interactions. Grand unified theories had in common that they did not conserve the baryon number, meaning that they allowed processes where the number of produced baryons, such as protons and neutrons, does not equal the number of antibaryons. It was this property that the Japanese physicist Motohiko Yoshimura exploited in 1978 to demonstrate how an initially symmetrical state could evolve into a state with a slight excess of nucleons, producing a world somewhat like the one we know. 'If my mechanism works, we may say that a fossil of early grand unification has remained in the form of the present composition of the universe,' he wrote, giving a new twist to the concept of

Table 5.1 Chronology of important events in the history of the Big Bang universe

Time after Big Bang	Temperature (K)	Energy	Events
10^{-43} s	10^{31}	10^{19} GeV	Quantum gravity. Planck time
10^{-33} s	10^{27}	10^{14} GeV	End of grand unification. Origin of matter–antimatter asymmetry. Inflation
10^{-2} s	10^{13}	1 GeV	Neutrons and protons form from quarks and antiquarks
10 s	10^9	1 MeV	Big Bang nucleosynthesis of light elements
10^6 y	10^3	0.1 eV	Formation of atoms. Origin of background radiation
10^{10} y	3	10^{-3} eV	Formation of solar system. Life

nuclear archaeology.[6] Theory predicts that very massive X particles ($m_X \cong 10^{15}$ GeV) will be produced in the GUT era. Although anti-X particles will be produced in the same amount as the X particles, they decay into quarks and antiquarks at different rates, thereby generating an excess of quarks over antiquarks and baryons over antibaryons.

Subsequent work by Yoshimura, Weinberg, Frank Wilczek, Leonard Susskind, and others gave a more detailed explanation of why there are a billion more photons than nucleons in our universe. The application of grand unified theory to cosmology led to successes, but also to problems. For example, a large number of primordial, very massive magnetic monopoles were expected to be formed, which would lead to an absurdly great mass density. Yet not a single magnetic monopole has ever been observed since Dirac predicted their existence back in 1931 (in a form different from the monopoles of the grand unified theories).

Although grand unified theories accounted for the preponderance of matter over antimatter, the first suggestion along these lines pre-dated grand unified theories by several years. In a prescient work of 1967, the Soviet physicist and famed political dissident Andrei Sakharov suggested that baryon number might not be conserved exactly and that this might have important cosmological implications. Sakharov's brief paper, published in a Russian journal, was not much noticed at the time but has later been recognized to be a pioneering contribution to the interface between cosmology and particle physics. These topics remained important to Sakharov during the rest of his life and he continued working on them, even after he was exiled to Gorky in 1980. His interest in cosmology reflected a belief in a cyclical universe which was not so much grounded in scientific data as related to metaphysical and emotional desires. A Nobel Prize *peace* lecture does not usually include references to physics and cosmology, but in Sakharov's case it did, and it was not accidental:

> I support the cosmological hypothesis that states that the development of the universe is repeated in its basic characteristics an infinite number of times. Further, other civilizations, including more 'successful' ones, should exist an infinite number of times on the 'preceding' and the 'subsequent' pages of the book of the universe. Nevertheless, this weltanschauung cannot in the least devalue our sacred inspirations in this world, into which, like a gleam in the darkness, we have appeared for an instant from the black nothingness of the ever-conscious matter, in order to make good the demands of reason and create a life worthy of ourselves and of the goal we only dimly perceive.[7]

But back to the cosmology–physics interface in the late 1970s. The dark-matter problem forged further links to particle physics when it turned out that a large part of the missing or dark matter was presumably in an unusual form, something very different from the nucleons and electrons of ordinary 'baryonic' matter.[8] Many cosmologists in the 1980s believed that the universe was critical ($\Omega = 1$), not because of observations but because the successful inflation theory predicted it. If so, this made the dark-matter problem more urgent since it then followed that most of the mass of the universe was probably not normal matter.

Observations and theoretical arguments based upon Big Bang nucleosynthesis indicated that visible matter (stars and the like) made up only about 10% of the amount of ordinary matter, but even if the dark matter was included, the baryonic part amounted to a mere 5% of the critical value. A density as small as $\Omega \cong 0.05$ could not account for the formation of structures in the universe, such as the clustering of galaxies. Many cosmologists therefore concluded that there must be a large component of dark, non-baryonic matter, a suggestion that found an enthusiastic response among particle physicists who had independently predicted exotic particles that might possibly serve as the stuff needed to make the universe critical.

The most obvious candidate was the neutrino, predicted by Wolfgang Pauli in 1930 and detected experimentally in 1956. Although the neutrino was well known, it was 'exotic' in so far that it is not part of ordinary matter (contrary to Pauli's original idea, according to which the neutrino was a constituent of the atomic nucleus). Of course, in order to be of any relevance to the dark-matter problem, it needed to have a mass, however small. Although the neutrino was traditionally taken to be massless, experiments could not rule out a very small mass, and some theories predicted that it did have a mass.[9] The question remained unsettled until 1998, when it was unexpectedly shown in the Super-Kamiokande neutrino observatory near Tokyo that cosmic-ray neutrinos 'oscillate' into some other kind of neutrino, which meant that at least one of the neutrinos involved must possess mass. The result caused problems for the standard model of elementary particles (which presupposed zero-mass neutrinos) and also implied that neutrinos must contribute to the non-baryonic matter density. Results from the Australian-based Two Degree Field (2dF) project showed in 2002 that the neutrino density was about the same as the density of ordinary matter ($\Omega_\nu \cong 0.05$) and therefore not of great significance to the dark-matter problem. The absolute masses of the neutrinos have not yet been determined, but recent experiments indicate that $m_\nu \leq 0.30$ eV for the three species taken together.

In the late 1980s a consensus emerged that the exotic part of the dark matter was 'cold', made up of relatively slowly moving particles unknown to the experimenters but predicted by physical theory. The particles of CDM (cold dark matter, a term introduced by Peebles in 1982) were collectively known as WIMPs, which stands for 'weakly interacting massive particles'. One of the CDM particle candidates was the axion, a particle Steven Weinberg had introduced in 1978 in connection with the theory of strong interactions (quantum chromodynamics), while others were predicted by supersymmetric theories, according to which all particles have partner particles. The most promising of these particles was the neutralino, a stable combination of supersymmetric partner particles called a photino, a zino, and a higgsino (partners to the photon, the Z-boson, and the elusive Higgs particle, respectively). There were, however, several other possibilities for dark-matter particles, including primordial black holes. Whatever the WIMPs are, they interact only by the weak and gravitational forces, and have in this respect much in common with the neutrinos (which adds to the difficulty of detecting them). So far, exotic dark matter, whether WIMPs or something else, has escaped detection, although there have been a few controversial claims of discovery.

High-energy physics made its most spectacular impact on cosmology in 1979–82, with the advent of inflation theory, a radically new conception of the earliest phase of the Big Bang universe. Several of the ingredients of the inflation scenario were first proposed by the Russian physicist Alexei Starobinsky, who in papers of 1979–80 developed a model in which the initial state of the universe was replaced by a brief but fiery expansion phase of the de Sitter type. Starobinsky's model was discussed by his mentor Zel'dovich and attracted much attention among Russian cosmologists, but it failed to make any impact on their Western counterparts.[10]

The effective invention of the inflation theory was due to Alan Guth, a young American particle theorist, who came to the idea while trying to understand the discrepancy between the abundance of magnetic monopoles predicted by grand unified theories and the fact that no such particle has been observed. Guth called attention to two other questions which he believed had been left unanswered by standard Big Bang cosmology. One was the horizon problem, which is essentially the problem that in the very early universe distant regions could not have been in causal contact, meaning that they could not communicate by means of light signals. The large-scale uniformity of the universe was hard to understand if it had developed from a huge number of causally separate regions. It could be ascribed to a Leibnizian pre-established harmony in the early universe, but such kinds of explanation did not appeal so much to physicists of the late twentieth century as they appealed to Leibniz. The horizon problem played an important role in the controversy over the steady-state theory in the 1950s, when it was discussed by Whitrow and Felix Pirani, among others. It was clarified by Wolfgang Rindler in a work of 1956 and seems to have been recognized for the first time in 1953 by Alpher, Herman, and Follin.[11]

Whereas the horizon problem was well known by 1980, the flatness problem was not. It was first described by Dicke in a popular lecture of 1969 and given further attention by Dicke and Peebles in 1979. A roughly flat space today ($\Omega \cong 1$) implies that the density must have been extremely close to the critical value at the beginning of the universe, and Guth emphasized that the standard model offered no explanation for this remarkable example of an apparent pre-established harmony: 'A universe can survive $\sim 10^{10}$ years only by extreme fine tuning of the initial values of ρ and H, so that ρ is very near ρ_{cr}. For the initial conditions taken at $T_0 = 10^{17}$ GeV, the value of H_0 must be fine tuned to an accuracy of one part in 10^{55}. In the standard model this incredibly precise initial relationship must be assumed without explanation.'[12]

The essence of Guth's revision of the standard model was that it introduced a brief phase in the life of the very early universe in which it expanded or 'inflated' by a truly gargantuan factor. This was assumed to take place by a kind of phase transition occurring shortly after the Planck time, the era $t_P = (hG/c^5)^{1/2} \cong 10^{-43}$ s, which marks the effective beginning of cosmological theory.[13] The phase transition, which was linked to certain properties of grand unified theories, produced a 'false vacuum', a temporary state of the lowest possible energy density, which could be described by particle theory but for which there was no experimental evidence. Guth explained that the false vacuum would lead to a gravitational repulsion, an expansion of the type proposed by de Sitter in 1917 and driven by the cosmological constant. One can think of the early universe, filled with a false vacuum, as expanding as $R(t) \sim \exp(t/\theta)$, where θ is an expansion time constant depending on the energy density of the false vacuum ρ_{fv}. The relationship follows roughly the expression

$$\theta^2 \cong \frac{3c^2}{8\pi G\rho_{fv}}.$$

The characteristic energy in the GUT era is 10^{14} GeV, from which ρ_{fv} follows, and then a value for the expansion constant of about 10^{-33} s. The corresponding Hubble parameter during the inflation phase is $H = \theta^{-1} \cong 10^{33}$ per second (or about 10^{14} in the standard units km/s/Mpc). Although the inflation period lasts only a split second, during this brief interval of time the radius of the universe will inflate by a factor of, say, 10^{40}. Since the energy density of the false vacuum (and not the energy itself) has the remarkable property that it remains constant, the inflation generates an enormous amount of latent energy.

The false vacuum can be conceived of as an excited vacuum state; like all excited quantum states it will be unstable, in this case decaying to the normal vacuum. When this happens, the explosive repulsion disappears, attractive gravity takes over, and the energy stored in the false vacuum is released. This will produce a universe filled with hot radiation energy and traces of matter at a temperature of about 10^{27} K, just as assumed in the initial conditions for the standard Big Bang model. From this point onward, the universe will continue to expand, but at the much slower rate of the standard theory.

Guth emphasized that the inflation model eliminated at least two fine-tuning problems of the standard theory. It provided a mechanism that drove the density Ω towards one during the inflation era, and in this sense solved the flatness problem, and it also predicted that the current value of the density parameter must be approximately one. The horizon problem was taken care of by the assumption that the universe—or the region of it corresponding to the observable universe—was incredibly small before inflation set in, so small that any part of it was causally connected with any other part and therefore at the same temperature. Starobinsky did not originally refer to the horizon problem, but in 1981 Zel'dovich pointed out that Starobinsky's theory solved this problem too. Finally, Guth's model also solved the monopole problem, not by ruling out the hypothetical particles but by making them extremely scarce because of the phenomenal expansion of space.

Inflation cosmology became an instant success in spite of various flaws in the original version, which Guth himself pointed out. One of the flaws was that the inflation did not leave the universe sufficiently homogeneous to be compatible with observations, a problem known as the 'graceful exit' problem. As early as 1982, Guth's theory was transformed into an improved version ('new inflation') by Andrei Linde in Russia and Paul Steinhardt and Andreas Albrecht in the United States. The new inflationary scenario was eagerly taken up by several research groups, who agreed that Guth's paper set the standards for any new theory of the early universe that was to be taken seriously. To speak of a paradigm is not much of an exaggeration, and physicists did indeed use the term. Between 1981 and the summer of 1996 about 3100 papers were published which referred to various aspects of inflation cosmology.

Of particular importance was the Nuffield Workshop on the very early universe that convened in Cambridge, England, in the summer of 1982 and was attended by, among others, Guth, Hawking, Starobinsky, Steinhardt, Sciama, Rees, and Michael Turner. One of the focal problems addressed in Cambridge was the calculation of the initial density inhomogeneities ('perturbations') that were needed to explain how structures had emerged in the universe. It was found that on the basis of the new inflation theory it was possible to understand the inhomogeneities as arising from quantum fluctuations, and calculations led to a definite prediction of the density inhomogeneities. The shape of the spectrum of these inhomogeneities agreed with what had been found by Edward Harrison in 1970 and, in a different way, by Zel'dovich two years later. The spectrum proposed by Harrison and Zel'dovich was motivated by attempts to account for the origin of large-scale structures, such as galaxies. Whereas the original Harrison–Zel'dovich spectrum was phenomenological, in 1982 it became based on physical theory. This was one more triumph for inflation theory, and reinforced its status as a promising new approach to the study of the very early universe.

The hugely successful inflation theory was not without its critics. Many observational astronomers found it as irrelevant as it was incomprehensible, and others wondered whether the success was on the social rather than the scientific level. Georges Ellis, an eminent South African theoretical cosmologist, wrote with Tony Rothman a paper where they sharply criticized the new inflation fashion. Their rhetoric brings to mind the earlier controversies related to Milne's rationalistic cosmology and the steady-state theory:

> A peculiar situation has arisen in cosmology. Over the last five years physicists have been hard at work on a theory that set out to resolve two problems that may not exist. This theory has no evidence to support it, and the one prediction it does make appears to be incorrect. . . . It is too early to make a conclusive judgment on inflation, which is, without argument, aesthetically pleasing. But there can also be no argument that cosmology is approaching the frontier where science is no longer based on experimental evidence and makes no testable predictions. Once this border is crossed, we have left the world of physics behind and have entered the realm of metaphysics.[14]

Other critics argued that, by the 1990s, there really was no inflation theory but only a wide range of inflationary models which included so many forms that, taken together, they could hardly be falsified observationally. Whatever the validity of such criticism, the inflation paradigm continued to prosper and to remain the preferred theory of the very early universe. Who could resist an idea which was, as Guth claimed, 'too good to be wrong'?

5.2 Observational surprises

The remarkable progress that occurred in cosmology in the late twentieth century owed as much to observations as it did to theory, if not more. The cosmic microwave background was the main empirical argument for the Big Bang theory, and work to obtain a more detailed picture of the radiation continued in the decades after its discovery in 1965. Observations showed that the background radiation was very uniform across the sky, in agreement with theory; but it must not be too uniform, for if there were no deviations from uniformity it seemed impossible to explain how the structures in the universe had been formed. Several experiments had looked for irregularities, or deviations from isotropy, but by the mid 1980s none had been found. In 1988 a team of astronomers from Nagoya University, Japan, and the University of California at Berkeley measured with a rocket-borne microwave detector the background radiation at three wavelengths. To everyone's surprise, they reported evidence for a major distortion of the background radiation relative to the black-body form. For two of the wavelengths they measured an excess flux, corresponding to a temperature significantly higher than the accepted 2.7 K, a result which was clearly inconsistent with the well-established Big Bang theory.

This controversial result was not confirmed by other experiments, and it was flatly contradicted by the superior data obtained from the COBE satellite experiment. The COBE, or Cosmic Background Explorer, project had been initiated as early as 1972, and the instrument-packed satellite was eventually launched into orbit 900 km above the Earth in the autumn of 1989. The first results from the satellite's FIRAS instrument, a specially designed spectrophotometer cooled to a temperature of 1.5 K, were announced in early 1990 by John Mather, and they showed that the background radiation fitted a black-body curve of temperature 2.735 ± 0.06 K most precisely. Only weeks after COBE had been launched, a rocket experiment reported a background temperature of 2.736 ± 0.017 K, in excellent agreement with that found by Mather and his FIRAS team. With these results, the Nagoya–Berkeley distortion was history, thought to be caused by some instrumental error.

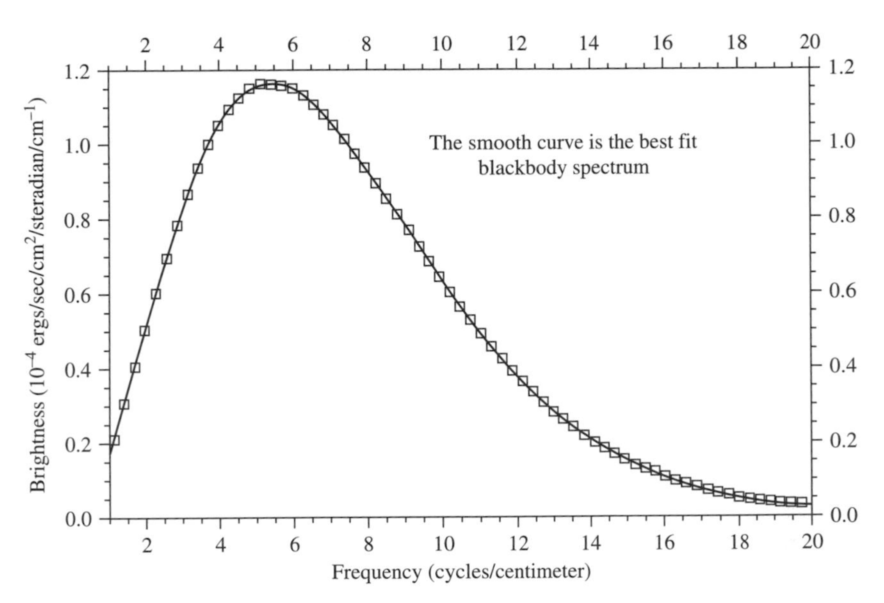

Fig. 5.1 The first COBE spectrum of the cosmic microwave background. The boxes are the measured data with their errors. From J. C. Mather *et al.*, 'A preliminary measurement of the cosmic microwave background spectrum by the Cosmic Background Explorer (COBE) satellite', *Astrophysical Journal* **354** (1990), L37–L40. Reproduced by permission of the American Astronomical Society. Credit to NASA Goddard Space Flight Center and the COBE Science Team.

Later COBE measurements gave an even more perfect fit, corresponding to a temperature which in 1994 was revised to 2.726 $\pm$ 0.010 K. (The precise temperature is not important by itself, though, as it slowly diminishes with the expansion of space.)

The COBE satellite also carried with it an instrument designed to detect variations in the microwave background by measuring simultaneously the radiation from two different directions in space; this instrument was known as the DMR (Differential Microwave Radiometer). After more than a year of operation, the DMR detector found what was hoped for, small variations which indicated density fluctuations in the early universe, some 300 000 years after the Big Bang.[15] What was measured was actually a temperature variation as small as $\Delta T/T \cong 5 \times 10^{-5}$, but this could be translated into a density variation of the same magnitude, approximately what was needed to provide the seeds from which galaxies had evolved. Some areas of the sky were slightly more dense and other areas slightly less dense than the average, the earliest indications of what would evolve into galactic clusters and void regions of space. This was good news, and it was also good news that the spectrum—the variation of the temperature difference with the angle of separation between the two directions—agreed nicely with predictions based upon the inflationary theory. Consequently, the COBE–DMR data were hailed not only as a success of the Big Bang theory but also of inflation cosmology. When George Smoot, the spokesman for this part of the experiment, presented the results in a TV-transmitted press conference in April 1992, he interpreted them as 'direct evidence of the birth of the universe'. As if this was not enough, he added that looking at the data was 'like seeing God'.[16]

The idea of constructing an orbiting astronomical observatory goes back to the early 1960s and was eventually, after many years of frustration and controversy, realized in April 1990 when the Hubble Space Telescope was launched into space on board the space shuttle Discovery. NASA's two-billion-dollar project was the most expensive pure-science project ever, astronomers' answer to the particle physicists' accelerator facilities.[17] ESA, the European Space Agency, contributed about 15% of the expenditure. After problems caused by misconstruction of one of the telescope's mirrors had been solved (by astronauts replacing some of the instruments), observations of very distant celestial objects started in 1993. The method was essentially the same as Hubble had used in the 1930s, but the Space Telescope could measure cepheids in galaxies much farther away than the Andromeda Galaxy, the most distant object that can be seen with the naked eye. The first results of 1994 were discomforting as they resulted in a Hubble constant of $H_0 = 73$ km/s/Mpc, which was taken to indicate an embarrassingly young universe. If a flat, matter-filled universe was assumed, the age of the universe came out as 8 billion years, about half the age estimated for certain clusters of galaxies.

Even with the excellent view provided by the Space Telescope, cepheids could not be seen at truly cosmological distances. The idea of using the much brighter—but also much rarer and ephemeral—supernovae as standard candles was first proposed by Baade and Zwicky in 1938, and half a century later, work was in progress to apply the method to measure the cosmic expansion. Tammann and his student Bruno Leibundgut had shown in 1990 that supernovae of a certain kind known as type Ia were extremely uniform, which made them ideally suited to the search for a more authoritative value of the Hubble constant.

It was primarily the competitive work of two international supernova research teams that led in the late 1990s to the surprising conclusion of an accelerating, yet critically dense universe.[18] The Supernova Cosmology Project (SCP) was formed in 1988 with the American physicist Saul Perlmutter as its leader, and the rival High-z Supernova Research Team (HZT) was established six years later by the Australian astronomer Martin Schmidt and others. The groups used ground-based telescopes and also, occasionally, the Hubble Space Telescope to look for rapidly receding type Ia supernovae. The SCP team found its first object in 1992 and in 1997 it reported data which indicated a universe with much more matter (in all forms) than the conventional value $\Omega_m \cong 0.3$ based on studies of galaxies and clusters. For a flat universe ($\Omega_m + \Omega_\Lambda = 1$), they got $\Omega_m = 0.94$, which did not leave much room for matter–energy associated with the cosmological constant. Perlmutter and his colleagues concluded that their results were inconsistent with a low-density, flat universe dominated by the cosmological constant.

By the end of 1997, the HZT scientists were convinced that Perlmutter's conclusion was wrong. From coordinated ground-based and Space Telescope observations of distant supernovae, they found a much smaller matter density and indications that the universe was either open or flat. Consequently, the dark energy associated with the cosmological constant had to play a significant role. At about this time the confusion began to disappear, with consensus emerging that the universe was in a state of acceleration and had been so for the last five billion years or so. What the two groups observed, to their surprise, was that the distant supernovae were dimmer than expected from conventional theory, the difference amounting to 20–25%. This might be due to local, more mundane effects, but it turned out

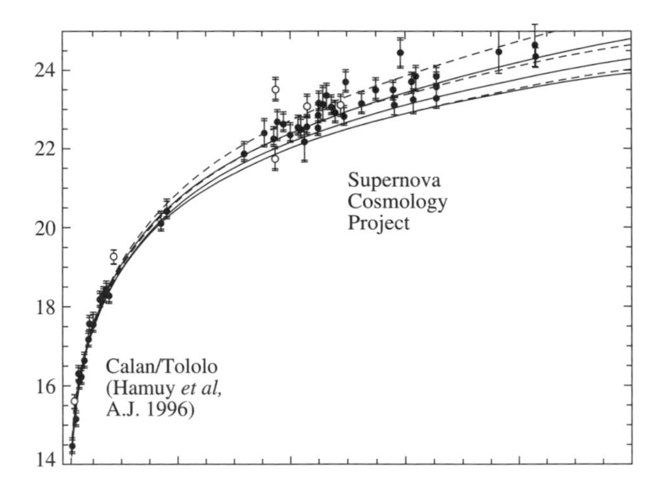

Fig. 5.2 Hubble diagram (apparent magnitude versus redshift) for 42 high-redshift type Ia supernovae measured by the Supernova Cosmology Project. Added to the plot are data from 18 low-redshift supernovae found by another group. From S. Perlmutter *et al*., 'Measurements of Ω and Λ from 42 high-redshift supernovae', *Astrophysical Journal* **517** (1999), 565–586. Reproduced by permission of the American Astronomical Society.

that these could be ruled out, which led the scientists to think that the faintness of the supernovae was caused by the very fabric of the cosmos.

In an article of 1998, two of the HZT scientists concluded that the universe was Λ-dominated and of an age of about 14.2 Gyr. Moreover, 'The derived value of q_0 is -0.75 ± 0.32, implying that the expansion of the Universe is accelerating. It appears that the Universe will expand eternally.'[19] When the SCP researchers published new data in 1999, they agreed essentially with the conclusion of HZT. During the next few years the new picture of the universe stabilized, at least as far as observations were concerned. Further work on high-z supernovae by the two teams confirmed that about 30% of the total energy density was mass and 70% was vacuum energy related to Λ, the cosmological constant. The new picture of the expansion of the universe, with a coasting point (rather than phase) marking the transition between deceleration and acceleration, was not unlike the one that Lemaître had proposed back in 1931, but few of the new generation of cosmologists would have known that.

The values of Ω_m and Ω_Λ could only be obtained by combining the supernova results with data on the anisotropy in the cosmic microwave background that were more precise than those provided by COBE. It was mainly for this purpose that the BOOMERANG project performed experiments with balloon-borne detectors in the Antarctic area (there was also another project of this kind, called MAXIMA, operating in the United States). These experiments resulted in 1999 in excellent data which showed a density of $\Omega_{\text{total}} = 1.00 \pm 0.04$. In 2001, a new satellite with instruments designed to measure variations in the background radiation was put into operation. The Wilkinson Microwave Anisotropy Probe (WMAP), named in honour of David Wilkinson, provided very precise data which, together with other evidence, resulted in an age of the universe of 13.4 ± 0.2 Gyr. As to the distribution

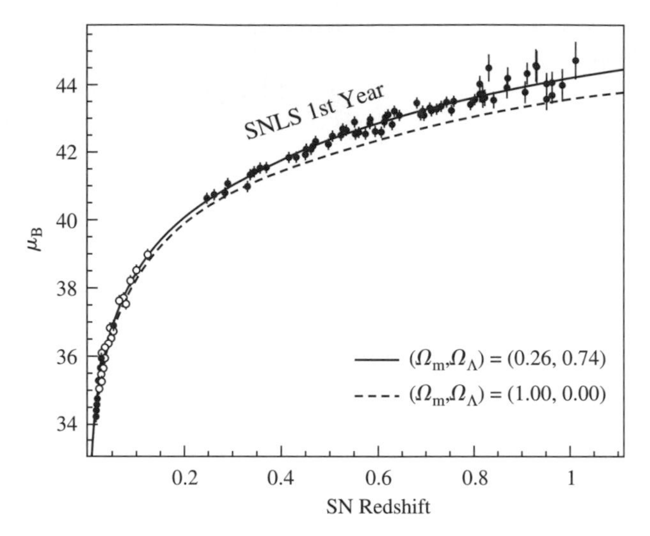

Fig. 5.3 Hubble diagram from the SNLS collaboration. The data, released in November 2005, gave strong support to a flat Λ cosmology. From P. Astier *et al.*, 'The Supernova Legacy Survey: Measurement of Ω_M, Ω_Λ and w from the first year data set', *Astronomy and Astrophysics* (to be published in 2006). Reproduced by permission of the Editor.

of energy, the result was $\Omega_m = 0.27$ and $\Omega_\Lambda = 0.73$. So far, this result has been fairly stable. It has received confirmation from several experiments, including a new Supernova Legacy Survey (SNLS), which presented its first measurements in the autumn of 2005. For a flat universe dominated by dark energy and cold dark matter, the SNLS scientists reported $\Omega_m = 0.263 \pm 0.042$ and a state of the dark energy that indicated that it was due to the cosmological constant.

With the new, accelerating universe there followed a new and mysterious form of energy (with relative density Ω_Λ), which many scientists identified with the vacuum energy associated with the cosmological constant.[20] As early as 1919, in an attempt to connect gravitation with electromagnetism, Einstein had suggested that the constant might play a role in atomic theory. In a later paper, of 1927, he considered a classical model of electrically charged particles with a negative pressure in the interior. This pressure he related to the cosmological constant. At that time, shortly after the introduction of quantum mechanics, a few physicists vaguely conceived that there might be some connection between the cosmological constant and the new quantum theory of atomic structure. On 3 February 1927, Weyl wrote to Einstein, 'all the properties I had so far attributed to matter by means of Λ are now to be taken over by quantum mechanics.'[21]

The first to point out that Einstein's constant of 1917 could be understood as a vacuum energy density was Lemaître in a paper of 1934, where he wrote: 'Everything happens as though the energy *in vacuo* would be different from zero. . . . We must associate a pressure $p = -\rho c^2$ to the energy density of vacuum. This is essentially the meaning of the cosmological constant λ which corresponds to a negative density of vacuum ρ_0 according to $\rho_0 = \lambda c^2/4\pi G \sim 10^{-27}$ gr/cm^3.'[22] Lemaître's consideration was not followed up, but similar ideas played a role in the version of steady-state cosmology McCrea developed in the 1950s. It was only when Zel'dovich connected the vacuum energy with ideas of quantum mechanics in 1968 that the energy interpretation of the cosmological constant became well known. And with Guth's inflation cosmology, it almost became fashionable, because the brief inflation phase can be ascribed to the repulsive effect of a cosmological constant.

According to quantum mechanics, the cosmological term does not belong to the left side of the general-relativistic field equations, the space–time part, but to the right side, together with other forms of energy (see p. 132). Although formally Einstein's cosmological constant and the vacuum energy constant are indistinguishable, conceptually they are entirely different. The suggestion that the major part of the critical density might consist of quantum vacuum energy was first made by Lawrence Krauss and Michael Turner in a paper titled 'The Cosmological Constant is Back' of 1995, that is, before observations had given support to an accelerating universe driven by dark energy. Krauss and Turner based their suggestion on the age of the universe, in particular, and argued that the best model would be one where the cosmological constant made up 60–70% of the critical density.[23]

The experiments of the 1990s convincingly demonstrated that there existed a 'dark energy' in addition to dark matter, and that the density of dark energy was about twice that of matter, mostly consisting of cold dark matter (CDM). Many cosmologists were persuaded by the 'ΛCDM paradigm', the belief that the universe consisted essentially of Λ-energy and cold dark matter. Some of the experiments in the early years of the twenty-first century, including WMAP and SNLS, were able to distinguish between different forms of dark energy, and they generally favoured Λ-energy or something acting like it. The WMAP measurements of 2003 did much to increase confidence in the ΛCDM model, which turned out to fit excellently with the data. The generally cautious Peebles concluded that the measurements provided 'strong evidence that the ΛCDM model is a good approximation to reality'.[24]

But not all physicists were comfortable with the intangible cosmological-constant, and there was no shortage of alternative explanations for the dark energy. One such alternative, which existed in several versions, carried the name 'quintessence', a fitting reference to Aristotle's semi-metaphysical, celestial element. Like vacuum energy, quintessence has negative pressure and acts antigravitationally at very large scales, but its pressure is less negative and can change with place and time. Another possibility, known as 'phantom energy', corresponds to an antigravity force increasing to infinity within a finite span of time. In this scenario, everything from atoms to galaxies will eventually be torn apart in a final 'big rip'.

Among the reasons for scepticism was that the cosmological-constant model seemed to require a few implausible coincidences. For example, since the end of inflation the mass density has decreased drastically, whereas the vacuum energy density has remained constant, and yet we happen to live in a world where the two energy densities are approximately the same. Why? Further, it was known from astronomical observations that Einstein's cosmological constant cannot be larger than 10^{-56} cm^{-2}, a bound which was translated into a vacuum energy density not greater than 10^{-29} g/cm^3. By contrast, the quantity as calculated by the quantum physicists was at least 10^{40} times greater! No wonder that Guth wrote, in his 1981 paper on the inflationary model, 'The reason Λ is so small is of course one of the *deep mysteries* of physics.'

Whether or not the accelerated expansion of the universe was caused by a cosmological constant, as Eddington and Lemaître believed in the 1930s, the new picture of a runaway universe caught the attention not only of the scientists, but also of the mass media and the general public. By the end of 1998, the matter-dominated, flat universe of the Einstein–de Sitter kind was buried and the accelerating universe was hailed as a fact, if still a controversial one. The historical significance of the 1998 'revolution' was emphasized by a public programme on the 'Nature of the Universe Debate: Cosmology Solved?' held at the Smithsonian National Museum of Natural History. Here, James Peebles and Michael Turner debated the new world picture in the same format and in the same auditorium as where Shapley and Curtis had had their 'Great Debate' nearly eighty years earlier. The whole session was arranged to create parallels to the historic 1920 event.[25]

The change in the world picture even inspired the literary world, as in the case of the American author John Updike, who in 2004 wrote a short story titled 'The Accelerating Expansion of the Universe'. Updike's story included a precise account of the new universe:

> But the fact, discovered by two independent teams of researchers, seemed to be that not only did deep space show no relenting in the speed of the farthest galaxies but instead a detectable acceleration, so that an eventual dispersion of everything into absolute cold and darkness could be confidently predicted. We are riding a pointless explosion to nowhere.

Like John Donne four hundred years earlier, Updike expressed concern about the implications of the revolution in the world picture: 'The accelerating expansion of the universe

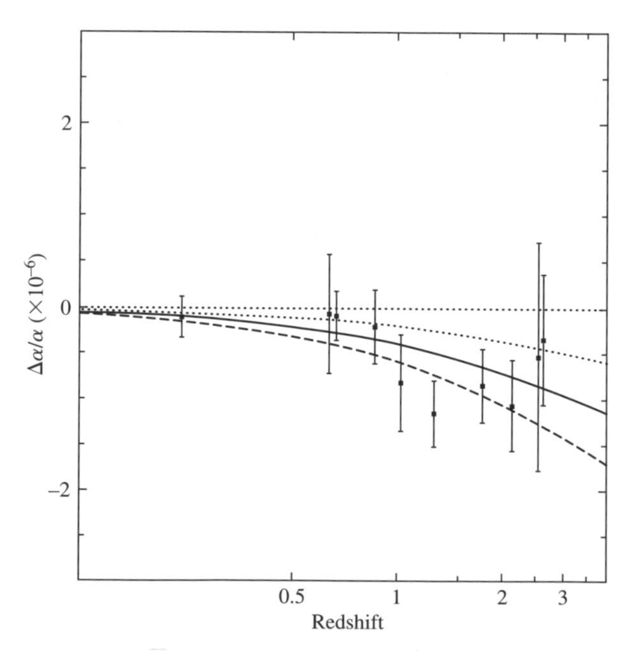

Fig. 5.4 Data points for the variation of the fine-structure constant with the redshift of quasars, based on measurements around 2000. The solid line depicts the best fit according to some theoretical $\alpha(z)$ models. Reproduced from Magueijo 2003, p. 2056, with the consent of the author and Institute of Physics Publishing.

imposed an ingnominious, cruelly diluted finitude on the enclosing vastness. The eternal hypothetical structures—God, Paradise, the moral law within—now had utterly no base to stand on. All would melt away.'[26]

The general feeling of optimism and excitement in the cosmological community received one more boost when it was announced in 2001 that analysis of absorption lines from quasars suggested that the fine-structure constant ($\alpha = 2\pi e^2/hc$) varied with time. The variation corresponded to a change of $(\Delta\alpha/\Delta t)/\alpha = 5 \times 10^{-16}$ per year, a tiny but measurable effect. If the result had been confirmed, it would have meant that at least one of the fundamental constants e (the elementary charge), h (Planck's constant), and c (the velocity of light) varied with time, which presumably would have important cosmological consequences. The announcement caused great excitement, and theoretical cosmologists immediately constructed cosmological models that accommodated a varying fine-structure constant. However, enthusiasm cooled when the much-publicized observations were contradicted by new, still more precise absorption data. According to a press release from the European Southern Observatory of May 2004, 'New quasar studies keep fundamental constant constant.'[27]

5.3 Anthropic and other speculations

Ideas of multiple universes (sometimes called 'multiverse') have a long history in speculative cosmology and exist in many forms, some contemplating temporally and others spatially separate universes.[28] For example, the latter concept of simultaneously existing sub-universes was entertained by Anaximenes in ancient Greece, and, much later, it was discussed by Boltzmann in the 1890s. In the early 1960s the idea was developed by Jaroslav Pachner and R. Giovanelli, among others. With inflation theory, it seemed to some cosmologists that such speculations might be turned into scientifically respectable hypotheses, indeed that they followed from the inflationary scenario. Shortly after Guth introduced his theory of the inflationary scenario in 1981 it was developed into hypotheses of multiple universes by J. Richard Gott, Katsuhiko Sato, Andrei Linde, and others, and within a few years many-worlds cosmology was established as a small but thriving cottage industry. In his survey of the new inflation theory at the Nuffield Workshop in 1982, Linde noted (tongue in cheek?) that 'it is not quite clear how one can describe creation of the universe from nothing', and went on to speculate that there might be no initial creation to worry about.[29] Perhaps the inflationary universe was born as a 'child universe' spatially disconnected from a 'mother universe', which was again the offspring of a 'grandmother universe', etc.

Linde would soon develop these ideas into what he called the chaotic inflation theory, a kind of theory which differed drastically from what just a few years earlier had been the standard Big Bang model of the universe. Whereas the 1982 Linde–Albrecht–Steinhardt theory still assumed an initial singularity and a hot early phase after the Planck time, Linde now envisaged a multitude of mini-universes, none of which was 'first'. These mini-universes produce numerous others ('baby universes'), and although some of them will eventually collapse, the process of universe generation will proceed endlessly. The essence of what he called 'a kind of Darwinian approach to cosmology' was that there was no single beginning of the universe as a whole and no end either. It seemed more likely, he argued,

> that the universe is an eternally existing, self-producing entity, that it is divided into many mini-universes much larger than our observable portion, and that the laws of low-energy physics and even the dimensionality

of space–time may be different in each of these mini-universes. This modification of our picture of the universe and of our place in it is one of the most important consequences of the inflationary scenario.[30]

According to Linde, we live in a four-dimensional, low-energy universe because this is the only kind of universe we can live in. The chaotic inflation model, as developed by Linde and others through the 1990s, described a stationary universe where the notion of *the* Big Bang lost its meaning and was removed to an indefinite past. As Linde and others noted, it had some similarity to the long-discarded steady-state model and even more to the quasi-steady-state cosmology advocated by Hoyle, Narlikar, and Burbidge.

The chaotic inflation theory was only one among several cosmological scenarios of a somewhat speculative and sometimes bizzare nature that were discussed at the end of the twentieth century. Some of these theories derived from attempted unifications of quantum mechanics and general relativity theory, and included cosmological ideas based on superstring theory and related many-dimensional space–time theories. Generally speaking, this area of research seems to be a mathematician's delight where connections to the world of empirical physics are either missing or thought to be unimportant.

Many of the early-universe scenarios were developments of the inflation theory, but there were also attempts to construct alternative cosmologies that avoided inflation. One such theory, based upon the hypothesis that the speed of light has changed with time, was first proposed by John Moffat in 1993 but only received attention after it was developed independently by João Magueijo and Andreas Albrecht in 1999. Whereas the earlier speculations of the 1930s had assumed a slow variation in the speed of light, the new VSL (varying-speed-of-light) cosmology predicted a drastic decrease in the speed of light at a time close to the Planck time. According to one version of VSL theory, in the very early universe light moved with a speed of, for example, 10^{38} km/s and then the speed suddenly fell to its present value of merely 300 000 km/s. By assuming such a gigantic speed of light in the early universe, the horizon problem could be explained without the inflation hypothesis, and VSL theory also accounted for some of the other cosmological problems that inflation explained.

The 1999 Magueijo–Albrecht theory was deliberately constructed as an alternative to inflation, but later versions turned VSL cosmology into a much broader and more general type of theory.[31] The change in the velocity of light means that the theory violates the fundamental principle behind Einstein's relativity theory, and it also follows from VSL theory that energy conservation is violated in the very early universe. Advocates of VSL theory believe that these drastic changes in physics can be justified. VSL cosmology leads to a number of predictions that distinguish it observationally from other theories of the early universe, but empirical evidence in favour of the theory has so far been largely missing. It is particularly problematical that it has not proved possible to provide a theory of structure formation based on the assumption of a varying speed of light.

Some of the cosmological scenarios discussed in the late twentieth century related to the old question of life in the universe and its fate in the far future. In a general sense, interest in life elsewhere in the universe was stimulated by the discovery of planets outside the solar system, called exoplanets, in the 1990s. The first such discovery, of a Jupiter-sized planet orbiting around the star 51 Pegasi, was announced in 1995 by two Swiss astronomers, Michel Mayor and Didier Queloz, and ten years later more than 170 exoplanets were known. Although exoplanets are of no direct cosmological significance, the possibility of life-sustaining conditions far away from the Earth plays some role in cosmologists' discussions. This is not a new phenomenon. For example, it was an important motive behind the stationary models proposed in the interwar period that they allowed life processes to go on indefinitely and everywhere in the universe. This could also be a reason to prefer the steady-state model over the Big Bang model, as was in the case for the young Dennis Sciama.[32]

While the possibility of life elsewhere in the universe has been used as an argument in favour of steady-state models, it has also been argued that a class of otherwise possible steady-state models can be ruled out if it turns out that there are no life-forms in the Milky Way other than what we know on Earth. It has even been suggested that life may be *essential* to the universe, in the sense that only life can prevent the universe from destroying itself.[33] Without life, no universe! These kinds of speculation exist in a variety of forms, many of them related to the popular idea of baby or bubble universes coming from the inflationary scenario. Linde concluded in 1994 that it was an appealing feature of his chaotic inflation model that it predicted endless life: 'One can draw some optimism from knowing that even if our civilization dies, there will be other places in the universe where life will emerge again and again, in all its possible forms.'[34] Optimism?

Lee Smolin and others have gone a step further. According to Smolin, the entire universe may be considered a living entity which evolves in a strict Darwinian sense. In a paper of 1992, Smolin proposed a new approach to the problem of the 'unnatural values' of the dimensionless constants of fundamental physics.[35] As an alternative to anthropic pseudo-explanation (as he considered it), he suggested that the constants evolved in a series of collapses and expansions of universes controlled by the production of black holes. Relying on ideas of quantum gravity, he developed a speculative 'mechanism for natural selection in cosmology', which he argued was of a nature similar to biological evolution, including the element of randomness and the element of natural selection. As biological metaphors and reasoning have inspired some cosmologists, so have theological metaphors and reasoning. At a conference in the Vatican Observatory in 1991, George Ellis argued that even the omnipotent God could not have created an infinite universe filled with matter and life-forms. If our universe is God's, as Ellis believed, it must be finite in size.[36]

The intense discussion about the very early universe that followed the invention of the inflation model was to some extent paralleled by an interest in the other extreme of the timescale, the state of the universe in the very far future. Such scenarios had been discussed since the mid nineteenth century in connection with the heat death, and from about 1980 they were reconsidered from the perspective of Big Bang cosmology.[37] The new wave of 'physical eschatology' started in the 1970s, when papers on the long-term future of the universe were published by John Barrow, Frank Tipler, Jamal Islam, Freeman Dyson, Paul C. W. Davies, and others. Dyson, a distinguished theoretical physicist and one of the fathers of modern quantum electrodynamics, wrote in 1979 an article in which he advocated that the study of the end of the universe should be developed into a scientific discipline of the same respectability as the study of the early universe.

The standard view was (and presumably still is) that 'The more the universe seems comprehensible, the more it also seems pointless,' as Weinberg wrote in his best-selling *The First Three Minutes*. Or, in the words of Bertrand Russell, 'The universe may have a purpose, but nothing that we know suggests that, if so, this purpose has any similarity to ours.'[38] Dyson, who wanted a universe with meaning and life, strongly disagreed. He argued that it would be possible for life and intelligence to survive indefinitely in an open universe and possibly also to maintain intergalactic communication in spite of the never-ending expansion. 'I have found a universe growing without limit in richness and complexity, a universe of life surviving forever and making itself known to its neighbors across unimaginable gulfs of space and time.'[39] Of course, other scientists concluded differently.

It goes without saying that eschatological studies of Dyson's kind rely on gross extrapolations and involve a good deal of speculation. The point to notice is that they were nonetheless taken seriously by many physicists and astronomers and continue to be so. Shortly after the discovery of the accelerating universe, several papers appeared with titles such as 'Can the Universe Escape Eternal Acceleration?' and 'The Ultimate Fate of Life in an Accelerating Universe'. Although most of the literature ignored references to theology, there was no shortage of papers and books cultivating the links between religious and physical eschatology. A quarter of a century after Dyson's pioneering paper, the eschatological aspects of cosmology have turned into a small industry. This is interesting from a sociological and psychological point of view, but it does not necessarily follow that physical eschatology has therefore become a proper science.

Life in general, and the existence of humans in particular, came to play a surprising role in cosmological thinking with the formulation of the anthropic principle in the 1970s. The general idea of the anthropic principle is that the observed universe is conditioned by the fact that there exist human observers. In any of its many versions, it claims that the present epoch is privileged in the sense that it is the epoch in which carbon-based life originated: life could not have originated in any other epoch, and this constrains the numerical values of the natural constants and fundamental parameters of physical theory. The world is as it is because we exist!

Although anthropic-like arguments can be found in the late nineteenth century, they first appeared in the context of the expanding universe about 1960, when they were introduced by Grigory Idlis in the Soviet Union and Robert Dicke in the United States. In a lecture of 1958 Dicke pointed out that the present epoch was not random, but was conditioned by the fact that the biological conditions for the existence of humans must be satisfied. Three years later, he published a brief paper in which he discussed several of the large dimensionless numbers that Dirac had called attention to in his cosmological theory of 1937–38. Whereas Dirac had inferred from the numerical coincidences that some of the constants of nature varied with time, to Dicke the coincidences could be explained by 'the existence of physicists now and the assumption of the validity of Mach's principle'. More specifically, he argued that 'with the assumption of an evolutionary universe, T [the age of the universe] is not permitted to take one of an enormous range of values, but is somewhat limited by the biological requirements to be met during the epoch of man'.[40] The British biologist Charles Pantin pointed out in 1965 that the fortuitous properties of substances that are necessary for life (such as carbon and water) could be explained if it was assumed that there was an indefinite number of universes, and that ours happened to be the one in which the the right conditions for life existed. Pantin spoke of 'a solution analogous to the principle of Natural Selection', another early reference to cosmo-Darwinism.

It was Brandon Carter, a student of Sciama, who coined the name 'anthropic principle' and elevated it to such a status that cosmologists began to take it seriously. Carter had for several years been occupied with trying to understand the role of microphysical parameters in cosmology, and in 1967 he wrote a manuscript on the subject which included a development of Dicke's line of reasoning. Although the manuscript was never published, its content was known to the small community of cosmologists and referred to in a 1973 paper by Hawking and Barry Collins with the title 'Why is the Universe Isotropic?' The two Cambridge cosmologists found that 'the most attractive answer' to the question was that 'the isotropy of the Universe and our existence are both results of the fact that the Universe is expanding at just about the critical rate. Since we could not observe the Universe to be different if we were not here, one can say, in a sense, that the isotropy of the Universe is a consequence of our existence.'[41]

When Carter published his ideas in 1974, he objected to what he found was an unwarranted extension of the Copernican principle that we do not occupy a privileged place in the universe. Although there is indeed no such privileged place, there is—contrary to the perfect cosmological principle—a privileged time, namely the epoch of life. Carter distinguished between a weak and a strong form of the anthropic principle, where the former states that 'our location in the Universe is *necessarily* privileged to the extent of being compatible with our existence as observers'. The strong anthropic principle goes further, stating that 'the Universe (and hence the fundamental parameters on which it depends) must be such as to admit the creation of observers within it at some stage'.[42]

Carter's formulation of 1974 was further developed by Bernard Carr and Martin Rees, who in 1979 published a review paper in *Nature* entitled 'The Anthropic Principle and the Structure of the Physical World'. With the works of Hawking and Collins, Carter, and Carr and Rees, the anthropic principle became established as an important ingredient of cosmological thinking, and these works served to make it legitimate, if by no means uncontroversial, to explain various aspects of the universe by appealing to the existence of human life. The number of articles involving anthropic reasoning exploded and the number of versions proliferated (in addition to Carter's weak and strong principles, there soon appeared 'final', 'participatory' and 'theological' versions, to mention but a few).[43] The subject was discussed not only by physicists and astronomers, but also by philosophers and theologians, who found a welcome opportunity to contribute to modern cosmology, a field from which they had long been banished.

In spite of its undeniable popularity, the epistemic nature of the anthropic principle is questionable. Is it merely a philosophical toy, or does it provide genuine explanations endowed with predictive power? P. C. W. Davies and Frank Tipler argued in about 1980 that the anthropic principle ruled out any cosmology with an infinite past, such as the Hoyle–Narlikar steady-state theory, which to anthropically minded cosmologists proved that the principle does have a kind of predictive power, if in this case only post hoc. Another often-quoted example is Fred Hoyle's 1953 prediction of a 7.7 MeV resonance level in the carbon-12 nucleus, a breakthrough in the understanding of stellar nucleosynthesis because it was necessary for the production of heavier elements. Since carbon evidently exists, it must have been produced in some way, which caused Hoyle to predict the resonance state necessary for the production of carbon.[44] Shortly after Hoyle's prediction, this was confirmed experimentally. However, it is far from clear that this is a genuine example of anthropic reasoning, and at any rate it does not relate to the existence of human life, only to carbon atoms. Although carbon is essential to humans, it is equally essential to limestone, ferns, and cockroaches.

5.4 The problem of creation

The Big Bang theory, as it emerged victoriously in the 1960s, assumed a hot, dense primordial state of the universe, but it did not offer an explanation of how this state had come into existence nor of what caused the bang. The only sort of explanation was that it was possibly the outcome of a previous, collapsing universe, which was no real explanation, as it left open the question of how this earlier universe was formed. The inflationary theory gave an account of the universe as early as, say, 10^{-30}s after $t = 0$, and it explained how all the particles (about 10^{90}) of the observed universe were produced. In this limited sense it was a creation theory, but it presupposed a small amount of matter to start inflation, and for this reason inflation theory did not come up with an explanation of the ultimate creation of the universe either. It starts shortly after the Planck time, but has nothing to say about earlier times, not to mention the magical moment of $t = 0$.

Cosmologists had traditionally avoided the problem of creation, a concept which might seem to be foreign to science and better suited to philosophical and theological discourses. If the universe had come into existence, was created, this could not be explained by reference to an earlier state. Neither did it help to say that the universe had been created from nothingness, or from an initial singularity, for neither of these concepts could be assigned physical meaning. And of course one could not appeal to a supernatural act, which would be to leave science altogether and enter theology. The somewhat timid attitude to the creation problem that characterized cosmologists in the era up to the 1970s changed markedly in the decades that followed.

Cosmologists in the modern period had different attitudes to the question of cosmic creation, depending in part on their favoured models and in part on their spontaneous philosophical preferences. According to some models, such as the Hoyle–Narlikar steady-state theory and Linde's self-reproducing inflation theory, there was no problem, since the universe at large had no beginning. Other cosmologists picked up the Milne–Misner idea of a redefinition of time to avoid the unpleasant Big Bang beginning. The no-beginning solution was also a feature of cyclical models with phases of expansion and contraction endlessly following one another. Such models, going back in a relativistic context to Friedmann and Tolman, were, however, thought to be ruled out observationally, and in addition they faced a number of theoretical problems. For example, no mechanism had been found to reprocess the matter–energy of the previous universe into the new one. Nonetheless, the general idea of a phoenix universe continued to fascinate a minority of cosmologists, and it reappeared in the early years of the twenty-first century. Whereas the classical cyclical models presupposed a closed universe, the new kind of phoenix universe advocated by Steinhardt and Neil Turok operated with an infinite, flat space, in agreement with observations.[45] In spite of the variety of no-beginning models, the standard view was that the Big Bang was real and ultimately in need of some kind of explanation.

Edward Tryon, a physicist at the City University of New York, suggested in 1973 that the universe had appeared from 'nowhere' without violating any conservation laws. According to Heisenberg's uncertainty relations, an energy fluctuation ΔE can occur only for a period of $\Delta t = h/\Delta E$ and thus be sustained for a very long time only if the net energy is nearly zero. By suggesting that the net energy of the universe was indeed zero, Tryon could produce a long-lived universe from a quantum fluctuation. The zero net energy came about because the positive mass–energy of a piece of matter mc^2 would be cancelled by its contribution to the negative gravitational energy:

$$mc^2 - \frac{GmM}{R} \approx 0,$$

where M denotes the mass of the universe contained within the Hubble radius $R = c/H$. Unknown to Tryon, the same idea had been entertained by Haas and Jordan in the 1930s. As to the mechanism of the origin of the universe, Tryon thus proposed that the entire universe (assumed to be closed) had originated spontaneously as a giant fluctuation in a quantum vacuum; and as to why it occurred, he offered 'the modest proposal that our Universe is simply one of those things which happen from time to time'.[46] It followed from Tryon's idea that the universe must be closed and also that there must be equal amounts of matter and antimatter in the universe, which presumably were two reasons why it was not received with enthusiasm. Another reason might have been that it failed to explain the creation itself, as it merely pushed back the creation scenario to a hypothetical quantum vacuum scenario which Tryon chose to call 'nowhere'.

In the wake of inflation theory, Alexander Vilenkin proposed in 1982 that the universe was created by a kind of tunnel effect, as known from quantum mechanics.[47] The title of Vilenkin's paper, 'Creation of Universes from Nothing', must have surprised many readers of *Physics Letters*, but it was only the beginning of a trend. Vilenkin's model did not have a Big Bang singularity and did not require any initial conditions. Although he claimed that his theory explained how 'the universe is spontaneously created from literally *nothing*', Vilenkin too presupposed a fuzzy quantum space–time background that the universe tunneled *from*. The creation of a universe out of vacuum was also studied by Robert Brout, Heinz Pagels, and others, who tried to find a mechanism that allowed for the generation of an open universe.

A different approach was suggested in a 1983 paper by James Hartle and Hawking, who took their point of departure as the Wheeler–DeWitt equation, a quantum-mechanical wave equation supposed to describe the entire universe, which Bryce DeWitt had suggested in 1967 (Wheeler's contribution was in 1968). Hartle and Hawking developed a wave function of the universe and argued that it represented the amplitude of the universe coming into existence from a finite quantum fuzz. There would be no problem of creation, because near $t = 0$, in the quantum fuzz, the very notion of space and time would lose its meaning. By being self-contained, the universe would be neither created nor destroyed. Although the Hartle–Hawking model described a universe with a finite past, it had no initial singularity or beginning, for there was no initial boundary at all. In his best-selling *A Brief History of Time*, Hawking popularized the 'no boundary' idea:

> The quantum theory of gravity has opened up a new possibility, in which there would be no boundary to space–time and so there would be no need to specify the behaviour at the boundary. . . . One could say: 'The boundary condition of the universe is that it has no boundary.' The universe would be completely self-contained and not affected by anything outside itself. It would neither be created nor destroyed. It would just BE.

Hawking claimed further, famously and controversially, that his no-boundary picture of the universe had profound theological implications, as it made God superfluous: 'So long as the universe had a beginning, we could suppose it had a creator. But if the universe is really completely self-contained, having no boundary or edge, it would have neither beginning nor end: it would simply be. What place, then, for a creator?'[48]

The Hartle–Hawking theory was much discussed, not because its scenario was thought to be realistic but because of its conceptual and mathematical novelty. It was followed by a number of other proposals based on quantum gravity, the kind of theory which aims at unifying general relativity and quantum mechanics and assumedly governs the earliest state of the Big Bang universe around or before the Planck time. Among the many theories of this kind, string theory (in one of its many versions) is the best known, and the one most developed since it was introduced as a theory of strong interactions about 1970. String theory operates with tiny many-dimensional objects (strings or branes) of a size close to the Planck length of 10^{-35} m, and string theorists believe that all particles and fields can be understood in terms of these hypothetical objects. It will come as no surprise that the ambitious and mathematically complex string theory has also been used as a foundation for cosmological theories.

Some of the cosmological scenarios proposed by string theorists have in common with the Hartle–Hawking model that they avoid the troublesome cosmic beginning. Quantum strings have a non-zero size and cannot collapse to an infinitesimal point, which provides a way to eliminate the initial singularity, although not in a very satisfactory way. If the universe started in a non-reducible string object, where did the string come from? According to Gabriele Veneziano, an Italian physicist who pioneered the string theory in the late 1960s, it makes sense to ask what happened before the Big Bang. In what is known as the pre-Big Bang scenario, there existed an accelerating universe before the Big Bang, out of which our initially decelerating universe emerged in a violent transition. Veneziano explains that 'the pre-big bang universe was almost a perfect mirror image of the post-bang one. If the universe is eternal into the future, its contents thinning to a meager gruel, it is also eternal into the past. Infinitely long ago it was nearly empty, filled only with a tenous, widely dispersed, chaotic gas of radiation and matter.'[49] It may appear unlikely that the string scenario can ever establish contact with empirical physics, but advocates of the pre-Big Bang theory think that it does have physical consequences and that these may be measurable as small temperature variations in the cosmic microwave background.

Let me end by quoting two leading American particle cosmologists, Edward Kolb and Michael Turner, who in 1990 published a monograph on the early universe:

> Whatever future cosmologists write about cosmology in the decades following the discovery of the CMBR [cosmic microwave background radiation], we can be certain that they will not criticize contemporary cosmologists for failure to take their theoretical ideas—and sometimes wild speculations—seriously enough. Perhaps future cosmologists will laugh at our naïvité. But, if they do, we can hope they will admire our courage and boldness in attacking problems once thought to be beyond the reach of human comprehension.[50]

This evaluation is as valid today as it was in 1990. Indeed, in their preface to the 1993 paperback edition, Kolb and Turner noted that 'We would be most surprised if the future did not include a revolutionary idea or unexpected discovery.' They were closer to the mark than they could have imagined.

5.5 Cosmology in perspective

Cosmology is the science of the universe and thus is based on a concept—the universe, world, or cosmos—that is immensely greater and more abstract than what can be directly perceived. During the course of history the very meaning of the universe has changed, and sometimes radically so, which is a major reason for the somewhat irregular development of cosmological thought. The Babylonian and Greek concepts of the universe have very little in common with that of modern cosmologists, so it is no wonder if the long-term history may appear to lack continuity. Yet, in spite of all the changes, the modern world picture does have a connection to the thoughts about the universe that emerged some 2500 years ago. It is the outcome of a long and complex historical process with roots going back to the era of the Presocratic philosophers, if not before. Aristotle noted that cosmology was derived from humans' elementary quest to know as much as possible about the observed universe. In his *Metaphysics*, he wrote:

> It is owing to their wonder that men both now begin and at first began to philosophize; they wondered originally at the obvious difficulties, then advanced little by little and stated difficulties about the greater matters, e.g., about the phenomena of the Moon and those of the Sun and the stars, and about the origin of the universe.[51]

The sense of wonder that Aristotle spoke about is an invariant in the history of cosmology, rooted as it is in elementary curiosity. When modern physicists wonder about the nature of dark energy or why there are so many more photons than baryons in the universe, they are engaging in a quest of the same character as when Eudoxus wondered about Mars's retrograde motions in its journey over the heavens. Although the answers to the questions have changed greatly, there has been a measure of permanence in the kind of questions people have wanted to know the answers to. Is the world finite or infinite? Does life exist elsewhere in the universe? Has the world always existed, or did it have an origin? Is it static, or in a state of evolution? Does the universe have a purpose? Grandiose questions like these were asked more than two thousand years ago, and they can be followed through history since then, including the modern era. We know the answers to some of the questions, but not to all. And there are questions that may be important, but may well be scientifically meaningless.

For long periods of time the universe was not just an object of study, it was conceived as being meaningful and being associated with the existence of humans and their place in the cosmos. To understand the universe necessarily involved understanding its meaning and purpose, concepts that could not be separated from a religious context. The references to gods, spirits, and intelligences as active agents in the dynamic cosmos continued well into the sixteenth century. Newton explained the heavenly phenomena in terms of his law of gravity, but at the same time he needed God—'very well skilled in Mechanicks and Geometry'—to account for the stability of the universe. A century later, astronomers might still refer to the creator, but he was not called upon to explain phenomena in the heavens. When Napoleon Bonaparte had perused Laplace's *Exposition du système du monde* he supposedly commented to its author that while Newton spoke of God in *Principia*, he did not find God mentioned in the *Exposition*. Laplace is to have responded, 'Citizen First Consul, I have no need for that hypothesis.'[52] Nor do modern cosmologists have need of that hypothesis; yet, as we have observed, flirtation with religious aspects is not uncommon in present cosmological theory. The vast majority of cosmologists, however, pursue their science with no reference to religious questions, quite in the manner of Laplace.

5.5.1 *Paradigms and traditions*

Generally speaking, either a particular branch of science may evolve continuously and cumulatively, or its development may be characterized by a series of revolutionary shifts involving changes in the very foundation of the science. In his influential *The Structure of Scientific Revolutions*, Thomas Kuhn advocated a dynamical view of science where periods of paradigm-governed 'normal science' end in a state of crisis, which may trigger a 'revolution' that leads to the old paradigm being replaced by a new one. According to Kuhn's original model of 1962, two rival paradigms may coexist only briefly, until one of them eradicates the other. Since they describe nature in different, untranslatable languages, they are incommensurable and there is no way to decide between them in purely rational terms. Whatever the virtues of Kuhn's attempt to reform the philosophy of science, it has long been recognized that revolutionary breaks of the strong kind he spoke of hardly ever occur in the real history of science. On the other hand, one can reasonably use the concepts of paradigm and revolution in a weaker sense and look at the historical development in the framework of those concepts. I believe that the history of cosmology yields some support for the notion of paradigm-governed science and even of revolutionary breaks, if not in the radical sense originally proposed by Kuhn.

We have in the 'circle paradigm' a remarkable example of a tenacious fundamental belief that survived for nearly two millennia and which shows some of the characteristics of a Kuhnian paradigm. According to this belief, going back to Plato and Eudoxus, the observed motions of the celestial bodies must necessarily be understood in terms of uniform circular movements. As Geminus pointed out, it was part of the definition of astronomy, and hence indispensable. If astronomy was the art of reducing observations to circular motions, how could Ptolemy's theories of the planets possibly be seriously wrong?

It took Kepler's genius to recognize that the belief was merely a working hypothesis, not a necessity, and it took several decades until Kepler's insight was accepted by the majority of astronomers. The belief in celestial circles was part of a larger complex of paradigmatic ideas, which included the idea that the universe was finite (and itself of spherical shape) and divided into two very different regions, the world beneath the Moon and the heavenly world above it. Also in later history, we meet examples of beliefs and traditions that were rarely questioned and which formed the framework of cosmological thinking, and hence had the character of paradigms. Until about 1910 it was generally believed that the stellar universe was limited to the Milky Way, and until 1930 the static nature of the universe as a whole was taken for granted. Current cosmology is founded on the theory of general relativity and some kind of Big Bang scenario, elements that are beyond discussion and conceived of as defining features of cosmological theory. Yet, although it may be tempting to characterize these beliefs as 'paradigmatical', they are so in a different sense from what Kuhn spoke of in his 1962 work.

As there have been periods in which astronomy and cosmology were governed by paradigmatic dogmas, so there have been a series of changes in which the accepted view of the cosmos broke down and was replaced by a new one. None of these breaks, dramatic as they may have appeared to contemporary scientists, have been revolutions in the radical sense of clashes between incommensurable world views. The much-discussed Copernican revolution was a long-term transitional process in which the heliocentric system was compared rationally and systematically with the traditional geocentric system. The outcome of the revolutionary process was a rejection of the traditional world view, but astronomers involved in the process did not need to commit themselves wholeheartedly to one of the rivals. As the existence of Tychonian and other hybrid systems demonstrates, it was perfectly possible to take a middle way, a least for a period. To a large degree, the Copernican 'revolution' is a myth. There are events in the history of cosmology that come closer to revolutionary breaks, such as the recognition in about 1930 that the universe is expanding. Here we have a major change in the conception of the world that occurred almost instantaneously.

Modern cosmologists are, like most scientists (not to mention science journalists), fond of the revolution metaphor, which they use repeatedly and somewhat indiscriminately. They routinely speak of 'the ΛCDM paradigm' and the current 'revolution' constituted by the discovery of the accelerating universe supposedly filled with dark matter and energy. In the spring of 1997, at a time when the perspective of the new world view was just being recognized, one cosmologist wrote, 'The field of cosmology is undergoing a revolution, driven by dramatic observational progress and by novel theoretical scenarios imported from particle physics.'[53] Dozens of similarly worded evaluations can be found in the literature. This is most interesting, but it does not imply that what happened in the late 1990s was really a 'revolution' in the proper sense of the term.

The undeniable element of continuity in scientific development is to a large extent caused by the almost permanent nature of good experimental and observational data. New theories are expected to explain reliable observations as well as or better than the older theory, a requirement which naturally limits how far the new theory can deviate empirically from the older one. Copernicus' heliocentric theory had to include an explanation of the retrograde movements of the planets, just as the steady-state theory had to account for Hubble's redshift–distance law. And Copernicus had to explain the non-observation of the stellar parallax, just as Ptolemy's theory had explained it.

Yet, data are not purely observational, but are in part the product of theory and expectations, as is illustrated by several cases in the history of cosmology. The Hubble constant is an observational quantity, if not purely so, but its relation to the age of the the universe is not. The finite age of the universe is a theoretical construct which is only meaningful within the framework of certain cosmological theories. Moreover, when is an observation *relevant* to cosmology? This is far from obvious, as illustrated by the the darkness of the night sky, a most ordinary phenomenon which Chésaux elevated to a cosmological problem in 1744, later to be known as Olbers' paradox. Similarly, the chemical composition of the universe became cosmologically interesting only with Gamow's nuclear-archaeological research programme of the 1940s.

The history of cosmology is full of cases where observational claims have been accepted too readily. For example, in the early part of the twentieth century, van Maanen's apparently authoritative measurements of proper motions in spiral nebulae provided strong support against the island universe theory. His data were believed to be hard facts, but by 1930 they were seen as spurious. The Stebbins–Whitford effect, believed to be authoritative and grounded on solid data, evaporated in the 1950s. Likewise, when Ryle concluded in 1955 from radio astronomical measurements that the steady-state model was contradicted by observations, the data were inadequate and the conclusion was coloured by Ryle's dislike of the steady state-theory. The theory was indeed wrong, but so were Ryle's data.

Observations are of course crucial to cosmology, but if they are believed too confidently they may hamper or prevent progress rather than further it. Again an example from twentieth-century cosmology: Hubble's value of the expansion parameter, wrong by a factor of seven or so, was accepted as unproblematically correct for more than two decades. The small value of the Hubble time caused problems for cosmologies of the Big Bang type, which might have been received more positively had astronomers not taken Hubble's result to be so authoritative. It was only after 1952, when Baade and others reconsidered the time scale, that the age problem was brought into the light and it was realized that the 'age paradox' was no real paradox.

The intricate question of the relationship between observation and theory in cosmology was examined masterfully and provocatively by Bondi in an article of 1955, whose general message was that a discrepancy between theory and observation might just as well be blamed on the observations as on the theory. With respect to so-called pure observations, he wrote:

> The purely factual part of the vast majority of observational papers is small. It is also important to realise that these basic facts are frequently obtained at the very limit of the power of the instruments used, and hence are of considerable uncertainty. To refer to observational results as 'facts' is an insult to the labours of the observer, a mistaken attempt to discredit theorists, a disservice to astronomy in general and exhibits a complete lack of critical sense. Indeed I would go so far as to say that this sort of irresponsible misuse of terminology is the curse of modern astronomy.[54]

5.5.2 *The status of cosmological models*

The distinction between *realism* and *instrumentalism* (or related concepts such as positivism and pragmatism) is a central topic in the philosophy of science, where it is discussed particularly in connection with microphysics and quantum mechanics. According to the realist, theories and models are about nature's objects and mechanisms; if they describe these correctly they are true, and if not, they are wrong. In short, science endeavours to understand how nature really is. The instrumentalist, on the other hand, considers a theory as a method or tool to reproduce observed regularities and to predict the outcome of future observations and experiments. A theory can be useful in this regard, and that is all we can hope for and are interested in. It is outside the possibilities of science to determine the true nature of reality, a problem which is claimed to be metaphysical and not physical.

The distinction can also relate to objects rather than theories. In this case, the realist will claim that 'the universe' designates an entity that exists independently of all cosmological enquiry. On the other hand, the instrumentalist considers 'the universe' to be a concept that can be ascribed a meaning only in a pragmatic sense, as it is a construct of cosmological theory.[55] Although questions concerning realism and anti-realism have been mostly discussed in microphysics—do quarks *really* exist?—they are of no less relevance to the study of the universe and its constituents (do black holes *really* exist?). The tension between the two opposite views can be followed through much of the history of cosmology.

The Babylonians produced long tables with positions of the celestial bodies. Their attitude to astronomy was instrumentalistic, as their theories, in the form of tables, were pure computational tools. By extrapolation, they were able to predict celestial phenomena, whereas they showed no interest in explaining why the phenomena occurred or why the heavenly bodies moved as they did. Eudoxus provided a geometrical model for the motion of the planets, but he did not claim that it offered an explanation or that the planets really moved on a series of concentric spheres. A much more realistic perspective was introduced by Aristotle. Not satisfied by models that merely worked, he aimed at a physical theory of the universe with explanatory power. The zenith of Greek mathematical astronomy, as represented by Ptolemy's *Almagest*, had a clear instrumentalist orientation, with its elaborate system of deferents, eccentrics, and epicycles. It reproduced the heavenly phenomena, but it is hard to believe that Ptolemy really thought that the celestial bodies moved as prescribed by his model. On the other hand, this traditional view must be modified by his *Planetary Hypotheses* which, as we have seen, has a very different character. The model of the *Planetary Hypotheses*, was not phenomenological, but a realistic attempt to understand the structure of the universe in terms of Aristotelian physics. Ptolemy seems to have thought of the model as a true representation of the heavens, whereas truth is a quality that is foreign to the mind of the instrumentalist.

The same theme appeared during the Middle Ages and the Renaissance, when world models were often seen as models only, with no claim of representing the structure of the real universe. We have a statement of this position in Moses Maimonides' *Guide to the Perplexed*, dating from about 1200: 'The object of that science [astronomy] is to suppose as a hypothesis an arrangement that renders it possible for the motion of the star [planet] to be uniform and circular with no acceleration or deceleration or change in it and to have the inferences necessarily following from the assumption of that motion agree with what is observed.'[56] The task of the astronomers was 'to save the phenomena'. This was also the message of Osiander in his infamous foreword to Copernicus' *De revolutionibus*, whereas Copernicus was in fact a realist, not an instrumentalist. The battle between the two world systems in the decades around 1600 was not limited to which of the systems provided the most precise and economical description of the universe. If that was all there was to it, Galileo would not have been put on trial in 1633.

To jump ahead in time, the theme re-emerged in early relativistic cosmology, where the models of Einstein, de Sitter, and Friedmann must be counted in the instrumentalist tradition, whereas Lemaître was careful to point out that his 1927 model was meant as an account of how the real universe evolves over time. We have met another example of the instrumentalist attitude in Hubble's interpretation of the expansion of the universe: instead of concluding that the universe does in fact expand, he preferred the safer ground of keeping to the observationally based redshift–distance relation. Again, the controversy between the Big Bang and the steady-state theories in the 1950s did not merely concern comparison between data and theory. Gamow believed that the Big Bang had happened, that it was real, Hoyle that it was not. Contrariwise, Milne's understanding of cosmology was instrumentalistic, as he believed that the origin of the universe was not an objective fact but a product of the kind of theory used to describe the universe.

With the proliferation of Big-Bang-related models after 1965, the situation has become more complex. On the one hand, cosmologists want to understand the real universe and how it has evolved, not only to provide as-if stories. On the other hand, it is obvious that many of the models of modern cosmology, especially those relating to the very early universe, do not claim to represent the real world. They are scenarios or sometimes just mathematical toy models. If they can account for observed features of the universe or make predictions, so much the better, but correspondence with observations is not necessarily given high priority. Stephen Hawking has in common with Osiander that he favours an anti-realistic perspective on cosmological theories: 'I'm a positivist . . . I don't demand that a theory correspond to reality because I don't know what it is. Reality is not a quality you can test with litmus paper.'[57]

5.5.3 *Is cosmology a science?*

During most of its long history, cosmology has been a small backyard placed behind the splendid mansions of astronomy and philosophy. Most scientists had better things to do than engage in unfruitful speculations about the universe at large, speculations that were considered harmless pastimes but occasionally attracted epistemically based criticism. Cosmology is necessarily based on extrapolations and uniformity assumptions, for example that the fundamental laws of physics are valid throughout the universe and perhaps, even more problematically, *for* the universe as a whole. During the second half of the nineteenth century, in connection with the discussions about the heat death, cosmology came under fire from scientists who found such assumptions to be plainly unscientific. One of them was Ernst Mach, and his critical attitude was followed by many others of a positivist inclination. The German–American philosopher John Stallo charged in 1882 that 'all cosmogonies which purport to be theories of the universe as an absolute whole, in the light of physical and dynamical laws, are fundamentally absurd.'[58] Yet this is precisely what cosmologists aim at, to provide theories of the universe as an absolute whole.

The new cosmology, founded on Einstein's general theory of relativity, did not at first attract much attention, but with the discovery of the expanding universe it became much more visible and subject to philosophical criticism. How could these cosmologists construct models of the entire universe and discuss them with such confidence? Were they engaged in science or in a mathematical game? The American physicist Percy W. Bridgman, a Nobel laureate of 1946, was an early critic of the new cosmology. Bridgman subscribed to operationalism, a version of the positivist philosophy of science, and from this position he doubted if cosmology would ever become a true science. 'To the untutored critic it must appear a trifle rash to peer 10^{16} years back into the past or even greater distances into the future on the basis of laws verified by not more than 300 years' observation,' he wrote in 1932. The following year he attacked relativist cosmologists for their 'metaphysical conviction' that 'the universe is run on exact mathematical principles, and . . . it is possible for human beings by a fortunate tour de force to formulate these principles.'[59]

Bridgman's criticism came from an outsider, but it was repeated in a much stronger form a few years later by Herbert Dingle, who was well acquainted with cosmological theory and had himself contributed to the field (in 1933, he published one of the first studies of anisotropic universes). Although Dingle's attack of 1937 was particularly aimed at the rationalistic cosmophysics of Milne, Dirac, and Eddington, it also included a broader criticism of contemporary cosmology, which he tended to see as esoteric, arrogant, and remote from sound scientific reasoning. What annoyed him most was cosmology's uncritical reliance on mathematics and general principles of an a priori nature. 'Instead of the induction of principles from phenomena we are given a pseudo-science of invertebrate cosmythology,' he thundered.[60] When Dingle resumed his attack on cosmology after the Second World War, now aimed at the steady-state theory in particular, his voice had not softened. He and other critics objected not only to the Bondi–Gold–Hoyle theory, but to relativistic evolution cosmology as well. In whatever version, contemporary cosmology dealt with the dubious question of creation by means of complex and opaque mathematics, which only confirmed the critics in their opinion that cosmology had degraded into a pseudoscientific state. Dingle's 1953 characterization of the steady-state theory included also Big Bang theory: 'It has no other basis than the fancy of a few mathematicians who think how nice it would be if the world were made that way.'

Although cosmologists in the 1950s paid little attention to the objections of Dingle and his allies, they were not unconcerned with cosmology's scientific status. Is cosmology a science like physics and chemistry? What are the criteria of truth, and how do they differ from those adopted by other sciences? How sound is the conceptual basis of cosmology? Such questions were openly discussed, for instance in the previously mentioned 1954 debate between Bondi and Whitrow. Cosmology entered a much more mature phase after 1965, with the general recognition of the standard Big Bang theory, but some uneasiness about the field's scientific status remained. Not all astronomers and physicists welcomed the entry of grand unified theory and quantum gravity in to early-universe cosmology, an area where it could sometimes be difficult to distinguish between mathematics and physics. 'Mathematicians revel in harmless sophisticated fantasies, and new-cosmologists buy them as real estate,' as one critic expressed it.[61] As mentioned before, Tony Rothman and George Ellis warned in 1987 that cosmology might be on its way to becoming 'metaphysical'.

This kind of criticism or worry has continued up to the present, as illustrated by an article of 2000 in the journal *General Relativity and Gravitation*. M. J. Disney, a British extragalactic observational astronomer, repeated many of the accusations of earlier critics, not only in substance but also in rhetoric. His basic concern was the gulf between observation and theory, a problem which has always haunted cosmology, and also the cosmologists' unrestrained willingness to extrapolate known physics over huge ranges in space and time. Cold dark matter, accepted by the majority of cosmologists, sounded to Disney 'like a religious liturgy which its adherents chant like a mantra in the mindless hope that it will spring into existence'. The comparison between cosmology and religion was not new, nor was it accidental: 'The most unhealthy aspect of cosmology is its unspoken parallel with religion. Both deal with big but probably unanswerable questions. The rapt audience, the media exposure, the big book-sale, tempt priests and rogues, as well as the gullible, like no other subject in science.'[62]

5.5.4 *Technology and the universe*

It may be hard to associate cosmology with something as mundane as engineering and technology (it surely would have shocked Aristotle), yet there can be no doubt that instrument technologies have been of no less importance in the progress of cosmology than advances in theory have been. Until the seventeenth century, astronomers had to make do with their naked eyes and relatively primitive instruments such as quadrants, sextants, and armillary spheres. The use of these traditional techniques reached a climax with Tycho Brahe's instruments in his Uraniborg observatory. With the invention of the telescope in the early seventeenth century, a new chapter in the history of astronomy had its beginning, if at first of limited cosmological significance. But with larger and better telescopes, reaching out to the faint nebulae, the situation changed. It was William Herschel's superb 40-foot reflector which led him to the cosmology he described in 'The Construction of the

Heavens'; and it was Lord Rosse's giant telescope which in 1846 revealed the existence of spiral nebulae, a discovery with unforeseen consequences.

Astronomical photometry—methods to measure the intensity of starlight—had its beginning in the 1830s, and over the next half century, much-improved photometrical techniques were introduced by Zöllner in Germany and Edward C. Pickering in the United States. Photoelectrical methods for astronomical use were pioneered around 1910 by the American astronomer Joel Stebbins, who by means of a selenium cell could record the light curve for the variable star Algol and measure differences in light intensity as small as corresponding to 0.01 magnitudes. With the development of electronic circuits and photomultipliers after the Second World War, sensitivity further increased.

The problem with the traditional photographic plates was that they responded to only about one per cent of the incoming light energy, which implied that long exposure times were necessary for faint sources. This problem was reduced in the early 1960s when an electronic imaging device called the image tube was used to amplify the incoming light (a similar technology was used in microscopy). While the image tube did not replace the photographic plate, such a replacement happened when digital detectors were brought into use from about 1970. With this technology, the light signals were transformed directly into electrical pulses, which were stored in a computer. A breakthrough occurred in 1970, when George E. Smith and Willard S. Boyle, two scientists at Bell Laboratories, announced their invention of a new semiconductor-based apparatus that generated electrical signals when exposed to light. The CCD (charge-coupled device) was not invented as a detector of feeble light, but within a decade the first CCD detectors were in use for astronomical purposes. The CCD quickly became an indispensable tool for astronomers and cosmologists, functioning in ground- as well as space-based observatories.[63]

CCDs and related imaging technologies to a large extent replaced traditional photography, which as an astronomical technique goes back to the 1840s. The first example of astrophotography dates from 1840, when the American chemist John W. Draper produced a daguerreotype of the Moon. The exposure time was twenty minutes. Astrophotography was continually improved and became of particular importance when combined with the spectroscope, another of the great instrument inventions of the nineteenth century. Hubble's famous discovery of the velocity–distance relation in 1929 relied crucially on three kinds of technology, the telecope, the spectroscope, and photographic techniques.

The instrument revolution has been of direct cosmological significance especially since the 1970s, although it is worth recalling that radio and microwave technology played a role even earlier. As one commentator has noted, 'It is a striking thought that 10 years of radio astronomy have taught humanity more about the creation and organisation of the universe than a thousand years of religion and philosophy.'[64] The discovery of the cosmic microwave background radiation, often hailed as the cosmological discovery of the century, was an engineering achievement as much as a scientific achievement. Other techno-cosmological stories could be told about satellite observations and astronomy outside the visible part of the spectrum.[65] What matters is that ever since Galileo in the summer of 1609 directed his telescope towards the sky, technological progress has been essential to cosmology, a process which has greatly accelerated since the 1950s and is today of overwhelming importance.

Notes

1. Schramm and Steigman 1988, p. 66.
2. Schramm 1996, p. xvii.
3. This alludes to a conference held in May 1984 with an attendance of more than 200 scientists (Kolb *et al.* 1986). According to the preface of the proceedings, the theme of the conference was 'the connection between physics of the microworld (Inner Space) and physics of the macroworld (Outer Space)'. There are many popular books on the subject, and also many technical works, but no good historical works. Schramm 1996 is a collection of primary sources, some of which are also reproduced in Bernstein and Feinberg 1986. It is customary to express the mass of an elementary particle in terms of its corresponding energy measured in electron volts (i.e. in eV rather than eV/c^2). The mass of an electron is 0.51 MeV (million electron volts).
4. The 1973 paper by Rogerson and York is reproduced in Lang and Gingerich 1979, pp. 75–77.
5. Schramm and Turner 1998.
6. Yoshimura 1978, p. 281.
7. Quoted in Drell and Okun 1990, p. 32. Sakharov was not allowed to go to Stockholm to accept his 1975 Nobel Prize, which was instead received by his wife, Elena Bonner.
8. The more recent developments in the search for dark matter are complex and outside the scope of the present work. They can be followed through review articles, Web sites, and popular books such as Seife 2004.
9. It has also been suggested that the photon has a tiny mass, but although photons are abundant in the universe they can be left out of consideration. Experiments prove that if photons have a mass, it is ridiculously small, namely less than 10^{-16} eV.
10. On Starobinsky's model, see Smeenk 2003, pp. 80–85. Smeenk's dissertation offers a philosophically oriented discussion of inflation theory and other aspects of early-universe cosmology. See also Smeenk 2005.
11. On the horizon problem, see the careful review in Tipler, Clarke, and Ellis 1980. For its role in the steady-state controversy, see Kragh 1996a, pp. 233–235.
12. Guth 1981, p. 348. Reproduced in Bernstein and Feinberg 1986, pp. 299–320. The initial temperature, given as 10^{17} GeV, corresponds to about 10^{30} K. On the invention of the inflation model and its further development, see Guth 1997 and Smeenk 2005. Its impact on the cosmological community is described in a series of interviews in Lightman and Brawer 1990. Guth's paper belonged to the realm of particle physics rather than astronomy: among the 79 papers cited in his bibliography, two belonged to the astronomy literature and the rest to the physics literature.
13. Before the Planck time, general relativity is not applicable and needs to be replaced by a theory of quantum gravity that does not yet exist.
14. Rothman and Ellis 1987, p. 22. See also the extensive criticism in Earman and Mosterin 1999.
15. Smoot *et al.* 1992.
16. Lemonick 1993, pp. 285–296. According to Smoot, he meant the much-quoted phrase as a metaphor, not in an apologetic sense.
17. The story of the Space Telescope is as much a story of money, politics, and bureaucracy as of science. For these aspects and the Space Telescope as an example of big science, see Smith 1989.
18. Concise accounts are given in Filippenko 2003 and Perlmutter 2003. Several of the scientists involved have written popular works on the projects and their role in them.
19. Filippenko and Riess 1998, p. 38.
20. For substantial reviews of the cosmological constant and its role in physics and cosmology, see Earman 2001 and Peebles and Ratra 2003.
21. Kerzberg 1989, p. 334.
22. Lemaître 1934, p. 13.
23. Krauss and Turner 1995.
24. Peebles and Ratra 2003, p. 596.
25. See the special March 1998 issue of *Publications of the Astronomical Society of the Pacific* and also http://antwrp.gsfc.nasa.gov/debate/debate98.html.
26. *Harper's Magazine*, October 2004, quoted here from an online version.
27. For a review and references, see Olive and Qian 2004.
28. Gale 1990 provides a review, covering both philosophical and scientific aspects.
29. Linde 1983, p. 246.
30. Linde 1987, p. 68.
31. Magueijo 2003.

32. In an interview of 1978, Sciama said that his devotion to the steady-state theory was in part rooted in his belief that it was 'the only model in which it seems evident that life will continue somewhere'. See Kragh 1996a, p. 254. Dirac similarly defended his $G(t)$ cosmology by the argument that it 'allows the possibility of endless life' (Kragh 1990, p. 236).
33. See Barrow and Tipler 1986, pp. 602 and 674.
34. Linde 1994, p. 39.
35. Smolin 1992.
36. Ellis 1993, pp. 394–395. Compare Kragh 2004, p. 227. See also Ellis and Brundrit 1979, where it is argued that a low-density, infinite, uniform universe leads to highly bizarre consequences, such as an infinite number of genetically identical beings in the universe at any time.
37. Barrow and Tipler 1986, pp. 613–682. A useful bibliography is given in Ćirković 2003b.
38. Russell 1957, p. 75.
39. Dyson 1979, p. 459.
40. Dicke 1961. Dirac replied that on Dicke's assumption life would exist only for a limited period of time, whereas his own cosmological hypothesis allowed life to continue indefinitely, which he considered a strong argument in favour of the large-number hypothesis.
41. Collins and Hawking 1973, p. 334.
42. Carter 1974, pp. 293 and 295. The articles by Dicke and Carter are reprinted in Leslie 1990.
43. Barrow and Tipler 1986 is a rich and comprehensive source on the anthropic principle. For other literature until 1991, see Balashov 1991.
44. Hoyle 1994, pp. 256–267.
45. On the Milne–Misner solution, see Lévy-Leblond 1990, and for cyclical models, Steinhardt and Turok 2002.
46. Tryon 1973, p. 397.
47. Vilenkin 1982.
48. Hawking 1988, pp. 143–144, 149; Hartle and Hawking 1983. Theologians and Christian cosmologists have had no problems with finding a place for God within the Hartle–Hawking cosmology. See, for instance, Russell 1994.
49. Veneziano 2004, p. 55.
50. Kolb and Turner 1990, p. 498.
51. Quoted here from Copan and Craig 2004, p. 219.
52. The authenticity of the quotation is questionable. See Crowe 1990, p. 78.
53. Turok 1997, p. ix.
54. Bondi 1955, p. 158. See also Kragh 1996a, pp. 237–240.
55. Munitz 1986 argues that 'The universe is what a cosmological model says it is' (p. 62).
56. Quoted here from Crowe 1990, p. 74, who gives an account of the history of 'save the phenomena'.
57. Hawking and Penrose 1996, p. 121.
58. Stallo 1882, p. 276.
59. See Kragh 2004, p. 155.
60. On Dingle's criticism, see Kragh 1996a, pp. 69–71, 224–226.
61. Carey 1988, p. 332.
62. Disney 2000, pp. 1131 and 1133. See also the reply in Ćirković 2002, who pointed out the similarity between Disney's criticism and the earlier tradition of Dingle and others.
63. Smith and Tatarewicz 1985.
64. Davies 1977, p. 211.
65. See the review in Longair 2001.

REFERENCES

Aaboe, Asger (2001). *Episodes from the Early History of Astronomy*. New York: Springer.

Aiton, E. J. (1981). 'Celestial spheres and circles', *History of Science* **19**, 75–114.

Alexander, H. G., ed. (1956). *The Leibniz–Clarke Correspondence*. Manchester: Manchester University Press.

Alpher, Ralph A., Hans Bethe, and George Gamow (1948). 'The origin of chemical elements', *Physical Review* **73**, 803–804.

Alpher, Ralph A., James Follin, and Robert C. Herman (1953). 'Physical conditions in the initial stages of the expanding universe', *Physical Review* **92**, 1347–1361.

Alpher, Ralph A., and Robert C. Herman (1948). 'Evolution of the universe', *Nature* **74**, 774–775.

—— (2001). *Genesis of the Big Bang*. New York: Oxford University Press.

Arrhenius, Svante (1908). *Worlds in the Making: The Evolution of the Universe*. London: Harper & Brothers.

—— (1909). 'Die Unendlichkeit der Welt', *Scientia* **5**, 217–229.

Assis, A. K. T., and M. C. D. Neves (1995). 'The redshift revisited', *Astrophysics and Space Science* **227**, 13–24.

Baigrie, Brian S. (1993). 'Descartes' mechanical cosmology', pp. 164–176 in Hetherington 1993.

Balashov, Yuri V. (1991). 'Resource letter AP-1: The anthropic principle', *American Journal of Physics* **59**, 1069–1076.

Barbour, Julian B. (2001). *The Discovery of Dynamics*. Oxford: Oxford University Press.

Barnes, Ernest (1933). *Scientific Theory and Religion*. Cambridge: Cambridge University Press.

Barrow, John D., and Frank J. Tipler (1986). *The Anthropic Cosmological Principle*. Oxford: Clarendon Press.

Basri, Gibor (2000). 'The discovery of brown dwarfs', *Scientific American* **282** (April), 57–63.

Becker, George F. (1908). 'Relations of radioactivity to cosmogony and geology', *Bulletin of the Geological Society of America* **19**, 113–146.

Becker, Barbara (2001). 'Visionary memories: William Huggins and the origins of astrophysics', *Journal for the History of Astronomy* **32**, 43–62.

Belkora, Leila (2003). *Minding the Heavens. The Story of Our Discovery of the Milky Way*. Bristol: Institute of Physics Publishing.

Berendzen, Richard, Richard Hart, and Daniel Seeley (1976). *Man Discovers the Galaxies*. New York: Science History Publications.

—— (1984). *Man Discovers the Galaxies*. New York: Science History Publications.

Bernstein, Jeremy (1984). *Three Degress Above Zero*. New York: Scribner's.

Bernstein, Jeremy, and Gerald Feinberg, eds (1986). *Cosmological Constants: Papers in Modern Cosmology*. New York: Columbia University Press.

Bertotti, Bruno, R. Balbinot, Silvio Bergia, and A. Messina, eds (1990). *Modern Cosmology in Retrospect*. Cambridge: Cambridge University Press.

Blacker, Carmen, and Michael Loewe, eds (1975). *Ancient Cosmologies*. London: George Allen & Unwin.

Blair, Ann (2000). 'Mosaic physics and the search for a pious natural philosophy in the late renaissance', *Isis* **91**, 32–58.

Boltzmann, Ludwig (1895). 'On certain questions of the theory of gases', *Nature* **51**, 483–485.

Bondi, Hermann (1948). 'Review of cosmology', *Monthly Notices of the Royal Astronomical Society* **108**, 104–120.

—— (1952). *Cosmology*. Cambridge: Cambridge University Press.

—— (1955). 'Facts and inference in theory and in observation', *Vistas in Astronomy* **1**, 155–162.

Bondi, Hermann, and Thomas Gold (1948). 'The steady-state theory of the expanding universe', *Monthly Notices of the Royal Astronomical Society* **108**, 252–270.

Bondi, Hermann, William B. Bonnor, Raymond A. Lyttleton, and Gerald J. Whitrow, eds (1960). *Rival Theories of Cosmology*. London: Oxford University Press.

Borel, Emile (1960). *Space and Time*. New York: Dover Publications.

Boscovich, Roger J. (1966). *A Theory of Natural Philosophy*. Cambridge, Mass.: MIT Press.

Brehaut, E. (1912). *An Encyclopedist of the Dark Ages: Isidore of Seville*. New York: Columbia University Press.
Bronstein, Matvei (1933). 'On the expanding universe', *Physikalische Zeitschrift der Sowjetunion* **3**, 73–82.
Brown, G. Burniston (1940). 'Why do Archimedes and Eddington both get 10^{79} for the total number of particles in the universe?', *Philosophy* **15**, 269–284.
Bruggencate, P. Ten (1937). 'Dehnt sich das Weltall aus?' *Die Naturwissenschaften* **25**, 561–566.
Brush, Stephen G. (1987). 'The nebular hypothesis and the evolutionary worldview', *History of Science* **25**, 245–278.
—— (1993). 'Prediction and theory evaluation: Cosmic microwaves and the revival of the big bang', *Perspectives on Science* **1**, 565–602.
Burbidge, E. Margaret, Geoffrey R. Burbidge, Fred Hoyle, and William A. Fowler (1957). 'Synthesis of the elements in stars', *Reviews of Modern Physics* **29**, 547–650.
Burtt, Edwin A. (1972). *The Metaphysical Foundations of Modern Physical Science*. London: Routledge and Kegan Paul.
Canuto, V., and J. Lodenquai (1977). 'Dirac cosmology', *Astrophysical Journal* **211**, 342–356.
Čapek, Milic (1976). *The Concepts of Space and Time: Their Structure and Development*. Dordrecht: Reidel.
Cappi, Alberto (1994). 'Edgar Allan Poe's physical cosmology', *Quarterly Journal of the Royal Astronomical Society* **35**, 177–192.
Carey, S. Warren (1988). *Theories of the Earth and Universe. A History of Dogma in the Earth Sciences*. Stanford: Stanford University Press.
Carroll, William (1998). 'Thomas Aquinas and big bang cosmology', *Sapientia* **53**, 73–95.
Carter, Brandon (1974). 'Large number coincidences and the anthropic principle in cosmology', pp. 291–298 in Malcolm S. Longair, ed., *Confrontation of Cosmological Theories with Observational Data*. Dordrecht: Reidel.
Charlier, Carl V. L. (1896). 'Ist die Welt endlich oder unendlich in Raum und Zeit', *Archiv für systematische Philosophie* **2**, 477–494.
—— (1908). 'Wie eine unendliche Welt aufgebaut sein kann', *Arkiv för Matematik, Astronomi och Fysik* **4**, 1–15.
Christianson, John R. (1968). 'Tycho Brahe's cosmology from the *Astrologia* of 1591', *Isis* **59**, 313–318.
Christianson, Gale E. (1995). *Edwin Hubble. Mariner of the Nebulae*. New York: Farrar, Straus and Giroux.
Ćirković, Milan M. (2002). 'Laudatores temporis acti, or why cosmology is alive and well—a reply to Disney', *General Relativity and Gravitation* **34**, 119–1130.
—— (2003a). 'The thermodynamical arrow of time: Reinterpreting the Boltzmann–Schuetz argument', *Foundations of Physics* **33**, 467–490.
—— (2003b). 'Resource letter: Pes-1: Physical eschatology', *American Journal of Physics* **71**, 122–133.
Clausius, Rudolf (1868). 'On the second fundamental theorem of the mechanical theory of heat', *Philosophical Magazine* **35**, 405–419.
Clerke, Agnes M. (1890). *The System of the Stars*. London: Longmans, Green and Co.
—— (1903). *Problems in Astrophysics*. London: Adam & Charles Black.
Cohen, I. Bernard, ed. (1978). *Isaac Newton's Papers & Letters on Natural Philosophy*. Cambridge, Mass.: Harvard University Press.
—— (1985). *The Birth of a New Physics*. New York: Norton & Company.
Cohen, Morris R., and I. E. Drabkin, eds (1958). *A Source Book in Greek Science*. Cambridge, Mass.: Harvard University Press.
Collins, C. Barry, and Stephen W. Hawking (1973). 'Why is the universe isotropic?', *Astrophysical Journal* **180**, 317–334.
Copan, Paul, and William L. Craig (2004). *Creation Out of Nothing. A Biblical, Philosophical, and Scientific Exploration*. Grand Rapids: Baker Academic.
Copernicus, Nicolaus (1995). *On the Revolutions of the Heavenly Spheres*. Translated by Charles G. Wallis. Amherst: Prometheus Books.
Copp, C. M. (1982). 'Relativistic cosmology, I: Paradigm commitment and rationality', *Astronomy Quarterly* **4**, 103–116.
—— (1983). 'Relativistic cosmology, II: Social structure, skepticism, and cynicism', *Astronomy Quarterly* **4**, 179–188.
Cornford, Francis M. (1956). *Plato's Cosmology*. London: Routledge & Kegan Paul.
Cosmas (1897). *The Christian Topography of Cosmas, an Egyptian Monk*. Translated by J. W. McCrindle. New York: Burt Franklin.
Crelinsten, Jeffrey (2006). *Einstein's Jury: The Race to Test Relativity*. Princeton: Princeton University Press.

Crombie, Alistair C. (1953). *Robert Grosseteste and the Origins of Experimental Science 1100–1700.* Oxford: Clarendon Press.
Crookes, William (1886). 'On the nature and origin of the so-called elements', *Report, British Association for the Advancement of Science*, 558–576.
Crowe, Michael J. (1990). *Theories of the World from Antiquity to the Copernican Revolution*. New York: Dover Publications.
—— (1994). *Modern Theories of the Universe: From Herschel to Hubble*. New York: Dover Publications.
Cusanus (1997). *Nicholas of Cusa: Selected Spiritual Writings*. Translated by H. Lawrence Bond. New York: Paulist Press.
Dalton, John (1808). *A New System of Chemical Philosophy*. Manchester: Bickerstaff.
Davies, Gordon L. (1966). 'The concept of denudation in seventeenth-century England', *Journal of the History of Ideas 27*, 278–284.
Davies, Paul C. W. (1977). *Space and Time in the Modern Universe*. Cambridge: Cambridge University Press.
Dean, Dennis R. (1981). 'The age of the earth controversy: Beginnings to Hutton', *Annals of Science* **38**, 435–456.
Debus, Allen G. (1977). *The Chemical Philosophy. Paracelsian Science and Medicine in the Sixteenth and Seventeenth Centuries*. Mineola, New York: Dover Publications.
Descartes, René (1983). *Principles of Philosophy*. Translated by V. R. Miller and R. P. Miller. Dordrecht: Reidel.
—— (1996). *Discourse on Method and Meditations on First Philosophy*. Edited by David Weissman. New Haven: Yale University Press.
de Sitter, Willem (1917). 'On Einstein's theory of gravitation, and its astronomical consequences. Third paper', *Monthly Notices of the Royal Astronomical Society* **78**, 3–28.
Detweiler, S., ed. (1982). *Black Holes: Selected Reprints*. Stony Brooks: American Association of Physics Teachers.
Dick, Steven J. (1982). *Plurality of Worlds: The Origins of the Extraterrestrial Life Debate from Democritus to Kant*. Cambridge: Cambridge University Press.
Dicke, Robert H. (1961). 'Dirac's cosmology and Mach's principle', *Nature* **192**, 440–441.
Dingle, Herbert (1924). *Modern Astrophysics*. London: Collins Sons.
Dirac, Paul (1937). 'The cosmological constants', *Nature* **139**, 323.
—— (1938). 'A new basis for cosmology', *Proceedings of the Royal Society A* **165**, 199–208.
—— (1939). 'The relation between mathematics and physics', *Proceedings of the Royal Society (Edinburgh)* **59**, 122–139.
Disney, M. J. (2000). 'The case against cosmology', *General Relativity and Gravitation* **32**, 1125–1134.
Drake, Stillman, ed. (1957). *Discoveries and Opinions of Galileo*. New York: Doubleday.
—— (1981). *Galileo at work: His Scientific Biography*. Chicago: University of Chicago Press.
Drell, Sidney, and Lev Okun (1990). 'Andrei Dmitrievich Sakharov', *Physics Today* **43** (August), 26–36.
Dreyer, J. L. E. (1953). *A History of Astronomy from Thales to Kepler*. New York: Dover Publications.
Dyson, Freeman J. (1979). 'Time without end: Physics and biology in an open universe', *Reviews of Modern Physics* **51**, 447–460.
Earman, John (1999). 'The Penrose–Hawking theorems: History and implications', pp. 235–270 in Goenner *et al.* 1999.
—— (2001). 'Lambda: The constant that refuses to die', *Archive for History of Exact Sciences* **55**, 189–220.
Earman, John, and Jean Eisenstaedt (1999). 'Einstein and singularities', *Studies in the History and Philosophy of Modern Physics* **30**, 185–235.
Earman, John, and Jesus Mosterin (1999). 'A critical look at inflationary cosmology', *Philosophy of Science* **66**, 1–49.
Easton, Cornelis (1900). 'A new theory of the Milky Way', *Astrophysical Journal* **12**, 136–158.
—— (1913). 'A photographic chart of the Milky Way and the spiral theory of the galactic system', *Astrophysical Journal* **37**, 105–118.
Eddington, Arthur S. (1920). *Space, Time and Gravitation: An Outline of the General Relativity Theory*. Cambridge: Cambridge University Press.
—— (1923a). *The Mathematical Theory of Relativity*. Cambridge: Cambridge University Press.
—— (1923b). 'The borderland of astronomy and geology', *Nature* ***111***, 18–21.
—— (1924). 'A comparison of Whitehead's and Einstein's formulae', *Nature* **113**, 192.
—— (1928). *The Nature of the Physical World*. Cambridge: Cambridge University Press.
—— (1930). 'On the instability of Einstein's spherical world', *Monthly Notices of the Royal Astronomical Society* **90**, 668–678.
—— (1931). 'The end of the world: from the standpoint of mathematical physics', *Nature* **127**, 447–453.

Eddington, Arthur S. (1933). *The Expanding Universe*. Cambridge: Cambridge University Press.

—— (1944). 'The recession-constant of the galaxies', *Monthly Notices of the Royal Astronomical Society* **104**, 200–204.

—— (1946) *Fundamental Theory*. Cambridge: Cambridge University Press.

Edge, David O., and Michael J. Mulkay (1976). *Astronomy Transformed: The Emergence of Radio Astronomy in Britain*. New York: Wiley and Sons.

Einstein, Albert (1916). 'Die Grundlage der allgemeinen Relativitätstheorie', *Annalen der Physik* **49**, 769–822.

—— (1917). 'Kosmologische Betrachtungen zur allgemeinen Relativitätstheorie', *Sitzungsberichte der Preussischen Akademie der Wissenschaften*, 142–152.

—— (1931). 'Zum kosmologischen Problem der allgemeinen Relativitätstheorie', *Sitzungsberichte der Preussischen Akademie der Wissenschaften*, 235–237.

—— (1945). *The Meaning of Relativity*. Princeton: Princeton University Press.

—— (1947). *Relativity. The Special and General Theory*. New York: Hartsdale House.

—— (1982). 'How I created the theory of relativity', *Physics Today* **35** (August), 45–47.

—— (1998). *The Collected Works of Albert Einstein*. Volume 8. Edited by Robert Schulmann, A. J. Kox, Michael Janssen, and Jószef Illy. Princeton: Princeton University Press.

Einstein, Albert, and Willem de Sitter (1932). 'On the relation between the expansion and the mean density of the universe', *Proceedings of the National Academy of Sciences* **18**, 213–214.

Einstein, Albert, Hendrik A. Lorentz, Hermann Weyl, and Hermann Minkowski, (1952). *The Principle of Relativity*. New York: Dover Publications.

Eisenstaedt, Jean (1989). 'The low water mark of general relativity, 1925–1955', pp. 277–292 in Don Howard and John Stachel, eds, *Einstein and the History of General Relativity*. Boston: Birkhäuser.

—— (1993). 'Dark bodies and black holes, magic circles and Montgolfiers: Light and gravitation from Newton to Einstein', *Science in Context* **6**, 83–106.

Eliade, Mircea (1974). *The Myth of the Eternal Return*. Princeton: Princeton University Press.

Ellis, George F. R. (1984). 'Alternatives to the big bang', *Annual Review of Astronomy and Astrophysics* **22**, 157–184.

—— (1993). 'The theology of the anthropic principle', pp. 363–400 in Robert J. Russell, Nancy Murphy, and C. J. Isham, eds, *Quantum Cosmology and the Laws of Nature: Scientific Perspectives on Divine Action*. Vatican City State: Vatican Observatory.

Ellis, George F. R., and G. B. Brundrit (1979). 'Life in the infinite universe', *Quarterly Journal of the Royal Astronomical Society* **20**, 37–41.

Evans, James (1993). 'Ptolemy's cosmology', pp. 528–544 in Hetherington 1993.

Farrington, Benjamin (1953). *Greek Science*. Melbourne: Penguin Books.

Felber, Hans-Joachim, ed. (1994). *Briefwechsel zwischen Alexander von Humboldt und Friedrich Wilhelm Bessel*. Berlin: Akademie Verlag.

Ferguson, James (1778). *Astronomy Explained upon Sir Isaac Newton's Principles*. London: W. Strathan.

Fernie, J. D. (1969). 'The period–luminosity relation: A historical review', *Publications of the Astronomical Society of the Pacific* **81**, 707–731.

Fick, Adolf (1869). *Die Naturkräfte in Ihrer Wechselbeziehung*. Würzburg: Stahel'schen Buchhandlung.

Filippenko, Alexei V. (2003). 'Einstein's biggest blunder? High-redshift supernovae and the accelerating universe', *Publications of the Astronomical Society of the Pacific* **113**, 1441–1448.

Filippenko, Alexei V., and Adam Riess (1998). 'Results from the High-*z* Supernova Search Team', *Physics Reports* **307**, 31–44.

Finlay-Freundlich, Erwin (1951). *Cosmology*. Chicago: University of Chicago Press.

Frankfort, Henri, H. A. Frankfort, John A. Wilson, and Thorkild Jacobsen, (1959). *Before Philosophy: The Intellectual Adventure of Ancient Man*. Harmondsworth: Penguin Books.

Frenkel, Victor (1994). 'George Gamow: World line 1904–1933', *Soviet Physics Uspekhi* **37**, 767–789.

Freudenthal, Gad (1983). 'Theory of matter and cosmology in William Gilbert's *De magnete'*, *Isis* **74**, 22–37.

—— (1991). '(Al-)Chemical foundations for cosmological ideas: Ibn Sîna on the geology of an eternal world', pp. 47–73 in Sabetai Unguru, ed., *Physics, Cosmology and Astronomy, 1300–1700: Tension and Accommodation*. Dordrecht: Kluwer.

Friedmann, Alexander A. (1922). 'Über die Krümmung des Raumes', *Zeitschrift der Physik* **10**, 377–386.

—— (2000). *Die Welt als Raum und Zeit*. Edited and translated by Georg Singer. Frankfurt am Main: Harri Deutsch.

Gale, George (1990). 'Cosmological fecundity: Theories of multiple universes', pp. 189–206 in John Leslie, ed., *Physical Cosmology and Philosophy*. New York: Macmillan.

Galilei, Galileo (1967). *Dialogue Concerning the Two Chief World Systems*. Translated by Stillman Drake. Berkeley: University of California Press.

Gamow, George (1952). *The Creation of the Universe*. New York: Viking Press.

Gamow, George, and J. A. Fleming (1942). 'Report on the eighth annual Washington conference of theoretical physics', *Science* **95**, 579–581.

Gaukroger, Stephen (1995). *Descartes. An Intellectual Biography*. Oxford: Clarendon Press.

Gheury de Bray, M. E. J. (1939). 'Interpretation of the red-shifts of the light from extra-galactic nebulae', *Nature* **144**, 285.

Gilbert, William (1958). *De Magnete*. Translated by P. Fleury Mottelay. New York: Dover Publications.

Gingerich, Owen (1975). ' "Crisis" versus aesthetic in the Copernican revolution', *Vistas in Astronomy* **17**, 85–93.

—— (1985). 'Did Copernicus owe a debt to Aristarchus?', *Journal for the History of Astronomy* **16**, 37–42.

Gingerich, Owen, and Robert S. Westman (1988). *The Wittich Connection: Conflict and Priority in Late Sixteenth-Century Cosmology*. Philadelphia: American Philosophical Society.

Godart, Odon, and Michael Heller (1985). *Cosmology of Lemaître*. Tucson: Pachart Publishing House.

Gödel, Kurt (1949). 'An example of a new type of cosmological solutions of Einstein's field equations of gravitation', *Reviews of Modern Physics* **21**, 447–450.

Goenner, Hubert (2001). 'Weyl's contributions to cosmology', pp. 105–137 in Erhard Scholz, ed., *Hermann Weyl's Raum-Zeit-Materie and a General Introduction to His Scientific Work*. Basel: Birkhäuser.

Goenner, Hubert, Jürgen Renn, Jim Ritter, and Tilman Saner, eds (1999). *The Expanding World of General Relativity*. Boston: Birkhäuser.

Goldschmidt, Victor M. (1937). 'Geochemische Verteilungsgesetze der Elemente, IX. Die Mengenverhältnisse der Elemente und der Atom-Arten', *Skrifter av det Norske Videnskabs-Akademien i Oslo, Skrifter*, No. 4.

Goldstein, Bernard R. (1967). 'The Arabic version of Ptolemy's planetary hypotheses', *Transactions of the American Philosophical Society* **57**, Part 4.

Gombrich, R. F. (1975). 'Ancient Indian cosmology', pp. 110–142 in Blacker and Loewe 1975.

Gorst, Martin (2002). *Aeons. The Search for the Beginning of Time*. London: Fourth Estate.

Graham, Loren R. (1972). *Science and Philosophy in the Soviet Union*. New York: Knopf.

Grant, Edward (1969). 'Medieval and seventeenth-century conceptions of an infinite void space beyond the cosmos', *Isis* **60**, 39–60.

—— ed. (1974). *A Source Book in Medieval Science*. Cambridge, Mass.: Harvard University Press.

—— (1994). *Planets, Stars, & Orbs: The Medieval Cosmos, 1200–1687*. Cambridge: Cambridge University Press.

Gray, George W. (1953). 'A larger and older universe', *Scientific American* ***188*** (June), 56–67.

Gray, Jeremy J. (1979). *Ideas of Space*: *Euclidean, Non-Euclidean, and Relativistic*. Oxford: Clarendon Press.

Gribbin, J. R. (1976). *Galaxy Formation. A Personal View*. New York: John Wiley and Sons.

Gunn, James E., and Beatrice M. Tinsley (1975). 'An accelerating universe', *Nature* **257**, 454–457.

Guth, Alan H. (1981). 'Inflationary universe: A possible solution to the horizon and flatness problems', *Physical Review D* **23**, 347–356.

—— (1997). *The Inflationary Universe*. Reading, Mass.: Addison-Wesley.

Haas, Arthur E. (1936). 'An attempt to a purely theoretical derivation of the mass of the universe', *Physical Review* **49**, 411–412.

Haber, Francis C. (1959). *The Age of the World*. Baltimore: Johns Hopkins University Press.

Hall, A. Rupert, and Marie Boas Hall, eds (1962). *Unpublished Scientific Papers of Isaac Newton*. Cambridge: Cambridge University Press.

Halley, Edmund (1720–21). 'Of the infinity of the sphere of fix'd stars', *Philosophical Transactions* **31**, 22–24.

Harper, Eamon (2001). 'George Gamow: Scientific amateur and polymath', *Physics in Perspective* ***3***, 335–372.

Harper, Eamon, W. C. Parke, and G. D Anderson, eds (1997). *The George Gamow Symposium*. San Francisco: Astronomical Society of the Pacific.

Harrison, Edward (1986). 'Newton and the infinite universe', *Physics Today* **39**:2, 24–32.

—— (1987). *Darkness at Night: A Riddle of the Universe*. Cambridge, Mass.: Harvard University Press.

Hartle, James B., and Stephen W. Hawking (1983). 'Wave function of the universe', *Physical Review* **28D**, 2960–2975.

Harwit, Martin (1981). *Cosmic Discovery: The Search, Scope, and Heritage of Astronomy*. New York: Basic Books.

Hawking, Stephen W. (1988). *A Brief History of Time: From the Big Bang to Black Holes*. New York: Bantam Books.

Hawking, Stephen W., and Roger Penrose (1996). *Nature of Space and Time*. Princeton: Princeton University Press.

Hearnshaw, J. B. (1986). *The Analysis of Starlight: One Hundred and Fifty Years of Astronomical Spectroscopy*. Cambridge: Cambridge University Press.
Heath, Thomas, ed. (1953). *The Works of Archimedes*. New York: Dover Publications.
—— (1959). *Aristarchus of Samo: The Ancient Copernicus*. Oxford: Clarendon Press.
Heckmann, Otto (1932). 'Die Ausdehnung der Welt in ihrer Abhängigkeit von der Zeit', *Nachrichten von der Gesellschaft der Wissenschaften zu Göttingen, Math.-Phys. Klasse*, 97–106.
—— (1942). *Theorien der Kosmologie*. Berlin: Springer-Verlag.
Heninger, S. K. (1977). *The Cosmographical Glass: Renaissance Diagrams of the Universe*. San Marino, Calif.: Huntingdon Library Press.
Hentschel, Klaus (1997). *The Einstein Tower. An Intertexture of Dynamic Construction, Relativity Theory, and Astronomy*. Stanford: Stanford University Press.
Hetherington, Norriss S. (1972). 'Adriaan van Maanen and internal motions in spiral nebulae: A historical review', *Quarterly Journal of the Royal Astronomical Society* **13**, 25–39.
—— (1982). 'Philosophical values and observation in Edwin Hubble's choice of a model of the universe', *Historical Studies in the Physical Sciences* **13**, 41–68.
—— (1988). *Science and Objectivity. Episodes in the History of Astronomy*. Ames, Iowa: Iowa State University Press.
—— ed. (1993). *Encyclopedia of Cosmology: Historical, Philosophical, and Scientific Foundations of Modern Cosmology*. New York: Garland Publishing.
Hewish, Anthony (1986). 'The pulsar era', *Quarterly Journal of the Royal Astronomical Society* **27**, 548–558.
Hirsh, Richard F. (1979). 'The riddle of the gaseous nebulae', *Isis* **70**, 197–212.
Hoefer, Carl (1994). 'Einstein's struggle for a Machian gravitation theory', *Studies in the History and Philosophy of Science* **25**, 287–335.
Holmberg, Gustav (1999). *Reaching for the Stars: Studies in the History of Swedish Stellar and Nebular Astronomy 1860–1940*. Lund: Ugglan.
Holmes, Arthur (1913). *The Age of the Earth*. New York: Harper.
Hooykaas, Reijer (1972). *Religion and the Rise of Modern Science*. Edinburgh: Scottish Academic Press.
Hoskin, Michael (1963). *William Herschel and the Construction of the Heavens*. London: Oldbourne.
—— (1976). 'The "Great Debate": What really happened', *Journal for the History of Astronomy* **7**, 169–182.
—— (1982). *Stellar Astronomy: Historical Studies*. New York: Science History Publications.
—— (1987). 'John Herschel's cosmology', *Journal for the History of Astronomy* **18**, 1–34.
—— ed. (1999). *The Cambridge Concise History of Astronomy*. Cambridge: Cambridge University Press.
—— (2003). *The Herschel Partnership*. Cambridge: Science History Publications.
Howell, Kenneth J. (1998). 'The role of Biblical interpretation in the cosmology of Tycho Brahe', *Studies in the History and Philosophy of Science* **29**, 515–537.
Hoyle, Fred (1948). 'A new model for the expanding universe', *Monthly Notices of the Royal Astronomical Society* **108**, 372–382.
—— (1965). *Galaxies, Nuclei, and Quasars*. New York: Harper and Row.
—— (1994). *Home is Where the Wind Blows: Chapters from a Cosmologist's Life*. Mill Valley, Calif.: University Science Books.
Hoyle, Fred, Geoffrey Burbidge, and Jayant Narlikar (2000). *A Different Approach to Cosmology: From a Static Universe to the Big Bang towards Reality*. Cambridge: Cambridge University Press.
Hoyle, Fred, and Allan Sandage (1956). 'The second-order term in the redshift–magnitude relation', *Publications of the Astronomical Society of the Pacific* **68**, 301–307.
Hu, Danian (2005). *China & Albert Einstein: The Reception of the Physicist and his Theory in China 1917–1979*. Cambridge, Mass.: Harvard University Press.
Hubble, Edwin (1926). 'Extra-galactic nebulae', *Astrophysical Journal* **64**, 321–369.
—— (1929). 'A relation between distance and radial velocity among extra-galactic nebulae', *Proceedings of the National Academy of Sciences* **15**, 168–173.
—— (1936). *The Realm of the Nebulae*. New Haven: Yale University Press.
—— (1937). *The Observational Approach to Cosmology*. Oxford: Clarendon Press.
—— (1942). 'The problem of the expanding universe', *American Scientist* **30**, 99–115.
Hubble, Edwin, and Milton Humason (1931). 'The velocity–distance relation among extra-galactic nebulae', *Astrophysical Journal* **74**, 43–80.
Hubble, Edwin, and Richard C. Tolman (1935). 'Two methods of investigating the nature of the nebular redshift', *Astrophysical Journal* **82**, 302–337.
Huggins, William, and Margaret Huggins (1909). *The Scientific Papers of Sir William Huggins*. London: Wesley and Son.

Humason, Milton (1929). 'The large radial velocity of N.G.C.7619', *Proceedings of the National Academy of Sciences* **15**, 167–168.

Huygens, Christiaan (1722). *The Celestial Worlds Discover'd*. London: James Knapton.

Israel, Werner (1987). 'Dark stars: the evolution of an idea', pp. 199–276 in S. Hawking and W. Israel, eds, *Three Hundred Years of Gravitation*. Cambridge: Cambridge University Press.

Jacobsen, Thorkild (1957). 'Enumah Elish—"the Babylonian Genesis" ', pp. 8–20 in Munitz, 1957.

Jaki, Stanley L. (1969). *The Paradox of Olbers' Paradox*. New York: Herder and Herder.

—— (1974). *Science and Creation*. Edinburgh: Scottish Academic Press.

—— (1978). 'Johann Georg von Soldner and the gravitational bending of light', *Foundations of Physics* **8**, 927–950.

—— (1990). *Cosmos in Transition. Studies in the History of Cosmology*. Tucson: Pachart Publishing House.

James, Frank A. J. L. (1982). 'Thermodynamics and sources of solar heat, 1846–1862', *British Journal for the History of Science* **15**, 155–181.

—— (1985). 'The discovery of line spectra', *Ambix* **32**, 53–70.

Jammer, Max (1993). *Concepts of Space: The History of Theories of Space in Physics*. New York: Dover Publications.

Jeans, James (1928). 'The physics of the universe', *Nature* **122**, 689–700.

Jordan, Pascual (1944). *Physics of the 20th Century*. New York: Philosophical Library.

—— (1952). *Schwerkraft und Weltall*. Braunschweig: Vieweg.

—— (1971). *The Expanding Earth: Some Consequences of Dirac's Gravitational Hypothesis*. Oxford: Pergamon Press.

Kaiser, David (1998). 'A ψ is just a ψ? Pedagogy, practice, and the reconstitution of general relativity, 1942–1975', *Studies in the History and Philosophy of Modern Physics* **29**, 321–338.

Kant, Immanuel (1981). *Universal Natural History and Theory of the Heavens*. Translated by S. L. Jaki. Edinburgh: Scottish Academic Press.

Kargon, Robert H. (1982). *The Rise of Robert Millikan: Portrait of a Life in American Science*. Ithaca, New York: Cornell University Press.

Kerzberg, Pierre (1989). *The Invented Universe. The Einstein–De Sitter Controversy (1916–17) and the Rise of Relativistic Cosmology*. Oxford: Clarendon Press.

Kilmister, Clive W. (1994). *Eddington's Search for a Fundamental Theory: A Key to the Universe*. Cambridge: Cambridge University Press.

Knopf, Otto (1914). 'Kosmogonie', pp. 977–989 in E. Korschelt et al., eds, *Handwörterbuch der Naturwissenschaften*, vol. 5. Jena: Gustav Fischer.

Knorr, Wilbur R. (1990). 'Plato and Eudoxus on the planetary motions', *Journal for the History of Astronomy* **21**, 313–329.

Kolb, Edward W., and Michael S. Turner (1990). *The Early Universe*. New York: Addison-Wesley.

Kolb, Edward W., Michael S. Turner, David Lindley, Keith Olive, and David Seckel, eds (1986). *Inner Space, Outer Space: The Interface Between Cosmology and Particle Physics*. Chicago: University of Chicago Press.

Kox, A. J., and Jean Eisenstaedt, eds (2005). *The Universe of General Relativity*. Boston: Birkhäuser.

Koyré, Alexandre (1965). *Newtonian Studies*. Chicago: University of Chicago Press.

—— (1968). *From the Closed World to the Infinite Universe*. Baltimore: Johns Hopkins Press.

Kragh, Helge (1982). 'Cosmo-physics in the thirties: Towards a history of Dirac cosmology', *Historical Studies in the Physical Sciences* **13**, 69–108.

—— (1990). *Dirac: A Scientific Biography*. Cambridge: Cambridge University Press.

—— (1995). 'Cosmology between the wars: The Nernst–MacMillan alternative', *Journal for History of Astronomy* **26**, 93–115.

—— (1996a). *Cosmology and Controversy: The Historical Development of Two Theories of the Universe*. Princeton: Princeton University Press.

—— (1996b). 'Gamow's game: The road to the hot big bang', *Centaurus* **38**, 335–361.

—— (1997). 'The electrical universe: Grand cosmological theory versus mundane experiments', *Perspectives on Science* **5**, 199–231.

—— (1999). 'Steady-state cosmology and general relativity: Reconciliation or conflict?' pp. 377–402 in Goenner *et al.* 1999.

—— (2000). 'The chemistry of the universe: Historical roots of modern cosmochemistry', *Annals of Science* **57**, 353–368.

—— (2001a). 'From geochemistry to cosmochemistry: The origin of a scientific discipline, 1915–1955', pp. 160–190 in Carsten Reinhardt, ed., *Chemical Sciences in the 20th Century*. Weinheim: Wiley-VCH.

Kragh, Helge (2001b). 'Nuclear archaeology and the early phase of physical cosmology', pp. 157–170 in Martínez, Trimble, and Pons-Bordería 2001.

—— (2003). 'Expansion and origination: Georges Lemaître and the big bang universe', pp. 275–294 in Patricia Radelet-de Grave and Brigitte van Tiggelen, eds, *Sedes Scientiae: L'Émergence de la Recherche á l'Université*. Turnhout: Brepols.

—— (2004). *Matter and Spirit: Scientific and Religious Preludes to Modern Cosmology*. London: Imperial College Press.

—— (2005). 'George Gamow and the "factual approach" to relativistic cosmology', pp. 175–188 in Kox and Eisenstaedt, 2005.

Kragh, Helge, and Simon Rebsdorf (2002). 'Before cosmophysics: E. A. Milne on mathematics and physics', *Studies in the History and Philosophy of Modern Physics* **33**, 35–50.

Kragh, Helge, and Robert Smith (2003). 'Who discovered the expanding universe?', *History of Science* **41**, 141–162.

Krauss, Lawrence M., and Michael S. Turner (1995). 'The cosmological constant is back', *General Relativity and Gravitation* **26**, 1137–1144.

Kubrin, David (1967). 'Newton and the cyclical cosmos: Providence and the mechanical philosophy', *Journal of the History of Ideas* **28**, 325–346.

Kuhn, Thomas S. (1957). *The Copernican Revolution: Planetary Astronomy in the Development of Western Thought*. Cambridge, Mass.: Harvard University Press.

Lactantius 1964. *The Divine Institutes*, translated by M. F. McDonald. Washington, DC: Catholic University of America Press.

Lambert, W. G. (1975). 'The cosmology of Sumer and Babylon', pp. 42–65 in Blacker and Loewe 1975.

Lambert, Johann H. (1976). *Cosmological Letters on the Arrangement of the World-Edifice*. Translated and annotated by S. L. Jaki. Edinburgh: Scottish Academic Press.

Lambert, Dominique (2000). *Un atome d'univers: La vie et l'oeuvre de Georges Lemaître*. Brussels: Éditions Racine.

Lanczos, Cornelius (1925). 'Über eine zeitlich periodische Welt und eine neue Behandlung des Problems der Ätherstrahlung', *Zeitschrift für Physik* **32**, 56–80.

Landsberg, Peter T. (1991). 'From entropy to God?', pp. 379–403 in K. Martinás, L. Ropolyi, and P. Szegedi, eds, *Thermodynamics: History and Philosophy*. Singapore: World Scientific.

Lang, Kenneth R., and Owen Gingerich, eds (1979). *A Source Book in Astronomy and Astrophysics, 1900–1975*. Cambridge, Mass.: Harvard University Press.

LeBon, Gustave (1907). *The Evolution of Matter*. New York: Walter Scott Publishing Co.

Lemaître, Georges (1925). 'Note on De Sitter's universe', *Journal of Mathematical Physics* **4**, 188–192.

—— (1927). 'Un univers homogène de masse constante et de rayon croissant rendant compte de la vitesse radiale des nébuleuses extra-galactiques', *Annales de Sociétés Scientifique de Bruxelles* **47**, 49–56.

—— (1931a). 'The beginning of the world from the point of view of quantum theory', *Nature* **127**, 706.

—— (1931b). 'L'expansion de l'espace', *Revue Questions Scientifiques* **17**, 391–410.

—— (1933). 'L'Univers en expansion', *Annales de Sociétés Scientifique de Bruxelles* **53**, 51–85.

—— (1934). 'Evolution of the expanding universe', *Proceedings of the National Academy of Sciences* **20**, 12–17.

—— (1949). 'The cosmological constant', pp. 437–456 in Paul A. Schilpp, ed., *Albert Einstein: Philosopher-Scientist*. Evanston, Ill.: Library of Living Philosophers.

—— (1958). 'Rencontres avec A. Einstein', *Revue des Questions Scientifiques* **129**, 129–133.

Lemonick, Michael (1993). *The Light at the Edge of the Universe*. Princeton: Princeton University Press.

Lenz, Wilhelm (1926). 'Das Gleichgewicht von Materie und Strahlung in Einsteins geschlossener Welt', *Physikalische Zeitschrift* **27**, 642–645.

Lerner, Lawrence S., and Edward A. Gosselin (1973). 'Giordano Bruno', *Scientific American* **228** (April), 86–94.

Leslie, John, ed. (1990). *Physical Cosmology and Philosophy*. New York: Macmillan.

Lévy-Leblond, Jean-Marc (1990). 'Did the big bang begin?', *American Journal of Physics* **58**, 156–159.

Lewis, Gilbert N. (1922). 'The chemistry of the stars and the evolution of radioactive substances', *Publications of the Astronomical Society of the Pacific* **34**, 309–319.

Lightman, Alan I., and Roberta Brawer (1990). *Origins: The Lives and Worlds of Modern Cosmologists*. Cambridge, Mass.: Harvard University Press.

Lindberg, David C. (1992). *The Beginnings of Western Science*. Chicago: University of Chicago Press.

—— (2002). 'Early Christian attitudes toward nature', pp. 47–56 in Gary B. Ferngren, ed., *Science & Religion. A Historical Introduction*. Baltimore: Johns Hopkins University.

Linde, Andrei (1983). 'The new inflationary universe scenario', pp. 205–250 in G. W. Gibbon, S. W. Hawking, and S. Siklos, eds, *The Very Early Universe*. Cambridge: Cambridge University Press.

—— (1987). 'Particle physics and inflationary cosmology', *Physics Today* **40** (September), 61–68.

—— (1994). 'The self-reproducing inflationary scenario', *Scientific American* **249** (November), 32–39.

Lizhi, Fang (1991). *Bringing Down the Great Wall: Writings on Science, Culture, and Democracy in China*. New York: Alfred A. Knopf.

Lodge, Oliver (1921). 'On the supposed weight and ultimate fate of radiation', *Philosophical Magazine* **41**, 549–557.

Longair, Malcolm S. (2001). 'The technology of cosmology', pp. 55–74 in Martínez, Trimble, and Pons-Bordería 2001.

Lovejoy, Arthur O. (1964). *The Great Chain of Being: A Study of the History of an Idea*. Cambridge, Mass.: Harvard University Press.

Lovell, Bernard (1973). *Out of the Zenith: Jodrell Bank 1957–1970*. London: Oxford University Press.

Lubbock, Constance A. (1933). *The Herschel Chronicle*. Cambridge: Cambridge University Press.

Lucretius (1997). *On the Nature of Things*. Translated by John S. Watson. Amherst, New York: Prometheus Books.

Luminet, Jean-Pierre, ed. (1997). *Alexandre Friedmann, Georges Lemaître. Essais de cosmologie*. Paris: Éditions du Seuil.

McGucken, William (1969). *Nineteenth-Century Spectroscopy: Development of the Understanding of Spectra*. Baltimore: Johns Hopkins Press.

Mach, Ernst (1909). *Die Geschichte und die Wurzel des Satzes von der Erhaltung der Arbeit*. Leipzig: Barth.

McKirahan, Richard D. (1994). *Philosophy Before Socrates*. Indianapolis: Hackett Publishing Company.

McLaughlin, P. J. (1957). *The Church and Modern Science*. New York: Philosophical Library.

MacMillan, William (1925). 'Some mathematical aspects of cosmology', *Science* **62**, 63–72, 96–99, 121–127.

McMullin, Ernan (1987). 'Bruno and Copernicus', *Isis* **78**, 55–74.

—— (1993). 'Indifference principle and anthropic principle in cosmology', *Studies in the History and Philosophy of Science* **24**, 359–389.

McVittie, George (1939). 'The cosmical constant and the structure of the universe', *Observatory* **62**, 192–194.

—— (1940). 'Kinematic relativity', *Observatory* **63**, 273–281.

—— (1961). 'Rationalism versus empiricism in cosmology', *Science* **133**, 1231–1236.

Magueijo, João (2003). 'New varying speed of light theories', *Reports on Progress in Physics* **66**, 2025.

Martínez, Vicent J., Virginia Trimble, and María J. Pons-Bordería, eds (2001). *Historical Development of Modern Cosmology*. San Francisco: Astronomical Society of the Pacific.

Mascall, Erich L. (1956). *Christian Theology and Natural Science*. London: Longmans, Green and Co.

Mason, Brian (1992). *Victor Moritz Goldschmidt: Father of Modern Geochemistry*. San Antonio, Texas: The Geochemical Society.

May, Gerhard (1994). *Creatio ex Nihilo: The Doctrine of 'Creation out of Nothing' in Early Christian Thought*. Edinburgh: T&T Clark.

Meadows, Arthur J. (1972). *Science and Controversy: A Biography of Sir Norman Lockyer*. London: Macmillan.

Merleau-Ponty, Jacques (1977). 'Laplace as a cosmologist', pp. 283–291 in Wolfgang Yourgrau and Allen Breck, eds, *Cosmology, History, and Theology*. New York: Plenum Press.

—— (1983). *La science de l'univers à l'âge du positivism*. Paris: Vrin.

Miller, Arthur I. (1972). 'The myth of Gauss' experiment on the Euclidean nature of physical space', *Isis* **63**, 345–348.

Mills, Bernard, and Slee, O. Bruce (1957). 'A preliminary survey of radio sources in a limited region of the sky at a wavelength of 3.5 m', *Australian Journal of Physics* **10**, 162–194.

Milne, Arthur E. (1935). *Relativity, Gravitation and World Structure*. Oxford: Clarendon Press.

—— (1938). 'On the equations of electromagnetism', *Proceedings of the Royal Society A* **165**, 313–357.

—— (1952). *Modern Cosmology and the Christian Idea of God*. Oxford: Clarendon Press.

Misner, Charles W. (1969). 'Absolute zero of time', *Physical Review* **186**, 1328–1333.

Mitton, Simon (2005). *Fred Hoyle: A Life in Science*. London: Aurum Press.

Munitz, Milton K., ed. (1957). *Theories of the Universe: From Babylonian Myth to Modern Science*. New York: The Free Press.

—— (1986). *Cosmic Understanding: Philosophy and Science of the Universe*. Princeton: Princeton University Press.

Needham, Joseph A., and Colin A. Ronan(1993). 'Chinese cosmology', pp. 63–70 in Hetherington 1993.

Newcomb, Simon (1906). *Side-Lights on Astronomy. Essays and Addresses*. New York: Harper and Brothers.

Newton, Isaac (1952). *Opticks*. New York: Dover Publications.
—— (1999). *The Principia. Mathematical Principles of Natural Philosophy*. Translated by I. Bernard Cohen and Anne Whitman. Berkeley: University of California Press.
Nicholson, John W. (1913). 'The physical interpretation of the spectrum of the corona', *Observatory* **36**, 103–112.
North, John (1975). 'The medieval background to Copernicus', *Vistas in Astronomy* **17**, 1–25.
—— (1990). *The Measure of the Universe: A History of Modern Cosmology*. New York: Dover Publications.
—— (1996). *Stonehenge: Neolithic Man and the Cosmos*. London: HarperCollins.
Norton, John D. (1999). 'The cosmological woes of Newtonian gravitation theory', pp. 271–324 in Goenner *et al.* 1999.
O'Brien, D. (1969). *Empedocles' Cosmic Cycle. A Reconstruction from the Fragments and Secondary Sources*. Cambridge: Cambridge University Press.
Olbers, H. Wilhelm (1826). 'Ueber die Durchsichtigkeit des Weltraums', *Astronomisches Jahrbuch* **51**, 110–121.
Olive, Keith A., and Yong-Zhong Qian (2004). 'Were fundamental constants different in the past?', *Physics Today* **57** (October), 40–46.
Öpik, Ernst (1922). 'An estimate of the distance of the Andromeda nebula', *Astrophysical Journal* **55**, 406–410.
Oppenheimer, J. Robert, and Hartland Snyder (1939). 'On continued gravitational contraction', *Physical Review* **56**, 455–459.
Orr, Mary A. (1956). *Dante and the Early Astronomers*. London: Wingate.
Osterbrock, Donald E. (2001). *Walter Baade. A Life in Astrophysics*. Princeton: Princeton University Press.
Paul, Erich P. (1993). *The Milky Way Galaxy and Statistical Cosmology 1890–1924*. Cambridge: Cambridge University Press.
Peebles, P. James E. (1971). *Physical Cosmology*. Princeton: Princeton University Press.
—— (1993). *Principles of Physical Cosmology*. Princeton: Princeton University Press.
Peebles, P. James E., and Bharat Ratra (2003). 'The cosmological constant and dark energy', *Reviews of Modern Physics* **75**, 559–606.
Peebles, P. James E., David N. Schram, E. L. Turner, and R. G. Kron, (1991). 'The case for the relativistic hot big bang cosmology', *Nature* **352**, 769–776.
Perlmutter, Saul (2003). 'Supernovae, dark energy, and the accelerating universe', *Physics Today* **56** (April), 53–60.
Plaskett, John S. (1933). 'The expansion of the universe', *Journal of the Royal Astronomical Society of Canada* **27**, 235–252.
Pliny (1958). *Natural History*. Vol. 1. Translated by H. Rackham. Cambridge, Mass.: Harvard University Press.
Plumley, J. M. (1975). 'The cosmology of ancient Egypt', pp. 17–41 in Blacker and Loewe 1975.
Poincaré, Henri (1911). *Leçons sur les hypothèses cosmogoniques*. Paris: A. Hermann et Fils.
—— (1982). *The Foundations of Science*. Edited by L. P. Williams. Washington, DC: University Press of America.
Priestley, Joseph (1772). *The History and Present Stage of Discoveries Relating to Vision, Light, and Colours*. London: J. Johnson.
Proctor, Richard A. (1896). *Other Worlds Than Ours*. New York: Appleton.
Ptolemy (1984). *Ptolemy's Almagest*. Tranlated by G. J. Toomer. London: Duckworth.
Rankama, Kaleva, and Ture Sahama (1950). *Geochemistry*. Chicago: University of Chicago Press.
Rankine, William J. M. (1881). *Miscellaneous Scientific Papers*. London: Charles Griffin and Company.
Renn, Jürgen, ed. (2005). *Albert Einstein: Chief Engineer of the Universe*. Weinheim: Wiley-VCH Verlag.
Renn, Jürgen, and Tilman Sauer (2003). 'Eclipses of the stars: Mandl, Einstein, and the early history of gravitational lensing', pp. 69–92 in Abhay Ashtekar, Robert S. Cohen, Don Howard, Jürgen Renn, Sahotra Sarkar, and Abner Shimony, eds, *Revisiting the Foundations of Relativistic Physics*. Dordrecht: Kluwer Academic.
Reynolds, Andrew (1996). 'Peirce's cosmology and the laws of thermodynamics', *Transactions of the Charles S. Peirce Society* **32**, 403–423.
Rice, James (1925). 'On Eddington's natural unit of the field', *Philosophical Magazine* **49**, 1056–1057.
Riedweg, Christoph (2002). *Pythagoras: His Life, Teaching, and Influence*. Ithaca, New York: Cornell University Press.
Riemann, Bernhard (1873). 'On the hypotheses which lie at the bases of geometry', *Nature* **8**, 14–17, 36–37.
Roberts, Francis (1694). 'Concerning the distance of the fixed stars', *Philosophical Transactions* **18**, 101–103.
Robertson, Howard P. (1928). 'On relativistic cosmology', *Philosophical Magazine* **5**, 835–848.
—— (1929). 'On the foundation of relativistic cosmology', *Proceedings of the National Academy of Sciences* **15**, 822–829.
—— (1933). 'Relativistic cosmology', *Reviews of Modern Physics* **5**, 62–90.

Robertson, Peter (1992) *Beyond Southern Skies: Radio Astronomy and the Parkes Telescope.* Cambridge: Cambridge University Press.
Rochberg-Halton, Francesca (1993). 'Mesopotamian cosmology', pp. 398–407 in Hetherington 1993.
Roscoe, Henry E. (1875). *The History of the Chemical Elements.* London: William Collins.
Rosen, Edward, ed. (1959). *Three Copernican Treatises.* New York: Dover Publications.
—— (1965). *Kepler's Conversations with Galileo's Sidereal Messenger.* New York: Johnson Reprint Corp.
Rosevear, N. T. (1982). *Mercury's Perihelion. From Le Verrier to Einstein.* Oxford: Clarendon Press.
Rothman, Tony, and George F. R. Ellis (1987). 'Has cosmology become metaphysical?', *Astronomy* **15**, 6–22.
Rubin, Vera C. (1983). 'Dark matter in spiral galaxies', *Scientific American* **248** (June), 88–101.
Rüger, Alexander (1988). 'Atomism from cosmology: Erwin Schrödinger's work on wave mechanics and space–time structure', *Historical Studies in the Physical Sciences* **18**, 377–401.
Russell, Bertrand (1957). *Why I am not a Christian.* London: Unwin.
Russell, Colin A., ed. (1973). *Science and Religious Belief: A Selection of Recent Historical Studies.* Sevenoaks, Kent: Open University.
Russell, Robert J. (1994). 'Cosmology from alpha to omega', *Zygon* **29**, 557–577.
Ryan, Michael P., and L. C. Shepley (1976). 'Resource letter RC-1: Cosmology', *American Journal of Physics* **44**, 223–230.
Sambursky, Samuel (1963). *The Physical World of the Greeks.* London: Routledge.
—— (1973). 'John Philoponus', pp. 134–139 in *Dictionary of Scientific Biography*, vol. 7. New York: Charles Scribner's Sons.
—— (1987). *The Physical World of Late Antiquity.* Princeton: Princeton University Press.
Sánchez-Ron, José (2005). 'George McVittie, the uncompromising empiricist', pp. 189–222 in Kox and Eisenstaedt 2005.
Sandage, Allan (1970). 'Cosmology: a search for two numbers', *Physics Today* **23** (February), 34–41.
—— (1998). 'Beginnings of observational cosmology in Hubble's time: Historical overview', pp. 1–26 in Mario Livio, S. Michael Fall, and Piero Madau, eds, *The Hubble Deep Field.* Cambridge University Press: Cambridge.
Schaffer, Simon (1978). 'The phoenix of nature: Fire and evolutionary cosmology in Wright and Kant', *Journal for the History of Astronomy* **9**, 180–200.
Scheiner, Julius (1899). 'On the spectrum of the great nebula in Andromeda', *Astrophysical Journal* **9**, 149–150.
Schemmel, Matthias (2005). 'An astronomical road to general relativity: The continuity between classical and relativistic cosmology in the work of Karl Schwarzschild', *Science in Context* **18**, 451–478.
Scheuer, Hans Günter (1997). *Der Glaube der Astronomen und die Gestalt des Universums: Kosmologie und Theologie im 18. und 19. Jahrhundert.* Aachen: Shaker-Verlag.
Schiaparelli, Giovanni (1905). *Astronomy in the Old Testament.* Oxford: Clarendon Press.
Schild, Alfred (1962). 'Gravitational theories of the Whitehead type and the principle of equivalence', pp. 69–115 in Christian Møller, ed., *Evidence for Gravitational Theories.* New York: Academic Press.
Schmidt, Maarten (1990). 'The discovery of quasars', pp. 347–354 in Bertotti *et al.* 1990.
Schofield, Christine J. (1981). *Tychonic and Semi-Tychonic World Systems.* New York: Arno Press.
Schramm, David N., ed. (1996). *The Big Bang and Other Explosions in Nuclear and Particle Astrophysics.* Singapore: World Scientific.
Schramm, David N., and Gary Steigman (1988). 'Particle accelerators test cosmological theory', *Scientific American* **262** (June), 66–72.
Schramm, David N., and Michael S. Turner (1998). 'Big-bang nucleosynthesis enters the precision era', *Reviews of Modern Physics* **70**, 303–318.
Schröder, Wilfried, and Hans-Jürgen Treder (1996). 'Hans Ertel and cosmology', *Foundations of Physics* **26**, 1081–1088.
Schrödinger, Erwin (1937). 'Sur la théorie du monde d'Eddington', *Nuovo Cimento* **15**, 246–254.
—— (1939). 'The proper vibrations of the expanding universe', *Physica* **6**, 899–912.
Schuster, Arthur (1898). 'Potential matter. A holiday dream', *Nature* **58**, 367.
Schwarzschild, Karl (1998). 'On the permissible curvature of space', *Classical and Quantum Gravity* **15**, 2539–2544.
Sciama, Dennis W., and Martin J. Rees (1966). 'Cosmological significance of the relation between red-shift and flux density for quasars', *Nature* **211**, 1283.
Seeley, David, and Richard Berendzen (1972). 'The development of research in interstellar absorption, c.1900–1930', *Journal for the History of Astronomy* **3**, 52–64, 75–86.
Seeliger, Hugo von (1895). 'Über das Newtonsche Gravitationsgesetz', *Astronomische Nachrichten* **137**, 129–136.

Seeliger, Hugo von (1897–98). 'On Newton's law of gravitation', *Popular Astronomy* **5**, 544–451.
Seife, Charles (2004). *Alpha and Omega: The Search for the Beginning and the End of the Universe*. London: Bantam Books.
Shapley, Harlow (1918). 'Studies based on the colors and magnitudes in stellar clusters, VI: On the determination of the distances of globular clusters', *Astrophysical Journal* **48**, 89–124.
Silberstein, Ludwik (1930). *The Size of the Universe*. Oxford: Oxford University Press.
Singer, Dorothea W. S. (1950). *Giordano Bruno: His Life and Thought*. New York: Henry Schuman.
Smeenk, Chris (2003). *Approaching the Absolute Zero of Time: Theory Development in Early Universe Cosmology*. PhD thesis, University of Pittsburgh.
—— (2005). 'False vacuum: Early universe cosmology and the development of inflation', pp. 223–257 in Kox and Eisenstaedt 2005.
Smith, Robert W. (1982). *The Expanding Universe: Astronomy's 'Great Debate' 1900–1931*. Cambridge: Cambridge University Press.
—— (1989) *The Space Telescope: A Study of NASA, Science, Technology, and Politics*. Cambridge: Cambridge University Press.
Smith, Robert W., and Joseph N. Tatarewicz (1985). 'Replacing the telescope: The large space telescope and CCDs', *Proceedings of the IEEE* **73** (July), 1221–1235.
Smolin, Lee (1992). 'Did the universe evolve?', *Classical and Quantum Gravity* **9**, 173–191.
Smoot, George F., C. L. Bennett, A. Kogut, E. L. Wright, J. Aymon, N. W. Boggess, *et al*., (1992). 'Structure in the COBE Differential Microwave Radiometer first-year map', *Astrophysical Journal Letters* **396**, L1–L5.
Soddy, Frederick (1909). *The Interpretation of Radium*. London: John Murray.
Söderqvist, Thomas, ed. (1997). *The Historiography of Contemporary Science and Technology*. Amsterdam: Harwood Academic.
Sovacool, Benjamin (2005). 'Falsification and demarcation in astronomy and cosmology', *Bulletin of Science, Technology & Society* **25**, 53–62.
Spencer Jones, Harold (1954). 'Continuous creation', *Science News* **32**, 19–32.
Stallo, John (1882). *The Concepts and Theories of Modern Physics*. London: Kegan Paul, Trench & Co.
Steinhardt, Paul J., and Neil Turok (2002). 'A cyclic model of the universe', *Science* **296**, 1436–1439.
Stoffel, Jean-François, ed. (1996). *Mgr. Georges Lemaître, savant et croyant*. Louvain-la-Neuve: Centre Interfacultaire d'Étude en Histoire des Sciences.
Stukeley, William (1936). *Memoirs of Sir Isaac Newton's Life*. Edited by A. H. White. London: Taylor and Francis.
Sullivan, Woodruff T., ed. (1982). *Classics in Radio Astronomy*. Dordrecht: Reidel.
—— (1990). 'The entry of radio astronomy into cosmology: radio stars and Martin Ryle's 2C survey', pp. 309–330 in Bertotti *et al*. 1990.
Tait, Peter G. (1871). [Address by President of the Mathematics and Physics Section], *Report, British Association for the Advancement of Science*, 1–8.
Tammann, Gustav, Allan Sandage, and Amos Yahil (1979). 'The determination of cosmological parameters', pp. 53–126 in Roger Balian, Jean Audouze, and David Schramm, eds, *Physical Cosmology*. Amsterdam: North-Holland.
Taylor, Joseph H. (1992). 'Pulsar timing and relativistic gravity', *Philosophical Transactions of the Royal Society, London, A* **341**, 117–134.
Terzian, Yervant, and Elisabeth Bilson (1982). *Cosmology and Astrophysics: Essays in Honor of Thomas Gold*. Ithaca: Cornell University Press.
Thomson, William (1882–1911). *Mathematical and Physical Papers*. 6 vols. Cambridge: Cambridge University Press.
—— (1891). *Popular Lectures and Addresses*, vol. 1. London: Macmillan.
—— (1901). 'On ether and gravitational matter through infinite space', *Philosophical Magazine* **2**, 160–177.
Thoren, Victor E. (1990). *The Lord of Uraniborg: A Biography of Tycho Brahe*. Cambridge: Cambridge University Press.
Thorne, Kip S. (1994). *Black Holes and Time Warps: Einstein's Outrageous Legacy*. New York: Norton.
Tipler, Frank J. (1988). 'Johann Mädler's resolution of Olbers' paradox', *Quarterly Journal of the Royal Astronomical Society* **29**, 313–325.
Tipler, Frank J., C. Clarke, and George F. R. Ellis (1980). 'Singularities and horizons, a review article', pp. 97–206 in A. Held, ed., *General Relativity and Gravitation*, vol. 2. New York: Plenum Press.
Tolman, Richard (1929). 'On the astronomical implications of the de Sitter line element for the universe', *Astrophysical Journal* **69**, 245–274.
—— (1931). 'On the problem of the entropy of the universe as a whole', *Physical Review* **37**, 1639–1660.

—— (1934). *Relativity, Thermodynamics, and Cosmology*. Oxford: Oxford University Press.
—— (1949). 'The age of the universe', *Reviews of Modern Physics* **21**, 374–378.
Torretti, Roberto (1978). *Philosophy of Geometry from Riemann to Poincaré*. Dordrecht: Reidel.
Toulmin, Stephen, and June Goodfield (1982). *The Discovery of Time*. Chicago: University of Chicago Press.
Trimble, Virginia (1988). 'Dark matter in the universe: Where, what, and why?', *Contemporary Physics* **29**, 373–392.
—— (1990). 'History of dark matter in the universe (1922–1974)', pp. 355–363 in Bertotti *et al.* 1990.
—— (1996). 'H_0: The incredible shrinking constant 1925–1975', *Publications of the Astronomical Society of the Pacific* **108**, 1073–1082.
Tropp, Eduard A., Viktor Ya. Frenkel, and Arthur D. Chernin (1993). *Alexander A. Friedmann: The Man Who Made the Universe Expand*. Cambridge: Cambridge University Press.
Trumpler, Robert J. (1930). 'Preliminary results on the distances, dimensions, and space distribution of open star clusters', *Lick Observatory Bulletin* **14**, No. 420, 154–188.
Tryon, Edward P. (1973). 'Is the universe a quantum fluctuation?', *Nature* **246**, 396–397.
Turok, Neil, ed. (1997). *Critical Dialogues in Cosmology*. Singapore: World Scientific.
Urani, John, and George Gale (1994). 'E. A. Milne and the origins of modern cosmology: An essential presence', pp. 390–419 in J. Earman, M. Janssen, and J. D. Norton, eds, *The Attraction of Gravitation: New Studies in the History of General Relativity*. Boston: Birkhäuser.
Usher, Peter (1999). 'Hamlet's transformation', *Elizabethan Review* **7**, 48–64.
Vailati, Ezio (1997). *Leibniz and Clarke: A Study in their Correspondence*. New York: Oxford University Press.
Van den Bergh, Sidney (2001). 'A short history of the missing mass and dark energy paradigms', pp. 75–84 in Martínez, Trimble, and Pons-Bordería 2001.
Van der Kruit, P. C., and K. van Berkel, eds (2000). *The Legacy of J. C. Kapteyn: Studies on Kapteyn and the Development of Modern Astronomy*. Dordrecht: Kluwer Academic.
Van Helden, Albert (1985). *Measuring the Universe: Cosmic Dimensions from Aristarchus to Halley*. Chicago: University of Chicago Press.
Veneziano, Gabriele (2004). 'The myth of the beginning of time', *Scientific American* **290** (May), 54–65.
Vilenkin, Alexander (1982). 'Creation of universes from nothing', *Physics Letters* **117B**, 25–28.
Wade, Nicholas (1975). 'Discovery of pulsars: A graduate student's story', *Science* **189**, 358–364.
Wagoner, Robert V. (1990). 'Deciphering the nuclear ashes of the early universe: a personal perspective', pp. 159–187 in Bertotti *et al.* 1990.
Webb, Stephen (1999). *Measuring the Universe: The Cosmological Distance Ladder*. London: Springer.
Weinberg, Steven (1977). *The First Three Minutes: A Modern View of the Origin of the Universe*. New York: Basic Books.
Weizsäcker, Carl Friedrich von (1938). 'Über Elementumwandlungen im Innern der Sterne, II', *Physikalische Zeitschrift* **39**, 633–646.
Wesson, Paul (1980). 'Does gravity change with time?', *Physics Today* **33** (July), 32–37.
Westfall, Richard S. (1980). *Never at Rest: A Biography of Isaac Newton*. Cambridge: Cambridge University Press.
Weyl, Hermann (1924). 'Observations on the note of Dr L. Silberstein', *Philosophical Magazine* **48**, 348–349.
Whitehead, Alfred N. (1922). *The Principle of Relativity*. Cambridge: Cambridge University Press.
Whitrow, Gerald J. (1998). *Time in History: Views of Time from Prehistory to the Present Day*. Oxford: Oxford University Press.
Wilkinson, David T., and Peebles, P. James E. (1990). 'Discovery of the 3K radiation', pp. 17–31 in N. Mandolesi and N. Vittorio, eds, *The Cosmic Microwave Background: 25 Years Later*. Dordrecht: Kluwer.
Williams, James W. (1999). 'Fang Lizshi's big bang: A physicist and the state in China', *Historical Studies in the Physical and Biological Sciences* **30**, 49–114.
Wright, Thomas (1971). *An Original Theory or New Hypothesis of the Universe*. With an introduction by M. Hoskin. London: MacDonald.
Wright, M. R. (1995). *Cosmology in Antiquity*. London: Routledge.
Yoshimura, Motohiko (1978). 'Unified gauge theories and the baryon number of the universe', *Physical Review Letters* **41**, 281–284.
Zel'dovich, Yakov B., and Igor D. Novikov (1983). *Relativistic Astrophysics, II: The Structure and Evolution of the Universe*. Chicago: University of Chicago Press.
Zöllner, K. Friedrich (1872). *Über die Natur der Cometen. Beiträge zur Geschichte und Theorie der Erkenntnis*. Leipzig: Engelmann.
Zwicky, Fritz (1933). 'Die rotverschiebung von extragalaktischen Nebeln', *Helvetica Physica Acta* **6**, 110–127.
—— (1935). 'Remarks on the redshift from nebulae', *Physical Review* **48**, 802–806.

INDEX

aberration of light 76–77
absorption of light, interstellar 84–85, 113–14
accelerating universe 216, 230–34, 238
Adams, Walter 118, 131
Adelard of Bath 37
age paradox 159–60, 190–92
age of universe 18, 33, 36, 83, 150, 160, 167, 231, 245; Kepler's value of 62
Alexander, Stephen 111
Albrecht, Andreas 227, 236
Alfvén, Hannes 217
Almagest 28–29, 37, 40, 76
Alpher, Ralph 178–82, 202, 226
Ambarzumian, Victor 199
Anaxagoras 15
Anaximander 14–15
Anaximenes 235
Andromeda Nebula 100, 116, 119, 190
anthropic principle 238–39
antimatter 107, 208, 223, 241
apeiron 14
Apianus, Petrus 1, 41
Apollonius 28
Aquinas, Thomas 42–43
Archimedes 26–27
Argelander, Friedrich 87
Aristarchus 24–27
Aristotelianism, revival of 37–41
Aristotle 3, 14, 16, 28–29, 38, 43–45, 243; eternal universe 23, 36; his world picture 19–24
Arrhenius, Svante 2, 106, 110
arrow of time 149, 153
Aslaksen, Cort 54–55
astrochemistry 92–95, 152
astrology 27, 36, 60
astronomy: conceptions of 21, 39, 53; and astrophysics 89–90
astrophysics 89–92; nuclear 161–62
astrospectroscopy 90–95
atomism 16–18, 57
Augustine 35
Averroes 40–41
axion 225

Baade, Walter 190–91, 207, 230, 246
Bacon, Francis 57
Bacon, Roger 218 (n. 5)
Barnes, Ernest 155–56
Barrow, John 237
Baum, William 193
Becker, George F. 95
Bede 36
beginning, cosmic. *See*: creation of the world; universe, beginning of
Behr, Albert 190
Bekenstein, Jacob 211
Bell, Jocelyn 207
Beltrami, Eugenio 126
Bentley, Richard 72–74, 108, 131
Bernard Sylvester 37–38
Bessel, F. Wilhelm 77, 89
Besso, Michele 130–31
Bethe, Hans 161, 218 (n. 4)
Bible 10, 12, 33–34, 38, 45, 52, 60, 82, 199
Big Bang universe 149, 152–55; Gamow's 177–80; responses to 156–59, 199–200
big squeeze 218 (n.13)
binary pulsar 207
Birkhoff, Garrett D. 139
black-body radiation 91, 202–03, 228–29
black holes 129, 210–12
Boethius of Dacia 41
Boltzmann, Ludwig 106–07, 235
Bolyai, János 126
Bondi, Hermann 185–89, 196–97, 246, 249
Bonnor, William 189, 195, 197
BOOMERANGproject 231
Borel, Emile 138
Boscovich, Roger 82, 127
Bouguer, Pierre 84
Bowen, Ira 99

Boyle, Willard 250
Bradley, Charles 152
Bradley, James 76–77, 86
Bradwardine, Thomas 44
Brahe, Tycho 37, 51–55, 63
Brans, Carl 217–18
Brewster, David 91, 122 (n. 58)
Bridgman, Percy W. 248
Brodie, Benjamin 93
Bronstein, Matvei 149, 219 (n. 46)
Brout, Robert 241
brown dwarfs 208
Bruno, Giordano 57–59
Buffon 83
Bunge, Mario 196
Bunsen, Wilhelm 91, 93
Burbidge, Geoffrey 190, 217, 236
Burbidge, Margaret 190
Buridan, Jean 42, 44–45
Burnet, Thomas 72

Calinon, Auguste 127–28
Cameron, Alastair 218 (n. 24)
Campanus of Novara 39
Canuto, Vittorio 169, 217
Capella, Martianus 35
Carey, Samuel Warren 198
Carr, Bernard 239
Carter, Brandon 238–39
Cassini, Jacques 76
CCD detectors 250
cepheid variables 115–16, 119
C-field theory 188–89
Chalcidius 32
Chandrasekhar, Subrahmanyan 168
Chardin, Teilhard de 2
Charlier, Carl 85, 105, 109–10
Chaucer, Geoffrey 40
chemistry: Paracelsian 53–54, 60; stellar 92–95
Chéseaux, Jean-Philippe 84
Chrysippus 19, 24
Chwolson, Orest 106, 209
circle paradigm 19–21, 28, 58, 244
Clark, George 208
Clarke, Samuel 72
Clausius, Rudolf 101–02, 104
Clavius, Christoph 51
Cleanthes 27
Clerke, Agnes 110–11, 213
Clifford, William K. 126
clockwork metaphor 45, 71
COBE experiment 228–29
cold dark matter 225, 232
Collins, Barry 239
comets 51, 58, 84
communism 199–200
Comte, Auguste 89–90
conventionalism 127, 172
Copernican revolution 46–50, 60, 244–45
Copernican system 3, 26–27, 47–50, 57, 67
Copernicus 26–27, 34, 47–51, 53, 245, 247
coronium 94–95, 122 (n. 68)
Cosmas 34–35
cosmic microwave background 201–05, 228–29; anisotropies 229, 231; confirmation 203; discovery 202; prediction 180–81, 202
cosmic rays 153, 155, 171
cosmogony 1, 11, 38, 72
cosmography 1
cosmological constant 132–35, 147, 154–55, 197, 216; and age paradox 160; dark energy 231–34; inflation scenario 226; varying in time 149
cosmological principle 46, 63, 170
cosmology 19; atomistic 16–18; Babylonian 8–11, 246; in book title 77; Cartesian 67–69; Chinese 11; Egyptian 7–8, 11; institutionalization 215; Jewish 10; Kant's 78–81; meaning of 1–2; medieval 32–46; Newton's 69–75; philosophical issues 196–97; pre-literar 6; scientific status of 196, 221–22, 248–49
cosmophysics 1–2
cosmos 1. *See also*: universe
counter-Earth 15–16, 213
counter-entropic processes 71, 104, 106
creation of matter 168–69, 184; continual 186, 188, 196, 199
creation of the world 23, 37, 42, 69, 72, 80–81, 240–42; entropic 104; *ex nihilo* 12, 33, 184; and Olbers' paradox 85
Crookes, William 92–93, 218 (n. 5)
Crowther, James 148
Cultural Revolution 199–200
Curtis, Heber 118–19
curvature parameter 192

Cusanus 45–46, 57, 170
cyclical universe 12, 19, 72–73, 102, 108, 240; Einstein 147; Friedmann 141; Sakharov 224

Dalton, John 89
Dante Alighieri 40–41
dark energy 230–33
dark matter 208, 210, 213–14, 224; non-baryonic 223–25 *See also*: cold dark matter
Darwinism: cosmological 235, 237–38; inorganic 93
Davies, Paul 237, 239
decay, cosmic 24, 71–72, 75, 81
deceleration parameter 192–93, 216
deferent 28–29, 31, 49
demiurge, Plato's 22–23
Democritus 17
density, cosmic 137–38, 158–59, 187, 213–14, 223; critical 156, 187, 213, 226, 233
density fluctuations, early universe 227, 229
Descartes, René 67–70, 72
de Sitter, Willem 131, 133–35, 146–47, 156–57, 160
de Sitter model 134–36, 139
deuterium 222–23
DeWitt, Bryce 241
dialectical materialism 199–200
Dicke, Robert 202, 205, 217–18, 226, 238
Digges, Thomas 57–58
dimensions, cosmic. *See*: distances, cosmic
dimensions of space 79, 139
Dingle, Herbert 150, 172, 196, 248–49
Diodorus 34
Dirac, Paul 166–69, 208, 217, 224, 238
Disney, M. J. 249
dissociation hypothesis 94–95
distances, cosmic 24–26; Andromeda 119; Copernican universe 48, 50; Kepler's 63; medieval 39; Ptolemy's 31–32; Sun and Moon 25, 51; stars 76–77. *See also*: universe, size of
Donne, John 47, 60, 234
Doppler, Christian 91
Doppler effect 91–92, 116, 146, 164
Doroshkevich, Andrei 201
Draper, John 122 (n. 58), 250
Droste, Johannes 211
Duhem, Pierre 105
Dyson, Frank 130
Dyson, Paul 237–38

Earth 16, 29; age of 82–83, 160, 191; evolution of 72; expanding 198; location of 22, 35; rotating 44–46; shape of 14, 35–36; spherical 15, 23, 36, 38. *See also*: universe, Copernican
Easton, Cornelis 111–12
Ecphantus 16
Eddington, Arthur S. 27, 130–31, 136, 143–44, 152, 157, 163, 209; cosmological numbers 138–40, 165–67
Einasto, Jaan 214
Einstein, Albert 91, 110, 125, 156–57, 160, 198; cosmological constant 155, 160, 232; cosmological equations 131–33; and Friedmann 142; general relativity 128–31, 209, 211–12; and Lemaître 143; singularities 211–12
Einstein-de Sitter model 156, 191, 213, 216, 234
Einstein equations 129
Einstein model 132–34, 139
Einstein-Rosen bridge 198
electrical universe 189
elements, chemical: distribution of 95, 151–52, 162, 178; formation in stars 189–90; origin of 93, 152, 161–62, 178–79, 205–06; in stars 92–95; the four classical 14–15, 21–22, 24, 28, 38, 46. *See also*: ether
Ellis, George 212, 228, 237, 249
Empedocles 14–15
empyrean heaven 38, 40–41
end, cosmic. *See*: decay, cosmic; universe, end of
energy conservation, violation of 149, 186, 236
entropic creation argument 104, 150
entropy 24, 75, 101, 104, 106, 153; black hole 211
entropy law 101–06
Enuma Elish 9
Epicurus 17
epicycle 28–29, 31, 49
equant 29, 40, 49
equivalence, principle of 128
Eratosthenes 24–25

Ertel, Hans 166
eschatology, physical 108, 237–38
eternity 24, 36, 42, 75.
ether 107; the fifth element 22, 29, 38, 54, 58
Eudoxus 3, 19–20, 28, 243, 246
exoplanets 237

Faber, Johann 62
Farkas, Ladislaus 152
Fermi, Enrico 181
Fick, Adolph 104
fifth element. *See*: ether
fine-structure constant 165–66, 234–35
Fizeau, Hippolyte 122 (n. 60)
Flamm, Ludwig 198
Flammarion, Camille 103
flatness problem 226–27
Fludd, Robert 60–61
Follin, James 181–82, 226
Fontenelle, Bernard 69
Foucault, Jean 122 (n. 58)
Fourier, J. B. Joseph 101
Fowler, William 190, 205–06
Fraunhofer, Joseph 90
Freundlich, Erwin 130, 165, 212
Friedmann, Alexander 134, 141–44
Friedmann equations 141, 143

galaxies: formation of 189; rotation of 214. *See also*: nebulae
Galileo 35, 53, 60–62
gamma bursts 208
Gamow, George 161, 189, 198; Big Bang theory 177–80, 183–84
Gassendi, Pierre 55, 57
Gauss, Carl Friedrich 126–27
Geiser, Carl Friedrich 129
Geminus 20–21, 32, 244
Genzel, Reinhard 211
geocentric system 27. *See also*: Ptolemaic system
geochemistry 163, 183
geo-heliocentric systems 37, 51
Gerard of Cremona 37
Gerasimovich, Boris 219 (n. 46)
Geroch, Robert 212
Gilbert, William 59–60
Giovanelli, R. 235
God 33, 37, 42–44, 68, 74, 104, 171, 196, 237, 241, 243
Gödel, Kurt 197–98
Gold, Thomas 185–89, 194, 207, 217
Goldhaber, Maurice 208
Goldschmidt, Victor 162, 178
Goodricke, John 115
Gore, George Ellard 85
Gott, J. Richard 235
grand unified theory 223–24, 226
gravitation 46, 60; Newton's law of 70, 108–10, 128
gravitation paradox 73, 104, 108–10
gravitation theories, non-Einsteinian 138–39
gravitational collapse 109, 207, 211
gravitational constant 129; varying 167–69, 198, 207, 217
gravitational lenses 209–10
Great Debate 4, 118–19, 234
Greenstein, Jesse 206
Grosseteste, Robert 38
Grossmann, Marcel 129
Guericke, Otto von 55
Gunn, James 216, 222
GUT. *See*: grand unified theory
Guth, Alan 225–28, 233, 235

Haas, Arthur 104–05, 166, 169, 241
Hale, George 129
Hale, Matthew 75
Halley, Edmund 74, 76, 83–84
Harkins, William 151–52
Harriot, Thomas 62
Harrison, Edward 227
Harteck, Paul 152
Hartle, James 241
Hartle-Hawking model 241–42
Hawking, Stephen 5, 211–12, 227, 239, 247
Hawking radiation 211
Hawkins, Gerald 6, 165
Hayashi, Chusiro 181, 183
heat death 101–06, 150, 187, 237, 248
Heckmann, Otto 148, 159, 192
heliocentric system. *See*: Copernican system
helium 222; calculated content of 181–83, 204–06; cosmic abundance 151; discovery 94
Helm, Georg 105
Helmholtz, Hermann von 101–02
Henderson, Thomas 121 (n. 23)
Heracleides of Pontus 29, 37

Herman, Robert 180–82, 202, 226
Herschel, Caroline 86
Herschel, John 84–85, 97, 99, 110, 112, 122 (n. 58)
Herschel, William 75, 86–88, 90; and nebulae 96
Hertzsprung, Ejnar 115, 117
Hesiod 12
Hewish, Anthony 207
Hey, James 193
High-*z* Supernova Research Team 230–31
Hipparchus 3, 19, 28–29
Hippolytes 33
Holmes, Arthur 150
homocentric model 21–22
horizon problem 226–27
Hoyle, Fred 195, 204–07, 239; and element formation 190; and steady-state theory 185–87, 216–17, 236; and Stonehenge 6
Hubble, Edwin 137–38, 157–58, 160; expanding universe 144–46; island universe 119–20; on Milne's theory 172
Hubble constant 144–47, 156, 191, 210, 213, 216, 230
Hubble law 144–46
Hubble time 190–91, 216. *See also*: Hubble constant
Huggins, William 92, 99–100
Hulse, Russell 207
Humason, Milton 145–46, 191, 193
Humboldt, Alexander von 89
Huygens, Christiaan 69
Hyle 37–38, 179, 218 (n. 5)

Ibn al-Haytam 40
Ibn Rush, Muhammad. *See*: Averroes.
Idlis, Grigory 238
India, cosmologies in 11, 13
indifference principle 120 (n. 4)
infinitism 42, 46, 105–06; Cartesian cosmology 68; in Greek philosophy 16–18, 23–24; Kant 80; Newton's universe 70, 73–74; seventeenth century 55–60, 62
inflation theory 225–29, 235, 240; chaotic 235–37
infrared light 90
inner space—outer space 222, 251 (n. 3)
instrumentalism 49, 70, 184, 246–47
Isidore 36
Islam, Jamal 237
Islamic astronomy 32, 37, 40
island universe theory 80, 100, 117–20

Jansky, Karl 193
Jeans, James 113, 148, 150
Johnson, Martin 197
Jordan, Pascual 167–69, 241
Joule, James 101

Kant, Immanuel 64 (n. 16), 87, 96, 125, 153; cosmology of 78–82
Kant-Laplace hypothesis 122 (n. 71)
Kapp, Reginald 185
Kapteyn, Jacobus C. 112–14
Kepler, Johannes 49, 62–64, 83, 244
Kepler's planetary laws 69
Khalatnikov, Isaac 212
kinematic relativity 172
King, Ivan 219 (n. 27)
Kirchhoff, G. Robert 91, 93, 104
Klein, Felix 130, 135–36
Klein, Oskar 197
Klügel, Georg Simon 86
Kobold, Hermann 123 (n. 113)
Koestler, Arthur 2
Kolb, Edward 242
Kraushaar, William 208
Krauss, Lawrence 233
Kuhn, Thomas 244
Kumar, Shiv Sharan 208

Lactantius 34–35, 66 (n. 73)
Lalande, Joseph-Jérôme 86
Lambert, Johann H. 81–82, 84
Lanczos, Cornelius 140, 150, 165
Laplace, Pierre-Simon 96, 109, 210, 213, 243
large number hypothesis 167–69, 217
laws, natural 68–69, 71, 80, 248
Leavitt, Henrietta Swan 115
LeBon, Gustave 107
Leibniz, Gottfried W. 71–72
Leibundgut, Bruno 230
Lemaître, Georges 134, 140, 148–49, 189, 196, 199, 211, 216; expanding model 142–44; primeval-atom hypothesis 152–57; vacuum energy 232
Lemaître model 154–55, 159

Lemaître-Eddington model 147, 152, 154, 157
Lenard, Philipp 173 (n. 11)
Lenz, Wilhelm 151
Levi-Civitá, Tullio 136
Lewis, Gilbert 152
Liceti, Fortunio 62
life, in other worlds 18, 46, 69, 236–38
Lifschitz, Evgeny 212
light: gravitational deflection 128–31; velocity of 75–76, 163, 236
light cosmogony 38
limited elongation 29, 49
Lindberg, David 40
Lindblad, Bertil 123 (n. 113)
Linde, Andrei 227, 235–37
Lizhi, Fang 199–200
Lobachevskii, Nikolai 126–27
Lockyer, Norman 6, 92, 94–95
Lodge, Oliver 108, 210
Lovell, A. C. Bernard 194–95
Lucretius 17–18, 23
Lundmark, Knut 124 (n. 123), 137
Luther, Martin 55
Lyttleton, Raymond 189

Mach, Ernst 105–06, 132, 248
MACHO 214–15
Mach's principle 132, 135, 238
MacMillan, William 164, 185
Macrobius, Ambrosius 35
Mädler, Johann 85
Magueijo, João 236
Maillet, Benoît de 82
Maimonides, Moses 39–40, 247
Mandl, Rudi 209
Mascall, Erich 219 (n. 37)
mass gap problem 181, 183, 190
Mather, John 228
Matthews, Thomas 206
Mattig, Walter 192
Maxwell, James C. 102, 104
Mayall, Nicholas 159, 191, 193, 207
Mayer, J. Robert 101
Mayer, J. Tobias 76, 86
Mayor, Michel 236
McCrea, William 148, 188–89, 232
McKellar, Andrew 180, 203
McVittie, George 148, 159, 172
Mercury anomaly 109, 130
Mersenne, Marin 67
Mesopotamian universe 8–10
meteorites 93, 151
Michell, John 210, 213
Middle Ages. *See*: cosmology, medieval
Milky Way 63, 77–78, 99, 110–14; density 113, 213; Galileo and 61; Wm. Herschel 87–88; Kant's theory 80; rotation 114; Shapley's model 117–18
Milky Way universe 85, 100, 111, 117–18
Millikan, Robert 155
Mills, Bernard 194
Milne, Edward Arthur 149, 163, 169–72
Mineur, Henri 190
Minkowski, Hermann 128
Misner, Charles 212–13
Moffat, John 236
Monod, Jacques 2
monopole, magnetic 224, 226–27
Mosaic physics 55
Müller, Johannes 2
multiple universes 16, 43, 72, 82, 216–17, 235
multiverse 235
Munitz, Milton 196
mythologies: Egyptian 7–8; Babylonian 8–9

Napoleon Bonaparte 243
Narlikar, Jayant 189, 204, 216–17, 236
nebulae 80, 82, 96–100, 110; spiral 97–98, 111–12, 118, 158
nebular hypothesis 96–97, 99–100
nebulium 94–95, 99
Ne'eman, Yuval 211
Neoplatonism 36
Nernst, Walther 164–65
neutrino 225; mass 222, 225; species of 222
neutron stars 207
Newcomb, Simon 111–12, 127
Newton, Isaac 69–74, 76–77, 108, 131
Nichol, John Pringle 96, 99
Nicholas of Cusa. *See*: Cusanus
Nishida, Minoru 183
non-Euclidean geometry 85, 125–29
Novikov, Igor 201, 211, 215
nuclear archaeology 161
Nuffield Workshop 235
number-magnitude relation 159

numbers, cosmological 166–67; Eddington's 65 (n. 36), 139–40, 165–67

observation, theory and 245–46
Öpik, Ernst 197, 218 (n. 24)
Ørsted, H. C. 92
Ohm, Edward 201–02
Olbers, H. Wilhelm 84, 126
Olbers' paradox 83–86, 104, 110, 112, 114, 126, 171, 245
Oort, Jan 114, 207, 214
Oppenheimer, J. Robert 207, 211
Oresme, Nicole 42–45
Origen 33
Osiander, Andreas 48–49, 247
Ostriker, Jeremiah 214
other worlds. *See*: multiple universes
Ouspensky, Peter D. 2

Pachner, Jaroslav 235
Pacini, Franco 207
Pagels, Heinz 241
Pantin, Charles 238
Paracelsus 60
paradigms 243–44
parallax, stellar 50–52, 57, 60, 76–77
Parsons, William *See*: Rosse
Patterson, Clair 191
Pauli, Wolfgang 225
Peebles, James 202–05, 214, 225–26, 233–34
Peirce, Charles Sanders 106
Penrose, Roger 212
Penzias, Arno 202–03
perfect cosmological principle 185–86, 196–97
period-magnitude relation 115–17, 190
Perlmutter, Saul 230
Petrosian, Vahe 216
Philolaus 15–16, 213
Philoponus, John 36
phoenix universe 73, 81, 240. *See also*: cyclical universe
photography, astronomical 250
photometry, astronomical 250
photons, mass of 251 (n. 9)
physical cosmology 149–51
Pickering, Edward C. 124 (n. 117), 250
Pirani, Felix 226
Pius XII 195–96
Planck length 165, 242
Planck time 226, 236, 240, 242
planets 28, 40; Ptolemy's system 29–32; retrograde motions of 20. *See also*: circle paradigm
Plaskett, Harry H. 120
Plaskett, John 157
plasma cosmology 217
Plato 19–20, 22–23
Pliny the Elder 27–28, 36
Plutarch 27
Poincaré, Henri 106, 127
polyhedra, regular 15, 63
Popper, Karl 176 (n. 113), 196–97
Poseidonius 24
positivism 105, 111
pre-Big Bang scenario 242
precession, equinoxes 28
Presocratic philosophers 13–18
Prévost, Pierre 87
Priestley, Joseph 220 (n. 64)
Proctor, Richard 99, 109–10
Prout, William 218 (n. 5)
Ptolemaic system 28–32
Ptolemy 1, 3, 28–31, 247
pulsars 207
Pythagoras 15
Pythagoreans 15–16

quantum gravity 242
quasars 206–07, 212, 235
quasi-steady-state theory 217, 236
Queloz, Didier 236
quintessence 233

radial velocities: stars 92; nebulae 116–17. *See also*: redshifts
radioactivity 24, 104–05, 107, 152
radio astronomy 193–95, 201
Ramsay, William 94
Rankama, Kaleva 183
Rankine, William 102, 104
rationalism 172, 184
Raychaudhuri, Amalkumar 212
Rayner, C. B. 138
realism 246–47
redshifts: cosmological 135, 144; galactic 116–17, 135–38; gravitational 128, 131; non-Doppler explanations 164–65

redshift-distance relation 136–38, 140, 144–46, 192–93. *See also*: Hubble law
redshift-magnitude relation 192–93
Rees, Martin 206, 227, 239
Refsdal, Sjur 210
relativity: general theory 128–31, 209–13; special theory 128
religion 55, 171, 195–96, 219 (n. 37), 238, 249. *See also*: Bible; God
retrograde motion, planetary 20, 29, 49
revolution metaphor 245
Rheticus, Georg 47, 49
Riccioli, Giovanni 55–56
Rice, James 165
Richardson, Owen 176 (n. 124)
Riemann, Bernhard 126, 129
Rindler, Wolfgang 226
Ritter, Johann Wilhelm 90
Roberts, Francis 75
Robertson, Howard Percy 140, 142, 146–48, 172
Robertson-Walker metric 142, 172
Rømer, Ole 76
Rogerson, John 223
Roll, Peter 202–03
Rosen, Nathan 198
Rosse 97–98, 111, 250
Rothman, Tony 228, 249
Rothmann, Christopher 52–53
Rubin, Vera 214
Russell, Bertrand 237
Russell, Henry N. 117, 119
Ryle, Martin 193–94, 201, 245

Sacrobosco 40
Sahama, Ture 183
Sakharov, Andrei 200, 224
Salpeter, Edwin 190
Sambursky, Samuel 165
Sandage, Allan 191–93, 216
Sato, Katsuhiko 235
Schatzmann, Evry 183
Scheiner, Julius 92, 100
Schiaparrelli, Giovanni 10, 64 (n. 21)
Schmidt, Maarten 206
Schmidt, Martin 230
Schönberg, Nicolaus von 47
Schramm, David 221–22
Schrödinger, Erwin 166–67
Schuster, Arthur 107
Schwarzschild, Karl 113, 127, 129–30, 210
Sciama, Dennis 189, 203–04, 206–07, 227, 237
Secchi, Angelo 92
Seeliger, Hugo von 108–10, 112–13, 127, 131, 133
Selety, Franz 110
Seleucus 27
Shakespeare, William 47, 57
Shapley, Harlow 86, 114–19, 136, 143, 145–46
Sherburne, Edward 57
Siger of Brabant 41–42
Silberstein, Ludwik 137–38, 146
simplicity 197
Simplicius 19–20
singularity 135, 155–56, 210–12
singularity theorem 212
Slee, O. Bruce 194
Slipher, Vesto Melvin 116–17, 135, 144
Smith, George E. 250
Smith, John 6
Smolin, Lee 237
Snyder, Hartland 211
Soddy, Frederick 108
Soldner, Johann Georg 129, 210
Solvay congress 194–95, 209
Soviet Union 199
space, curved 85, 125–28. *See also*: non-Euclidean geometry
Space Telescope, Hubble 230
spectroscopy 4, 91. *See also*: astrospectroscopy
Spencer Jones, Harold 188
spheres, celestial 38–41
Stallo, John 105, 248
Stark, Johannes 92
stars, fixed 32, 50; age of 160; distances to 76–77; energy production 161; proper motions of 76, 86, 112; radial velocities 92
star streaming 113
Starobinsky, Alexei 225, 227
statistical cosmology 112
steady-state theory 184–90, 192, 194–96, 201, 217, 239
Stebbins, Joel 187, 250
Steigman, Gary 221–22
Steinhardt, Paul 227, 240
Stewart, John 164

Stoic philosophers 19, 23–24
Stonehenge 6
string theory 236, 242
Strömberg, Gustaf 144
structure formation 224, 227–28, 236
Struve, Friedrich 84
Stukeley, William 83
Sun 27, 49, 64, 112; heat 101; spectrum of 90–91
Supernova Cosmology Project 230–31
Supernova Legacy Survey 232
supernovae 207; distance indicators 230–31
supersymmetry 225
Susskind, Leonard 224
Suzuki, Seitaro 151
Synge, John 138

Tait, Peter G. 102, 104
Takeuchi, Tokio 163
Talbot, William Fox 122 (n. 58)
Tamann, Gustav 216, 230
Taub, Abraham 173 (n. 31)
Tayler, Roger 205
Taylor, Joseph 207
technology, cosmology and 249–50
telescopes 76, 249–50; Galileo's 61–63; Herschel's 87, 250
Teller, Edward 177
Tempier, Etienne 42
Tertullian 33–34
Thâbit ibn Quarra 32
Thales of Miletus 13
Theon of Smyrna 31
thermodynamics 91, 100–02, 104, 106–07, 151. *See also*: entropy law
Thierry of Chartres 37
Thomson, J. J. 95
Thomson, William 85, 101–02, 105, 109
Thorne, Kip 199
time, scales of 171, 212, 240
time travels 198–99
Tinsley, Beatrice 216
Tipler, Frank 237, 239
tired-light explanations 164
Tolman, Richard 146–48, 151, 158, 160, 216
Tolman-Bondi model 155, 220 (n. 82)
Toulmin, Stephen 2
Trimble, Virginia 216
Trumpler, Robert 114
Tryon, Edward 240–41
Turkevich, Anthony 181
Turner, Michael 227, 233–34, 242
Turok, Neil 240

ultraviolet light 90
universe: Aristotle's 22–23; beginning of 18, 43, 72, 150, 153; chemical composition of 93; concept of 1, 19, 81, 197, 242; Einstein's static 131–34; end of 18, 237; energy of 166, 169, 240–41; eternal 23–24, 42; expanding 139–49; hierarchical 85, 109–10; infinite size of 17–18, 23, 55–61; isotropy 239; Milne's 169–72; pyrocentric 16; size of 138, 140, 158. *See also*: accelerating universe; Big Bang universe; cyclical universe; Copernican system; Island universe theory; Milky Way universe; Newton; Ptolemaic system; rotating universe 198; steady-state theory
Updike, John 234
Uranus, discovery of 86
Urey, Harold 152

vacuum 55, 70; false 226–27
vacuum energy 149, 151, 232–33
Vallarta, Manuel 155
van Maanen, Adriaan 118–20, 245
van Rhijn, Pieter 113
varying speed of light 236
Vaucouleurs, Gérard de 216
velocity-distance relation 136–38. *See also*: redshift-distance relation
Veneziano, Gabriele 242
Venus, phases of 62
Vespucci, Amerigo 66 (n. 99)
Vilenkin, Alexander 241
Vogel, Hermann 92
void, cosmic 23–24, 29, 43–44, 52. *See also*: vacuum
Volkoff, George 207
vortices, celestial 68–70

Wagoner, Robert 205–06
Walker, Arthur G. 142, 172
Wallace, Alfred Russell 112
Walsh, Dennis 210
Waterston, John J. 101
Weinberg, Steven 195, 224–25, 237

Weizsäcker, Carl Friedrich 152, 161–63, 177
Wesson, Paul 217
Weyl, Hermann 136, 150, 167, 232
Weymann, Raymond 210
Wheeler, John 198, 211, 241
Whiston, William 72
white holes 211–12
Whitehead, Alfred North 2, 138–39
Whitford, Alfred 187
Whitrow, Gerald 196, 226, 249
Wickramasinghe, Chandra 204
Wilczek, Frank 224
Wilkins, John 65 (n. 73)
Wilkinson, David 202–03, 231
Wilkinson Microwave Anisotropy Probe 231
William of Conches 37
Wilson, Robert 202–03
WIMP 225
Wirtz, Carl Wilhelm 136
Wittich, Paul 51
Wolff, Christian 77
Wollaston, William 90
wormholes 198
Wright, Thomas 77–78, 87

Xenarchus 22

Yahil, Amos 214
York, Donald 223
Yoshimura, Motohiko 223–24

Zel'dovich, Yakov 199, 202, 215, 223, 225, 227, 233
Zeno of Citium 24
Zhdanov, Andrei 199
Zöllner, Karl Friedrich 85, 250
Zwicky, Fritz 151, 164, 207, 209–10, 213–14, 230